SCM with AI for Managers

Strategies and Best Practices

Nada R. Sanders

Northeastern University

Prospect Press

Founded in 2014, Prospect Press serves the academic disciplines of Information Systems and Business Analytics by publishing essential and innovative textbooks across the curriculum, including introductory, emerging, and upper-level courses. Based in Burlington, Vermont, Prospect Press distributes titles worldwide. We welcome new authors to send proposals or inquiries to Beth.golub@prospectpressvt.com.

Editor: Beth Lang Golub
Associate Editor: Dave Williams
Publishing Coordinator: Giulia Bronzi
Production Management: Peter Holm, Sterling Hill Productions
Copyeditor: Theresa Sandifer
Cover Design: Annie Clark
Illustrations: Marcia Blanco

eTextbook:
ISBN: 978-1-958303-41-2
Available from VitalSource, Kortext, Perusall, and BibliU

Printed Paperback:
ISBN: 978-1-958303-42-9

For more information, visit https://www.prospectpressvt.com/textbooks/sanders-scm-with-ai.

CONTENTS

Preface

We are living through an era of profound transformation—from global disruptions to the digital acceleration of every business function. Supply chains have moved from behind-the-scenes enablers to strategic front lines. They are no longer merely systems of production and delivery but instead are dynamic networks that reflect the pace, complexity, and volatility of the world itself. Supply-chain management (SCM) is now at the core of success for virtually all companies.

At the heart of this transformation is artificial intelligence (AI).

The development of AI, advanced analytics, and emerging technologies has created a new era of digitally driven supply chains that has changed the nature of SCM competitiveness. Today's SCM cannot happen without AI decision-making tools and business leaders who know how to use them.

This textbook was born from a recognition that supply-chain management (SCM) must evolve—not incrementally, but fundamentally. Traditional models and methods, while still essential, are increasingly insufficient in the face of today's hyper-connected, data-saturated environment. AI, with its ability to sense patterns, predict outcomes, optimize decisions, and learn from complexity, offers a powerful set of tools to reimagine how supply chains operate.

But this is not just a book about algorithms. It is about strategy, leadership, and systems thinking. It is about equipping future supply-chain professionals, analysts, and executives with the mind-set and capabilities needed to thrive in uncertainty. Whether it is forecasting demand in the face of climate volatility, personalizing products at scale, managing global logistics with real-time data, or ensuring resilience amid geopolitical shifts—AI is redefining the art and science of supply-chain management.

In this text, I have sought to capture the cutting edge. You will find not only foundational principles of SCM but also emerging practices, real-world case studies, and forward-looking frameworks that show how AI is being deployed across industries—from retail and manufacturing to healthcare and entertainment. My aim is to make the material rigorous yet accessible, technical yet strategic, and always grounded in practical impact.

I hope this book serves as both a guide and a challenge—to think beyond traditional silos, to embrace innovation, and to lead supply chains into the future.

Why This Book Stands Out

This book is designed to provide students and practitioners with the basics of SCM **within the context of AI, analytics, and latest technological capabilities**. It is different from all other SCM and analytics books; it is specifically written as a simple yet comprehensive SCM text providing integration with AI and analytics capabilities that are state-of-the-art tools in SCM decision making.

Stories and examples showcase exactly

- what decisions need to be made,
- capabilities of AI and technology that can be leveraged, and
- how to make supply-chain decisions with these technologies.

This is not a traditional academic text in that it is a highly readable, story-driven exploration of SCM's modern realities. Designed for both students and professionals, it allows readers to engage with any chapter independently, picking up key insights without needing to follow a linear path. Through compelling narratives, it illustrates how AI and other technologies empower supply chains—from marketing and operations to purchasing and distribution.

What to Expect

Stories That Inspire

- Each chapter is rich with real-world examples, illustrating how organizations harness AI to achieve strategic goals and deliver value.
- You'll read about how generative AI transforms inventory forecasting, how the IoT redefines logistics, and how blockchain ensures transparency in global sourcing.

Practical, Human-Centric Insights

- The focus is not on complex equations but rather on practical decision making.
- Learn how AI serves as a copilot for business leaders, helping navigate the challenges of modern SCM.

Comprehensive Yet Accessible

- This book is as much about today as it is about tomorrow.
- It explores the latest disruptive technologies while remaining grounded in actionable insights.

Flexible Learning

- Whether you are a student, an executive, or a curious professional, the book's modular structure lets you dive in to the topics most relevant to you.

Discussion Questions, Exercises, and Cases

- This book offers numerous tested pedagogical tools for the classroom.
- The Instructor's Manual is written by the author.

A Practical Guide to Ai-Driven SCM

This book is not mathematical or quantitative. There are plenty of excellent books on the market for that. Rather, this is an accessible book that looks at the state-of-the-art in AI, analytics, and emerging technology capabilities and **how they drive supply chains, from marketing to sourcing.** It is designed to foster human decision making in SCM with AI.

The book can be used alone or as a companion to a quantitative analytics textbook. Each segment of an end-to-end supply chain is specifically presented—from marketing to purchasing—and then brought together in the cohesive framework of a connected supply chain.

The book focuses heavily on industry best practices and how AI can provide a competitive advantage across supply chains. This text addresses SCM within a realistic global and technologically driven business environment, in a complete and comprehensive manner.

- It is written in an **accessible** manner, enabling students to easily grasp the material, and then extend and elevate discussion in the classroom.
- Each chapter ends with numerous **exercises** that use generative AI as a copilot for decision making and **business case discussions** that integrate AI.
- It is intended to provide the **foundational concepts** for undergraduate and graduate-level classes in SCM in the era of AI.
- In addition, the book is an excellent resource for **executive education** and training seminars.

A New Lens on Decision Making

In every chapter, AI takes center stage—not as a replacement for human judgment but as an essential partner in making smarter, faster decisions.

- The key is understanding **which SCM decisions** need to be made.
- Each **section of the supply chain** is examined through the lens of AI capabilities, showing how to transform challenges into opportunities.
- The end-to-end perspective presented **integrates** AI and best practices, empowering readers to **think critically** and **innovate**.

A Journey into Ai-Driven SCM Decisions

This book equips readers to thrive in a global, tech-driven supply-chain environment. Each chapter ends with deep discussion questions and exercises for the classroom that incorporate generative AI thinking and are designed to challenge the status quo. They serve as both part of a foundational text and a springboard for deeper discussion.

Whether used in undergraduate classrooms, executive training seminars, or personal study, this book is a guide to navigating the future of SCM with AI and analytics.

Table of Contents

Part I: AI-Driven Supply Chains

Chapter 1: AI—The Game Changer

This chapter explains how global disruptions revealed the critical importance—and fragility—of supply chains, making disruption the new normal. It introduces supply-chain management as an integrated system of flows and decisions, outlines the range of AI technologies enabling modern supply chains, and shows how core supply chain principles apply across physical, financial, and digital domains. The chapter establishes AI as a fundamental game changer for building responsive and resilient supply chains.

Chapter 2: Strategy, Competitiveness & AI "Fit"

This chapter explains why AI is not a one-size-fits-all solution and must be selected to support a firm's unique business and supply-chain strategy. It examines how strategic priorities and competitive objectives shape AI technology choices and identifies the key organizational and contextual factors that influence technology selection. The chapter provides a framework for ensuring that AI investments have the right strategic fit and deliver meaningful competitive advantage.

Chapter 3: Forecasting in the Age of AI

Chapter 3 introduces forecasting as the foundation of supply-chain decision making and risk identification. It explains the basic principles and processes that guide effective forecasting, reviews traditional and AI-enabled forecasting methods, and highlights the importance of collaboration across the supply chain. The chapter shows how improved forecasting accuracy and alignment enable more proactive, resilient supply-chain decisions.

Part II: AI's Impact Across the Supply Chain

Chapter 4: Impact on Marketing, or "Sell"

Chapter 4 explains how AI drives supply chains through the marketing, or Sell, function by reshaping customer-facing decisions. It examines how AI-enabled tools transform the four Ps of marketing—product, price, place, and promotion—enabling personalization at scale, advanced pricing and merchandising, and location-based marketing. The chapter shows how integrating these capabilities creates a customer-centric marketing mix that drives demand and improves overall supply-chain performance.

Chapter 5: Impact on Operations, or "Make"

This chapter explains the role of operations in creating the goods and services firms sell and why close coordination between Make and Sell is essential. It examines the key operational decisions shaped by AI, including production planning, inventory, and quality management, and introduces the concept of the digital factory. The chapter shows how AI-enabled operations connect and synchronize the entire value chain to improve efficiency, responsiveness, and resilience.

Chapter 6: Impact on Logistics, or "Move"

Chapter 6 explains the role of logistics in moving goods through the supply chain and the decisions required to deliver the perfect order. It examines the cost structure of transportation and logistics infrastructure and shows how AI optimizes movement across inventories, warehousing, transportation, and reverse logistics. The chapter highlights how AI-enabled logistics integrates activities across the supply chain to improve speed, reliability, and cost performance.

Chapter 7: Impact on Sourcing, or "Buy"

Chapter 7 explains why sourcing and purchasing are strategic functions that extend well beyond transactional buying. It examines how AI enables modern sourcing by transforming activities from order processing to supplier collaboration and co-creation and by supporting complex tradeoffs in sourcing structure. The chapter highlights how AI informs decisions around single versus multiple sourcing, global versus domestic sourcing, and outsourcing, while strengthening supplier risk management across the supply chain.

Part III: Building an AI-Powered Supply Chain

Chapter 8: The AI Roadmap

This chapter explains why an AI roadmap is essential for translating strategy into effective execution and highlights the risks of implementing AI without a clear plan. It introduces the DEFINE framework as a structured approach for guiding AI initiatives and emphasizes the role of measurement in evaluating performance. The chapter concludes with tools for assessing AI readiness and tracking metrics that ensure that AI investments deliver sustained value.

Chapter 9: Implementation & Scaling

Chapter 9 explains why implementation and scaling represent dual—and often underestimated—challenges in enterprise AI adoption. It introduces the three pillars of successful implementation—people, processes, and technology—and shows why they must be managed simultaneously as organizations move along the AI implementation maturity curve. The chapter highlights practical strategies for scaling AI and positions Sales and Operations Planning (S&OP) as a proven entry point for enterprise-wide AI implementation.

Chapter 10: AI Governance, Culture & Leadership

This chapter explains why effective AI governance is essential for sustainable AI success and how organizational culture shapes outcomes. It outlines the management steps required to implement AI and examines the critical role of leadership in guiding adoption. The chapter concludes with practical guidance on what leaders must do to ensure that AI initiatives are responsible, aligned, and value-creating across the organization.

Instructor Resources

Instructor resources include

- PowerPoint slides for all chapters,
- Test Bank in Word and QTI formats, and
- instructor guide, including suggested teaching approaches for the course; detailed lecture outlines for individual chapters with timing guidance; suggested course syllabi; recommended chapter sequences and tips for hybrid, online, and asynchronous delivery; case teaching notes and discussion question teaching notes; and ideas for class projects.

To access the Instructor Resources, visit https://www.prospectpressvt.com/textbooks/sanders-scm-with-ai:

- Scroll to the red horizontal menu bar.
- Click on Instructor Resources.
- Click on the yellow box with Resources Login.
- Click on Instructor Resources Access Request Form and fill out the form. *Only verified faculty will receive access information.*

Acknowledgments

I would like to thank Beth Lang Golub and the team at Prospect Press for their effort and help in working on this textbook. I also thank all the reviewers, listed below, who provided detailed feedback. Your suggestions and ideas significantly improved the text.

Triss Ashton, Tarleton State University
Dr. Ryan Atkins, PhD, Duquesne University
Gokul Bhandari, PhD - Area Chair, University of Windsor, Odette School of Business
Ronnie Brannon, Associate Professor, Palo Alto College
David Butler, Associate Professor, Grove City College, Supply Chain Management & International Business
Jen-Yi (Jay) Chen, Cleveland State University
Mark A. Jacobs, PhD, University of Dayton
Collin Johnson, Professor of Practice, Wittenberg University, Business and Economics Department
Michael Kane, Asbury University
Ata Karbasi, Assistant Professor of Information and Decision Science, Valparaiso University
John Larsen, Florida State University
Huiling Liu, Arkansas State University
Charles Luo, Metro State University
Kyle Moninger, Bowling Green State University
Jaideep Motwani, Grand Valley State University
Dr. Sharon Mills Ramp, APUS
Surender Reddy, Saginaw Valley State University
Wendy M. Serapiglia, Duquesne University
Andrew Zeiser, John Carroll University
Joseph H. Zhang, Department of Accounting and Finance, Bowling Green State University

Nada R. Sanders
March 2026

Chapter 1

AI—The Game Changer

In This Chapter You Will Learn

1. how the world learned why supply chains matter,
2. that disruptions have become the new normal and adoption of AI has become necessary for resilience,
3. about the complex nature of supply-chain management as a system of integrated functions,
4. the different types of AI that fall under the "umbrella" of enabling technologies
5. that supply-chain principles can apply to universal flows—from cartons to payment flows to music, and
6. how AI is a game changer for creating responsive and resilient supply chains.

Imagine this. You ordered a last-minute birthday gift—a smartwatch—from Amazon, expecting it might take a few days. But to your surprise, it arrives within hours, complete with gift wrapping and personalized recommendations for accessories. Just a few years earlier, that same order would have taken two to five days.

This experience was made possible by AI managing Amazon's supply chain behind the scenes. Amazon uses AI-driven predictive analytics to anticipate what products people in a region are likely to buy, often before they even place an order. It is because of this that the smartwatch you ordered was already stocked in a local fulfillment center near you. Then, their robotic systems and route optimization algorithms ensured that the package was picked, packed, and delivered rapidly. It feels like magic. But this is an AI-enabled supply chain—from forecasting and inventory placement to last-mile delivery—that made speed and convenience possible. The result? A happy friend and a more loyal customer.

AI has made supply chains faster and smarter. Thanks to AI, the world now moves with precision. However, the details of this process can seem mysterious. In this book, we will see behind the curtain and learn how AI governs every aspect of supply-chain management (SCM). First, in this initial chapter we begin by looking at the complexity of the supply-chain function, the volatility of the current global environment, and how AI has become an essential tool for responsiveness and resilience.

Let's get started.

1.1 Supply Chains & the New Global Disorder

Today our lives are shaped by AI in ways we barely notice. You can order groceries online and have them delivered in under two hours—because AI predicts what your local store needs before you even click "buy."

AI-powered apps like Google Maps adjust your route in real-time based on traffic patterns, accidents, or weather, saving you time and frustration. Platforms like Netflix and Spotify recommend movies and music that match your mood or preferences, thanks to machine-learning models trained on your behavior patterns. In healthcare, virtual assistants and AI triage tools now help patients get faster diagnoses or schedule appointments without calling a doctor's office. Instead of waiting on hold, you often chat with AI bots that answer questions instantly—booking flights, processing refunds, or solving tech issues on demand.

Someone recently said, *"I used to browse for toothpaste. Now it shows up at my door before I realize I'm out."*

But what happens when all this suddenly stops?

1.1.1 Why Supply Chains Matter

The year 2020 is often referred to as "the year the earth stood still." It was also the year the world learned why supply chains matter.

That was the year the world economy, as we know it, came to a grinding halt. It was the result of the world implementing shut-downs in response to the COVID-19 global pandemic. In the space of just a few frantic weeks in March 2020, "just-in-time" (JIT) became "not-any-time." Borders slammed shut, factories went dark, and the intricate choreography that keeps goods moving simply froze. The world discovered—painfully—that supply chains are not background plumbing. They are the heartbeat of the global economy. And when that heartbeat stops, everything from cereal to surgery is at risk.

All but essential service sectors simply stopped functioning. In early 2020, as the virus spread rapidly across the world, many countries imposed lockdown orders on communities and, to contain the outbreak, placed restrictions on virtually every economic activity. Global supply chains that companies and consumers depend upon just stopped.

The result was a shattering impact on businesses and industries worldwide. Many companies shut down, some temporarily and many permanently, due to lack of demand, unavailable workers, inaccessible supplies, unmanufactured parts, lack of materials, cumbersome compliance guidelines like social distancing, and dire financial constraints.

Global supply chains, those interconnected networks that facilitate the production and distribution of goods and services, for all practical purposes, collapsed. As countries went into lockdown, businesses were unable to purchase the raw materials and supplies needed to produce their products. Companies that relied heavily on one country or one supplier were hit the hardest. Borders were closed, disrupting shipping and transportation, resulting in massive shortages of critical supplies and raw materials as well as delays and disruptions in the production and distribution of goods. The results, among other things, were global shortages and stockouts of everything from toilet paper and pharmaceutical items to household appliances and semiconductors.

Many companies did not weather the storm, and in 2020, in the US alone, corporate bankruptcies reached levels higher than seen in more than a decade. The year of 2020 was called "the year of bankruptcy." Businesses of every type were in peril. Companies in every industry sector urgently began redesigning their business models to secure business continuity and build resilience. Companies that survived the pandemic began making rapid changes to their businesses to prevent ever being in that situation again.

Let's look at an example in the apparel industry and the changes companies made.

COMPANY EXAMPLE: Bangladesh's Apparel Industry

Few systems illustrate the shock more starkly than the fashion pipeline that links Western brands to Bangladeshi garment factories.

Before the COVID-19 pandemic, Bangladesh was one of the world's leading apparel manufacturing hubs, supplying a significant portion of the global retail fashion market. In fact, Bangladesh was the second-largest garment exporter in the world, after China. In 2019, apparel exports from Bangladesh totaled over $34 billion, accounting for more than 84 percent of the country's total exports. The industry employed over four million workers, in more than 4,500 garment factories.

Many of the world's largest fashion and retail brands sourced garments from Bangladesh due to its competitive labor costs, established factory base, and large production capacity. This included well-known retail brands such as H&M, Zara, Walmart, Gap, Primark, and Target.

Why Bangladesh?

Bangladesh had among the lowest labor costs globally, even compared to other Southeast Asian nations specializing in woven and knit garments. It had also established an export infrastructure, such as ports, logistics, and regulatory alignment with Western buyers. Last, it met all compliance with global retail standards. This was an important criterion as many global retailers had invested in auditing and upgrading safety standards, making Bangladesh more reliable.

In 2020, however, orders vanished overnight. By late April 2020, global retailers had canceled $3.8 billion in apparel orders from Bangladesh alone. Factories closed en masse. More than 1,100 factories shut for nearly a month during the first national lockdown, losing about $3.18 billion in lost revenue and leaving hundreds of thousands of workers without pay.

Consequences were cascading. There was a cash-flow collapse for many suppliers, mass furloughs, and unpaid wages. As suppliers defaulted on loans, raw-cotton shipments piled up in ports, and Western brands faced empty shelves months later. Even after factories reopened, staggered shifts, transport limits, and renewed lockdowns in 2021 kept capacity well below normal, forcing brands to rethink single-country sourcing strategies.

Brands quickly realized that single-hub sourcing and long lead times create a systemic risk. There was a strategic response to diversify geographies of suppliers. For example, US and EU buyers cut China's share from 30 percent to 22 percent and raised South Asia's, plus nearshoring to Turkey and Mexico. Companies also realized that they needed to shorten the supply-chain cycle. For example, Zara rolled out demand-sensing AI that recalibrates buys twice a week, cutting forecast error by 30 percent. Companies also engaged in predictive planning. For example, H&M used machine-learning models for size and color forecasting, cutting excess stock by 12 percent. Risk was now spread across multiple hubs, and tech investments were used to turn data into early warning signals.

As a result of these efforts, in 2024, exports rebounded to roughly $40 billion. Retail brands learned to diversify their supply chains, shortened more than cycles, and invested in demand-sensing AI to cut forecast error—actions that continue to reshape the industry today. The apparel sector remains vital, but the market has now shifted toward diversified, tech-enabled buyers.

What made this crash unique?

There were numerous lessons learned, but there were two significant problems. *First*, there was a high concentration of risk in one geographic area. In fact, more than 50 percent of some fast-fashion chains' woven tops all came from a 300-km radius around Dhaka. *Second*, there was high focus on a single source of supply coupled with Lean logistics, resulting in little inventory in the pipeline. This left brands without ready alternates when Bangladesh idled.

QUESTIONS

1. List the risks associated with having a high concentration of suppliers in one geographic area. Now explain the risks with using a single source of supply. Do these risks magnify each other?
2. How should companies balance cost efficiency with resilience when choosing suppliers across different geographies?
3. How could contract structures be redesigned to distribute risk more equitably between brands and suppliers?
4. In what ways did the lack of supply-chain visibility contribute to the severity of the disruption in Bangladesh? How can digital tools address this gap?
5. What ethical responsibilities do global brands have toward supplier workers during crises, especially when canceling contracts or halting orders?

The world learned that 2020 did not just pause the economy but rewrote the playbook for how global supply chains operate. The pandemic revealed vulnerabilities of global supply chains. Companies suddenly learned that they must be creative in reinventing their supply-chain models. What worked in the past was not going to work in the future. Companies that internalized these lessons—diversify sourcing, create data-rich visibility and agile, digitally enabled fulfillment—are better positioned for whatever shocks come next.

1.1.2 Uncertainty & Global Disruptions

Before the pandemic, "consumers had enjoyed the luxury of low prices for imported goods for a long time, but it was built on something that was very fragile."* The statement highlights that pre-pandemic global supply chains were not structured to withstand disruptions. Although the pandemic has long passed, it ushered in a new era—one of broad global turbulence and disruption. Everything had changed, including the rules for companies to survive and thrive.

The disruptions and breakages of global supply chains during the pandemic left a weak and vulnerable economic landscape. Historically, most businesses designed their operations for what are called "blue sky" days—when everything is functioning normally. When a disruption happened, companies invoked contingency plans—until the situation was resolved. Then they resumed normal operations. However, the long duration and global sweep of the pandemic's disruption collapsed businesses and entire supply chains. Businesses that survived were left weak and weary. Yet, global economic disruptions continued.

Disruptions have now become the "new normal." What had previously been isolated disruptions now occur simultaneously and repeatedly. Let's look at some of these.

First, geopolitics is now a major global risk. There is the Russia-Ukraine conflict, tensions between the United States and China, strenuous relationships among countries across the globe, and competing interests across Europe. In early 2024, attacks on commercial vessels in the Red Sea slashed container traffic in the corridor by 75 percent and forced carriers to add 10-to-15-day detours around Africa. This spiked freight rates and emissions. Impact on supply chains was huge—both Tesla's Berlin plant and Volvo's Ghent factory halted production as shipments failed to arrive on time.

Environmental risks are also rapidly mounting with failure to mitigate and adapt to climate change. A historic drought on the Panama Canal cut daily transits in half for months, forcing bulk carriers to reroute and adding weeks to Asia–US East Coast voyages. Canadian wildfires in 2025 blanketed North America in smoke, stalled rail lines, and curtailed lumber output, pushing building-material costs higher just as interest-rate hikes cooled demand. West African cocoa crops failed under extreme heat and disease, sending futures above $10,000 per ton and triggering a "chocolate shrinkflation" wave across consumer brands in 2024-2025. Even niche products felt these effects recurring; for instance, Sriracha shortages traced back to drought-stressed chili-pepper harvests in Mexico and California.

There is also a return of "older" risks—trade wars, inflation, cost-of-living crises, capital outflows from emerging markets, and widespread social unrest. Then there are societal risks, many that are a follow-up to the protracted pandemic, including employment crisis, severe mental health deterioration, erosion of social cohesion that kept society intact as well as increasing spread of misinformation and disinformation.

These simultaneous crises have created an environment of interlocking risk or what is referred to as "poly-crisis"—multiple crises at the same time. Climate shocks tighten food supply, food inflation fuels social tension, geopolitical conflict redirects shipping, and longer routes raise emissions that worsen climate risk, creating a feedback loop with no cooldown period.

What does this mean for supply chains?

1. Resilience is now more important than efficiency. Single-supplier dependencies and bare-bones buffers are liabilities, not virtues.
2. Visibility through AI is essential. This means ability to see upstream supplier data, real-time tracking of orders, and creation of early-warning systems, as opposed to static dashboards.

---

* Milken Institute Conference, June 2024.

3. Optionality is strategic. This means that it is strategic to have choices. Dual sourcing, nearshoring, and modular product designs reduce exposure to any one choke point.
4. Policy risk is supply-chain risk. Trade bans, sanctions, and tariffs can shut factories faster than any hurricane.

Surviving "all-at-once" disruption demands predictive, self-learning, supply-chain intelligence. In short, the post-pandemic world rewards companies that treat disruption as a baseline condition and deploy AI as an always-on radar, not a side project.

1.1.3 AI to the Rescue

The pandemic served as a global stress test for businesses, pushing many over what can now be seen as a technology tipping point. What had once been gradual, cautious adoption of digital technologies quickly became an urgent strategic imperative. Organizations that had already invested in artificial intelligence, or AI (technologies that enable machines to mimic human intelligence), automation, and digital infrastructure were far better equipped to navigate the crisis. These businesses weathered talent shortages more effectively, maintained greater visibility over their supply chains, and continued to meet customer demands with greater agility than their less-prepared counterparts.

According to Maria Villablanca, CEO of Future Insights Network, "Pre-COVID, AI was this aspirational thing . . . then businesses realized they must get a competitive advantage. AI offers businesses resilience and is no longer a future project. AI is now part of everyday operations and business."*

In the post-pandemic era, disruption has become a defining feature of the global business environment. Companies that survived the pandemic emerged with a sharpened awareness: *long-term resilience depends on technological readiness*. As such, AI has moved to the center of corporate strategy. In 2024, digital transformation moved from "optional upgrade" to a "requirement" for competitiveness. Numerous surveys, including those conducted by McKinsey and Deloitte, reveal that executives across all industries view AI as a critical investment.

One of the most profound shifts ushered in by the pandemic is the redefinition of technology's role in business. Once viewed primarily as a cost-cutting tool, technology is now seen as a driver of strategic value and organizational agility. Business leaders increasingly recognize that competitive advantage hinges on digital connectivity—not just within the enterprise but also across the broader ecosystem of suppliers, partners, and customers. The importance of this interconnectivity was made starkly clear during the early months of the pandemic, when more than half of companies were unable to identify the geographic location of their suppliers. This lack of visibility contributed to delayed deliveries, severe shortages, and widespread stockouts.

Today, companies are turning to AI-powered solutions to close these visibility gaps. Technologies such as cloud-based analytics, Internet of Things (IoT) sensors, and predictive algorithms enable real-time tracking of inventory, supplier performance, and demand fluctuations. According to a 2025 MIT Sloan Management Review report, more than 60 percent of companies have implemented AI in at least one business function, and for many, it represents their most significant strategic shift. Applications include intelligent forecasting, automated procurement, robotic process automation, and customer service chatbots—all designed to reduce costs and increase decision-making speed.

In fact, the cost of not adopting digital tools is becoming prohibitive. A recent IBM study estimated that companies incur an average of over $1 million annually in labor costs just to manage and correct data manually—a burden that is often passed on to consumers in the form of higher prices or poorer service. By contrast, digital systems can streamline these tasks, freeing up resources and enabling faster, more personalized customer experiences.

* Maria P. Villablanca, Co-Founder & CEO, Future Insights Network, personal interview, November 2024.

AI technology is also proving essential in business model innovation. AI and digital manufacturing tools are enabling mass customization, allowing companies to deliver personalized products at scale. For example, eyeglass manufacturers now use facial scanning and 3D-printing to create custom-fit frames in days. Footwear brands like Asics have launched 3D-printed sandals tailored to individual foot shapes, while LEGO has begun experimenting with 3D-printed toys designed to customer specifications.

In sum, the pandemic did not just accelerate the adoption of AI—it permanently altered the trajectory of business strategy. Companies now understand that supply-chain resilience, innovation, and competitiveness are deeply intertwined with AI technology.

COMPANY EXAMPLE: Walmart

When the pandemic struck, global supply chains experienced unprecedented disruption. Amid the chaos, Walmart emerged as a standout example of how AI can be leveraged to enhance supply-chain resilience, agility, and customer responsiveness during a crisis.

Walmart's supply chain is huge. It spans more than 100,000 suppliers and supports over 10,000 stores globally. Prior to the pandemic, they had already begun investing in AI to improve inventory management, logistics, and demand forecasting. The pandemic tested these systems at an entirely new scale. Demand for certain categories—such as cleaning supplies, toilet paper, and packaged foods—spiked dramatically, while other items saw sharp declines. Traditional forecasting methods, based on historical trends, suddenly became unreliable.

To adapt, Walmart accelerated its deployment of AI-driven demand sensing and inventory optimization tools. These systems processed massive amounts of real-time data—including point-of-sale transactions, e-commerce behavior, social media sentiment, and public health data—to generate more accurate, location-specific forecasts. AI models helped Walmart determine which products were surging in demand at which locations and automatically triggered replenishment orders to redirect inventory from distribution centers accordingly.

One of the most critical tools in this response was Walmart's AI-powered supply-chain visibility platform. This system enabled the company to track goods from suppliers to store shelves in near real time. When suppliers faced production or shipping delays—especially from overseas—Walmart's AI systems helped identify alternate sourcing options or reallocate inventory from other regions. This agility allowed the retailer to maintain stock availability while many competitors were still scrambling to understand where their goods were stuck.

In parallel, Walmart expanded its use of robotics and automation at distribution centers to minimize human contact and maintain operations amid labor shortages and health concerns. AI was also embedded into workforce scheduling, ensuring safe staffing levels while balancing peak delivery periods. In addition, Walmart's online grocery business, which surged during the pandemic, relied heavily on AI to manage order picking, route optimization for home deliveries, and curbside pickup logistics.

The results were tangible. Walmart maintained high service levels, grew e-commerce sales by over 79 percent in 2020, and reinforced its position as a supply-chain leader. The crisis validated their long-term investment in AI.

QUESTIONS

1. How did Walmart's prior investment in AI and digital infrastructure give it a competitive advantage during the COVID-19 crisis? What specific capabilities allowed it to respond faster than competitors?
2. In what ways did AI improve Walmart's demand forecasting and inventory management during the pandemic? How did this differ from traditional methods of forecasting?
3. What risks might Walmart have faced if it had not already begun digitizing its supply chain before the pandemic? What vulnerabilities did other companies face that Walmart was able to avoid?
4. How did AI contribute to both customer service and employee safety during the pandemic? Can AI be designed to address operational and human-centered outcomes at the same time?

5. What lessons from Walmart's use of AI during COVID-19 can be applied to building more resilient supply chains in the future? Should AI now be considered a requirement for global supply-chain competitiveness?

1.2 Supply Chains Are Complex

1.2.1 Supply-Chain Networks

Most people think a supply chain is a linear flow of products, as shown in Figure 1.1. However, this is a simplified view. Fundamentally, yes—a supply chain links a series of companies that make, transport, refine, and deliver the finished product to you as the consumer.

Consider a cup of coffee from Starbucks. Your coffee might begin as a pile of coffee beans grown and picked by a farmer in Guatemala. They are then shipped to a coffee roaster, say, in Seattle, who sends them on to a distributor near where you live, who sells them to your local Starbucks. A shutdown anywhere along the supply chain in any of these locations stops this flow and could prevent you from enjoying your morning brew.

However, a supply chain *is not* just a linear flow of products. It also involves flows of information, money, and returned products—all flowing in many directions. Therefore, a **supply chain** is the comprehensive network of entities, processes, and technologies involved in the creation and delivery of a product or service from its origin to the final customer. This network spans multiple stages, including the sourcing of raw materials, manufacturing and assembly, warehousing, order processing, inventory management, distribution, and, ultimately, delivery to the end customer.

While a coffee supply chain may seem relatively simple and linear, a supply chain can quickly get complicated for products that have many parts. A supply chain is more like a complex hyperconnected network. Many companies often share the same supplier, so a disruption in one link in the supply chain can have ripple effects across companies around the world.

At its core, the supply chain is a coordinated flow of materials, information, and finances. The physical flow begins with external suppliers who provide raw materials and components, which are transported to manufacturers. These manufacturers then transform the materials into finished products. Once complete, products are shipped to distribution centers, wholesalers, or retailers, depending on the model. From there, goods move to the end customers via various delivery channels. This is shown in Figure 1.2.

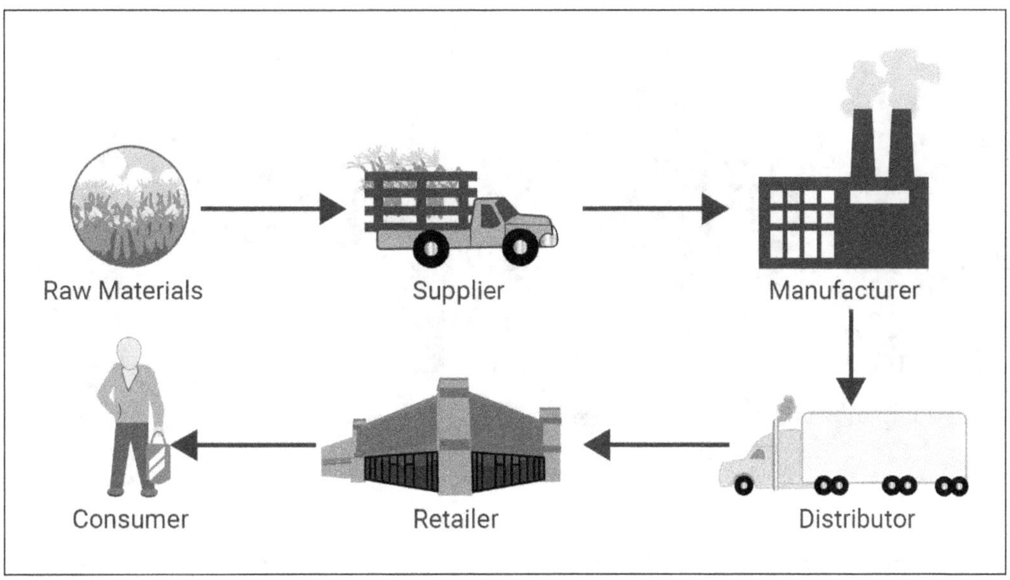

Figure 1.1 A Simplified Supply-Chain View

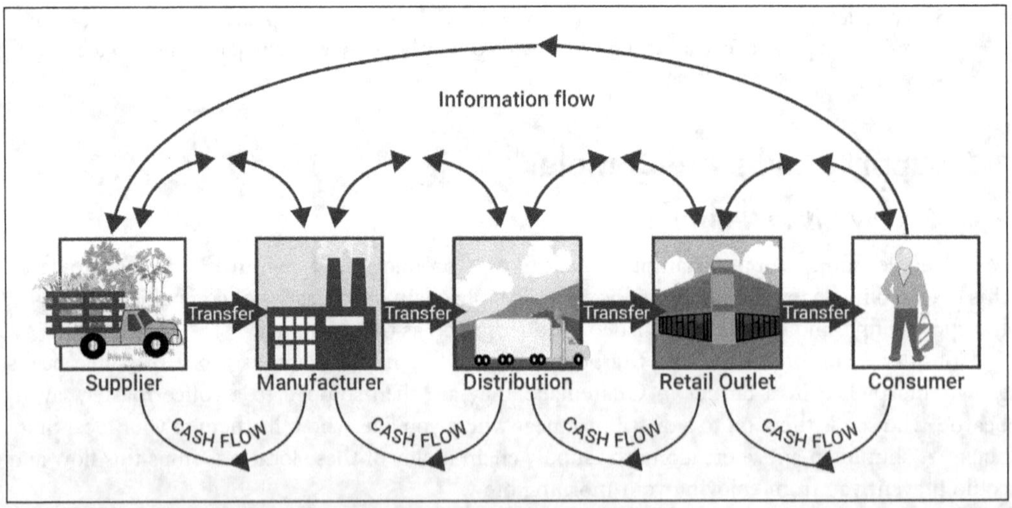

Figure 1.2 A Supply Chain Involves Multiple Flows

Simultaneously, a flow of information travels in both directions. Critical data—such as demand forecasts, inventory levels, customer orders, and promotional plans—must be shared among all partners. This flow of data is enabled by integrated information systems that support visibility, collaboration, and decision making across the supply chain.

In reality, a supply chain is a multi-flow network as shown in Figure 1.3. **Supply-chain management** in turn is the strategic function responsible for planning, coordinating, and optimizing this entire supply chain, from end-to-end.

SCM aims to ensure that the right product reaches the right customer at the right time, cost, and quality level. That means that SCM requires coordinating the movement of goods and services from suppliers to manufacturers, then to distributors, and ultimately to the final consumer. It also involves synchronizing activities across internal departments (e.g., procurement, production, logistics, sales) and with external stakeholders (e.g., suppliers, logistics providers, retail partners).

This process requires sharing and integrating information such as real-time sales data, forecasts,

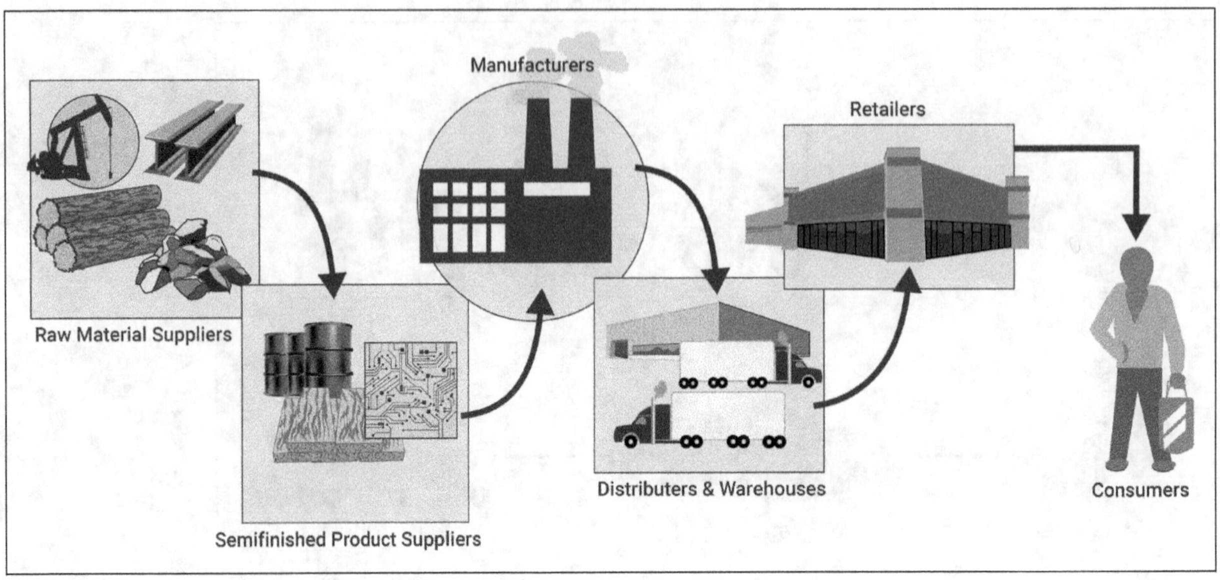

Figure 1.3 A Multi-Flow Network

inventory status, and marketing promotions to align supply and demand. At the same time, companies have to minimize costs—including transportation and production—and reduce lead times all while maximizing efficiency, customer satisfaction, and resilience. An effective supply chain must be agile, responsive, and capable of adapting to changes in demand, disruptions, or shifts in the global marketplace.

Let's look at a case for an Apple iPhone.

COMPANY EXAMPLE: Apple

The Apple iPhone is one of the most iconic consumer products in the world, and one of the most complex, from a supply-chain perspective. Far from being manufactured in a single location, the iPhone is the result of a deeply globalized production network involving more than 200 suppliers across 43 countries on six continents. Each iPhone integrates thousands of components, many of which traverse multiple borders before final assembly.

1. Design and Engineering (United States). Apple's supply chain begins in Cupertino, California, where the core product architecture, hardware design, and proprietary chips (such as the A-series processors) are conceived. While Apple does not manufacture these chips itself, it exerts tight control over their design specifications, performance, and security features.

2. Semiconductor Production (Taiwan, South Korea, United States). The custom chips designed by Apple are manufactured by Taiwan Semiconductor Manufacturing Company, headquartered in Taiwan. This chipmaking process uses leading-edge lithography, requiring billions of dollars in capital-intensive facilities. Some fabrication also occurs in South Korea, where Samsung produces other critical components like memory and OLED displays. The semiconductors may then be shipped to the Philippines or Malaysia for testing and packaging, which ensures they function properly under different thermal and electrical conditions.

3. Component Sourcing Across Continents. Each iPhone includes more than 1,600 individual parts. These parts originate from a vast array of suppliers. Camera modules, for example, are supplied by Sony (Japan), using precision lenses and sensors made in factories in Southeast Asia. Screens are supplied by Samsung Display (South Korea) and LG Display and BOE Technology (China). Batteries are assembled in China, using lithium sourced from Australia, cobalt from the Democratic Republic of Congo, and graphite from China. Then, the aluminum frame is sourced from Alcoa in the United States and Rio Tinto in Canada and Australia, then machined in China.

4. Logistics, Warehousing, and Flow Coordination. Before final assembly, components travel through complex logistics hubs. Raw materials and subcomponents are transported via air freight and cargo ships to various suppliers. Apple relies on sophisticated supply-chain visibility tools and forecasting models to manage inventory flow, avoid overstocking, and quickly respond to demand shifts. Apple also maintains a global network of hubs in Singapore, China, Ireland, and the United States to receive, consolidate, and reroute products globally.

5. Assembly and Final Production (China, India, Vietnam). The final assembly of the iPhone happens primarily in Zhengzhou, China, at the massive Foxconn facility sometimes dubbed "iPhone City." More than 300,000 workers operate around the clock on precision assembly lines.

Some iPhone models are also assembled in India and Vietnam, as Apple seeks to diversify its production to reduce geopolitical risks. During final assembly, chips from Taiwan, sensors from Europe, and screens from South Korea converge to become a single, unified product.

6. Distribution and Retail (Global). After assembly, finished iPhones are packaged and sent to regional distribution centers, and transported to Apple Stores and resellers in more than 175 countries. Apple also runs direct-to-consumer logistics through its online store, with fulfillment centers located strategically across North America, Europe, and Asia-Pacific.

As you can see, the Apple iPhone supply chain illustrates the interdependent nature of modern manufacturing, where materials and components cross multiple borders — often several times — before the final product is delivered. From rare earth elements mined in Africa and electronics made in Asia, to assembly in China and design in California, the iPhone is more than a product — it is a global orchestration of design, technology, labor, and logistics.

QUESTIONS

1. How does Apple's global supply chain contribute to both its competitive advantage and its vulnerability? Consider efficiency, specialization, cost, and risks such as geopolitical tension or natural disasters.
2. What are the implications of Apple relying heavily on one country (like China) for final assembly? Should Apple diversify further?
3. In what ways does Apple maintain control and visibility over such a vast and complex supply chain? What technologies enable this oversight?
4. How do ethical sourcing and sustainability concerns factor into Apple's supply-chain decisions—especially with components like cobalt or rare earth minerals?
5. If a major disruption were to occur in Taiwan, how might it affect Apple's global operations? What contingency strategies could Apple implement?

1.2.2 Supply-Chain Core Functions

A supply chain is more than a sequence of business activities; it is a dynamic system of interconnected functions that work in concert to deliver value to customers. At its core, the supply chain can be broken down into four essential functions: **purchasing, operations, logistics, and marketing**—commonly remembered as "buy, make, move, and sell." These four functions form the backbone of how raw materials are transformed into finished goods and delivered to end users. Understanding each of these areas is key to managing supply chains effectively in today's global economy.

1. Purchasing ("Buy"). Purchasing (also called Sourcing) refers to the procurement of raw materials, components, products, and services from suppliers. This function links the organization to suppliers and connects the supply chain upstream. It involves identifying reliable vendors, negotiating contracts, ensuring quality standards, and managing costs. The function has evolved from a simple buying role to strategic function, which is reflected in the evolution of the term from *purchasing* to *sourcing*, as well as *strategic sourcing*. It plays a critical role in ensuring continuity, especially in times of disruption, geopolitical shifts, or natural disasters. Modern supply chains prioritize diversified sourcing, supplier collaboration, and sustainability in purchasing to reduce risk and improve resilience.

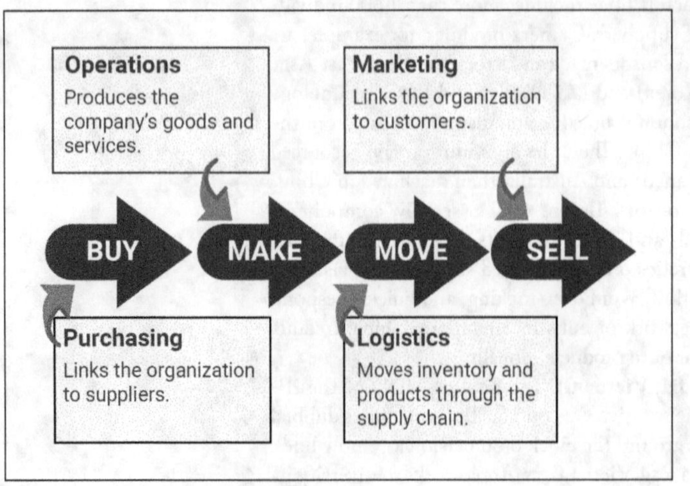

Figure 1.4 Supply-Chain Functions

Example: Apple sources components like semiconductors and display panels from a global network of suppliers. Managing these relationships and ensuring timely delivery is a critical part of its sourcing strategy.

2. Operations ("Make"). Operations refer to the processes involved in transforming raw materials into finished products. This includes manufacturing, assembly, quality control, and inventory management. Efficient operations are central to maintaining cost competitiveness and meeting customer expectations for quality and speed. Increasingly, companies are turning to AI, automation, and predictive analytics to enhance productivity and reduce waste in their operations.

Example: Toyota's Lean manufacturing system is a hallmark of efficient operations, minimizing inventory and maximizing quality through just-in-time production and continuous improvement.

3. Logistics ("Move"). Logistics covers the movement of goods from suppliers to manufacturing sites, between internal facilities, and finally to customers. It includes inbound logistics, transportation, warehousing, internal transfers, and outbound delivery. Efficient transportation management affects both cost and service levels. Companies must optimize routes, manage carriers, comply with regulations, and respond quickly to disruptions in shipping lanes or fuel price volatility.

Example: Amazon uses a combination of air, truck, and last-mile delivery networks—including its own fleet—to ensure fast, flexible deliveries and real-time tracking for millions of customers.

4. Marketing ("Sell"). While marketing is traditionally seen as separate from the supply chain, it is an essential function within it. This is where demand is generated and fulfilled. Sales forecasts, promotional campaigns, pricing strategies, and customer feedback loops directly inform sourcing, production, and distribution decisions. A responsive supply chain integrates marketing insights to align supply with customer preferences and market demand.

Example: Nike's direct-to-consumer strategy uses digital platforms and predictive analytics to align product launches and inventory levels with consumer buying behavior.

By viewing the supply chain as a system of four interdependent functions—purchasing (buy), operations (make), logistics (move), and marketing (sell)—organizations can better coordinate their strategies and improve responsiveness, efficiency, and customer satisfaction. Success in today's business environment requires mastering not just one link but also the entire chain.

1.2.3 An Interconnected System

One important understanding of supply chains is that they represent an interconnected system. As a result, they require end-to-end thinking. In today's fast-moving global environment, a company's supply chain is no longer a back-office function—it is a strategic enabler of value, resilience, and customer satisfaction.

At its core, a **supply chain is a system:** a set of interdependent activities—purchasing, operations, logistics, and marketing—that must work in coordinated harmony. Yet, many organizations continue to manage these functions in silos, resulting in inefficiencies, misalignment, and increased vulnerability to disruption. Treating the supply chain as a unified system rather than a series of disconnected parts is key for responsiveness.

The Interconnected Nature of the Supply Chain. Each function in the supply chain—buying, making, moving, and selling—is deeply interdependent. A decision in one area has cascading effects on the others. For example, if marketing overestimates demand, operations may produce excess inventory, transportation costs may spike, and purchasing may lock in unneeded raw materials. Conversely, if purchasing is delayed due to supplier issues, it can throw off production schedules, delay shipments, and disappoint customers. These are not isolated problems; they are systemic failures.

Only by viewing the supply chain holistically can businesses understand these cause-and-effect relationships and make decisions that *optimize the system* rather than optimizing individual functions.

The Cost of Siloed Thinking. When supply-chain functions operate independently, companies encounter several common disruptions. *First*, there is poor forecast accuracy. Without integrating marketing data with purchasing and operations, forecasts are less reliable, leading to stockouts or overproduction. *Second*, we see a mismatch between inventory and demand. Inventory may pile up in the wrong place at the wrong time because logistics was not coordinated with demand signals. *Third*, we see delayed decision making. Without real-time information flow among functions, decisions are made based on outdated or incomplete data. *Fourth*, it creates increased vulnerability to external shocks. During global disruptions—like the COVID-19 pandemic—companies that lacked end-to-end visibility could not adjust quickly, leading to supply shortages, production halts, and customer dissatisfaction.

A well-known example is the 2021 global chip shortage. Automakers who treated their supply chains as linear transactions found themselves with stalled assembly lines because upstream component suppliers were not integrated into their planning systems. In contrast, companies like Tesla, which takes a systems approach and maintains close collaboration across the entire chain, were more agile in reallocating resources and redesigning components.

Benefits of System-Level Thinking. Adopting a system-wide perspective brings several strategic advantages. An important benefit is *resilience*. Companies can model scenarios and anticipate the ripple effects of disruptions across the chain. Yet another is *agility*, or the ability to respond quickly. Integrated systems enable faster responses to demand shifts, supply constraints, or external shocks. Another benefit is *efficiency*. Data sharing and coordination eliminate redundancies and reduce lead times.

A unified supply chain aligns all the functions to better serve customer needs. These benefits are amplified by AI technologies and advanced analytics, which are most powerful when fed by data from the entire supply chain system, not just isolated segments.

COMPANY EXAMPLE: Target Canada

Target Canada offers an important lesson in the consequences of siloed supply-chain thinking. When US-based retailer Target expanded into Canada in 2013, it was one of the most anticipated retail launches in Canadian history. However, within two years Target Canada shut down all 133 stores and exited the market—losing more than $2 billion. One of the central reasons for the failure was siloed thinking across supply-chain functions, which led to a cascade of operational breakdowns.

What happened?

Disconnected Sourcing and Inventory Systems: The company's teams responsible for sourcing and procurement worked on aggressive timelines to secure products and build assortments. However, the inventory management system was incomplete and untested, and many products never made it to shelves. Some stores opened with empty shelves or large quantities of the wrong items.

Isolated IT Decisions: The IT systems supporting operations were adapted from Target's US operations, but they were not properly localized or integrated with Canadian suppliers. Errors in the data entry system (many items had incorrect product dimensions or barcodes) meant that automated replenishment systems failed, and stores did not receive the right products.

Lack of Coordination Between Operations and Store Teams: Logistics and transportation teams worked under the assumption that products were flowing as expected, while in-store employees were manually correcting errors, delaying operations, and struggling with unscannable merchandise.

Marketing & Operations Out of Sync: Despite operational issues, marketing continued with national ad campaigns promoting products that were unavailable in stores, frustrating customers and damaging brand trust. This is a very common problem with large consequences.

The most significant problem was customer dissatisfaction. Consumers arrived at stores to find empty shelves or wrong products on shelves. Despite interest and foot traffic, customers quickly lost confidence in the brand. At the same time, there was excess inventory in the warehouse. While store shelves were empty, distribution centers were overflowing with unsellable or misplaced inventory—highlighting the disconnect between logistics, inventory, and demand planning. Despite these obvious problems, there was failure to adapt. Because functions were not coordinated, Target Canada could not quickly diagnose or resolve the root causes. Operational teams, IT, logistics, and marketing were all responding separately, and often too late.

Target Canada's collapse serves as a cautionary tale of what happens when companies treat supply-chain functions as independent silos. Without integration between purchasing, operations, logistics, and marketing, even a major retailer with deep pockets and brand recognition can fail spectacularly. It underscores the critical importance of treating the supply chain as a unified system—especially during high-stakes expansions or transformations.

QUESTIONS

1. What specific supply-chain functions at Target Canada were managed in silos, and how did this contribute to the company's operational failures? Consider purchasing, logistics, store operations, and marketing.
2. How might an integrated supply-chain system have prevented or reduced the severity of Target Canada's challenges? Explain how real-time visibility or cross-functional data sharing could have helped.

3. In what ways did poor data quality and disconnected systems lead to both empty shelves in stores and excess inventory in warehouses? What about the relationship between inventory management and in-store execution?
4. Why is it important for marketing teams to be closely aligned with supply-chain operations, especially during new market entries or product launches? Discuss the risks of promoting products that are not available.
5. What role did leadership and organizational structure play in reinforcing siloed decision making at Target Canada? How could leadership have better supported cross-functional integration?

1.3 AI-Driven Supply Chains

AI is now reshaping every component of SCM by automating processes, uncovering insights from large data sets, and enabling faster, smarter decisions. However, remember that a supply chain is an integrated system. This means that the functions have to communicate with on another through real-time data and intelligence sharing.

Often companies make the mistake of investing in one functional application. But remember that the foundation of SCM is the efficient coordination of goods, information, and relationships. The key is to optimize the entire supply-chain system.

There are three tenets to remember:

- Every supply chain is a system connected by AI technology.
- The entire system must be optimized, not individual functions.
- Information at any one function must be conveyed to other functions—informing them and coordinating action.

Next, let's look at individual AI applications and then see how to connect them.

1.3.1 Functional Applications of AI

We described the core supply-chain functions and their roles. Today AI applications exist along every one of these supply-chain function—from purchasing to marketing. In later chapters we will go into the details of each of these functions and AI applications. For now, however, let's do a brief overview of individual functional AI capabilities.

1. Purchasing (Buy). Purchasing—or sourcing—is the function that connects the company to their suppliers and looks at the upstream part of the supply chain. Many companies spend a large part of their budgets—as much as 80 percent—on purchased materials. As a result, this function is critical. AI enhances purchasing by making procurement smarter, faster, and more resilient.

AI capabilities here include **monitor supplier risks.** For example, AI can scan news feeds, financial data, geopolitical trends, and weather patterns and aggregate all of these to identify at-risk suppliers in real time. AI can also assist in **predictive sourcing**. This is where machine-learning models forecast demand and recommend optimal reorder points, contract terms, or vendor shifts based on historical performance. AI-driven platforms can also **evaluate supplier**

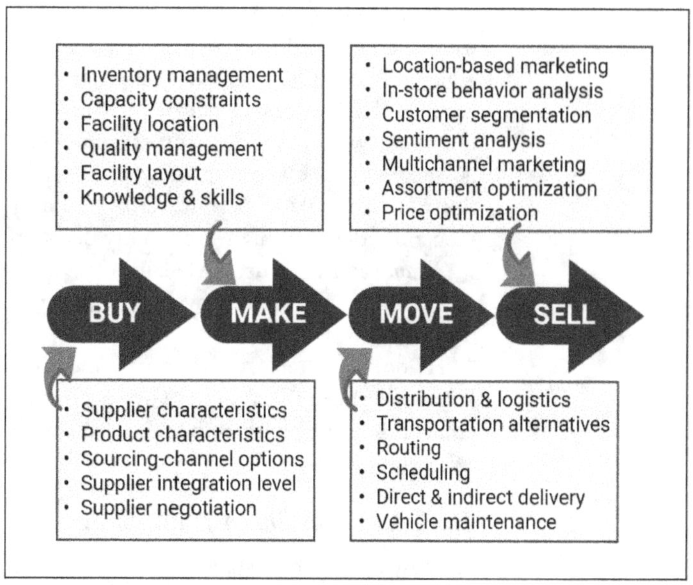

Figure 1.5 AI Capabilities

capabilities, costs, and certifications to suggest new, qualified vendors across the globe. In addition, AI can help in **contract negotiation**. For example, natural language processing tools analyze procurement contracts to identify pricing anomalies, compliance risks, and negotiation opportunities.

Example: IBM's "Supply Chain Intelligence Suite" is a platform that uses AI to predict supplier delays and recommend alternative sourcing strategies. As such it improves purchasing agility and therefore the entire supply chain.

2. Operations (Make). Remember that operations is the function that makes the company's goods and services. AI transforms operations by optimizing manufacturing processes and improving efficiency. AI capabilities include **predictive maintenance**, where AI sensors detect early signs of machinery failure and schedule maintenance proactively to avoid costly downtimes. They also include **production optimization**. Here AI algorithms balance workloads, adjust production schedules, and optimize plant capacity in real time. Then there is **quality control** with computer vision. AI detects defects in products through high-speed image recognition, increasing accuracy and reducing waste. There is also the creation of **digital twins**. These are AI-powered simulations of factory environments that allow managers to test changes in production lines virtually before physical implementation.

Example: Siemens uses AI to power its Digital Factory initiative, where machine learning is applied to optimize robotic assembly and prevent quality defects in real time.

3. Logistics (Move). Logistics involves moving, transporting, and storing goods. AI enhances logistics with better visibility, efficiency, and responsiveness. AI capabilities are enormous as logistics is a heavily automated function. Capabilities include **route optimization**, where machine-learning algorithms consider weather, traffic, fuel prices, and delivery urgency to suggest optimal routes. There is **dynamic fleet management**, where AI assigns delivery vehicles based on capacity, location, and predicted delays to improve asset utilization. There is also **real-time shipment tracking** something all consumers expect today. Here AI integrates sensor data and external signals to provide end-to-end visibility and proactive alerts for delays or disruptions. Then there is the use of **autonomous vehicles & drones**. AI enables self-driving trucks, warehouse robots, and delivery drones to perform physical movement tasks with minimal human oversight.

Example: DHL uses AI in its Smart Logistics Network platform to optimize routes, forecast delivery times, and adjust operations in response to real-time disruptions.

4. Marketing (Sell). AI makes marketing and demand generation more predictive, personalized, and agile. AI capabilities include **demand forecasting**. Here, AI models analyze point-of -sale (POS) data, social media trends, weather, and macroeconomic indicators to improve forecast accuracy. Then there is **dynamic pricing**. AI adjusts prices in real time based on competitor behavior, inventory levels, and customer demand elasticity. There is also **customer personalization**. This is where machine learning analyzes buyer behavior to tailor promotions, product recommendations, and messaging at scale. **Sales planning** is yet another capability, where AI generates sales scenarios, identifies high-probability deals, and prioritizes customers based on data patterns.

Example: Target uses AI-driven predictive analytics to anticipate customer needs, optimize promotions, and localize inventory by store, thus driving higher conversions and reduced markdowns.

AI does not replace the four foundational supply-chain functions—it elevates them. From smarter supplier selection to hyper-personalized marketing, AI injects speed, accuracy, and adaptability into every phase of the supply chain. By deploying AI across all four—buy, make, move, and sell—companies can build more resilient, responsive, and customer-centric supply chains.

1.3.2 An Umbrella of Technologies

AI, however, is not a stand-alone tool. Rather, AI acts as the intelligence layer integrated into a range or an umbrella of enabling technologies. There are four categories.

First are **Analytics & Decision Technologies** that enhance decision making by powering advanced analytics, forecasting engines, and digital twin simulations. They enable both predictive and prescriptive decision making.

Second are **Processing Technologies** involved in automated operations and physical processes. These AI-driven technologies enable robots and smart machines to make adaptive, high-precision movements on factory floors. We are so accustomed to Generative AI, driven by large language models (LLMs), that we may overlook AI-driven automation processes that automate the physical world.

Third are AI **Communications Technologies** that interpret and act on real-time signals from sensors, satellites, and global networks. AI-powered dashboards unify data from across the supply chain, giving managers real-time, actionable insights. This integrated visibility enhances collaboration and enables synchronized decisions across functions.

Fourth are **Integrative & Platform Technologies** in which AI harmonizes the flow of data across systems—driving seamless coordination among enterprise resource planning (ERP), IoT, and cloud platforms.

Table 1.1 shows the enabling technologies behind this transformation, each powered in part by AI. Together, they form the digital backbone of future-ready supply chains. Companies now use AI to anticipate demand, optimize production schedules, reduce waste, personalize customer experiences, and respond to disruptions faster than ever before. This is integration of AI across analytics, automation, communication, and platforms enabling supply chains to be intelligent, proactive, and resilient.

Let's look at examples of these technologies:

1. **Analytics & Decision Technologies.** Unilever uses AI-powered digital twins (a virtual representation of an object or system) and predictive analytics to optimize its end-to-end supply chain. Through platforms like o9 Solutions, Unilever can run what-if simulations, which can improve forecast accuracy by 20 percent and make faster planning decisions across global markets. These analytics tools support scenario planning for disruptions such as raw material shortages or shifts in demand.
2. **Processing Technologies.** Tesla leverages AI-enhanced robots and flexible manufacturing systems to automate its Gigafactories. Robots equipped with computer vision handle precision assembly, welding, and inspection, enabling 24/7 production.

Table 1.1 Categories of AI Technology Types

Type of Technology	Capabilities	Examples
Analytics / Decision Technologies	Provide computing power, AI intelligence, and real-time data orchestration, enabling predictive and prescriptive insights for faster, better decisions	• Advanced planning & optimization (APO) • AI-driven supply-chain digital twins • Generative AI for scenario planning • Machine learning for forecasting • Autonomous decision agents
Processing Technologies	Automate operations and physical processes with 24/7 availability, improving speed, quality, and cost-efficiency through intelligent automation	• Intelligent robots with computer vision • Smart manufacturing (Industry 5.0) • 3D printing with AI-enhanced designs • Autonomous mobile robots (AMRs) • AI-enabled quality-control systems
Communications Technologies	Enable real-time connectivity, visibility, and data sharing across ecosystems with edge computing and IoT orchestration	• 5G and edge networks for latency-sensitive operations • Smart sensors and AI-powered IoT platforms • Satellite logistics tracking • Blockchain-enabled visibility
Integrative / Platform Technologies	Orchestrate data, decisions, and digital twins across platforms; unify physical and digital assets with cloud-to-edge AI integration	• Multi-enterprise supply-chain platforms (e.g., Kinaxis, o9, Blue Yonder) • AI-integrated ERP and PLM systems • Cloud-native digital twin platforms • Blockchain + AI traceability systems

AI-driven automation has reduced defect rates and shortened vehicle production cycles, especially in the high-volume Model Y production line.

3. **Communications Technologies.** Maersk uses AI-enabled IoT and satellite connectivity to track shipping containers in real time. Smart sensors on containers monitor location, temperature, and delays, transmitting data via satellite links. The system supports dynamic route adjustments and real-time customer updates, thus reducing lost cargo incidents and improving visibility across global ocean logistics.
4. **Integrative & Platform Technologies.** Schneider Electric has integrated cloud-based ERP, AI, and IoT systems using a digital twin of its supply chain. The company uses its EcoStruxure platform to combine production, inventory, and customer data across global sites. AI-driven insights from the platform support demand planning, carbon tracking, and supplier collaboration, thus improving both responsiveness and sustainability performance.

1.3.3 Agentic AI

In the rapidly evolving landscape of AI, a new frontier has emerged: **Agentic AI**. Agentic AI refers to autonomous systems that can perceive their environment, formulate plans, make decisions, and take actions to achieve defined goals while continuously learning from outcomes and interactions.

Traditional AI systems have amazing capabilities but function largely as tools executing predefined tasks or making predictions within narrow domains. By contrast, Agentic AI systems act with greater autonomy, initiative, and adaptability. They function more like digital agents than passive software. This shift has profound implications, particularly for SCM, where responsiveness, complexity, and interdependence define success.

Agentic AI are systems that demonstrate goal-oriented behavior, make context-aware decisions, and operate independently in dynamic environments. These agents are capable of planning, reasoning, learning from feedback, and even negotiating or collaborating with other systems or humans to fulfill objectives. Unlike traditional AI, which often relies on fixed models and static inputs, Agentic AI systems are designed to adapt and operate across multiple steps or tasks with minimal human intervention. Traditional AI might tell you **what will happen**; Agentic AI may decide **what should be done about it**, and then do it.

These differences are shown in Table 1.2.

Table 1.2 Agentic AI versus Traditional AI

Feature	Traditional AI	Agentic AI
Scope	Narrow, task-specific (e.g., demand forecasting)	Broad, multi-step tasks (e.g., end-to-end inventory optimization)
Decision Making	Follows predefined rules or supervised learning outputs	Makes autonomous decisions, often through planning and feedback loops
Goal Orientation	Task execution based on human-defined triggers	Goal-driven behavior with self-initiated actions
Adaptability	Static models that require retraining	Continuously learns and adapts to context
Interaction	Input-output systems, often passive	Interactive agents that can collaborate, negotiate, and respond dynamically

While Agentic AI has emerged just recently—over the past few years—there are already numerous examples in SCM. Here are a few:

1. **Autonomous Procurement Agents.** Instead of simply flagging supplier risks or recommending vendors, an agentic system can autonomously source materials, assess risk scores, compare delivery times, and negotiate contracts through e-auctions or direct

interfaces with suppliers. For example, a system might detect a disruption in a tier-2 supplier and automatically reroute orders to an alternative vendor based on price, lead time, and compliance criteria.

2. **Inventory Rebalancing Agents.** Retailers like Walmart or Zara deal with inventory imbalances across regions. An Agentic AI can identify surplus stock in one region, forecast local demand, and trigger inter-store transfers or markdowns, without waiting for centralized human decision making. The agent can continuously monitor inventory flows and autonomously adjust decisions in near real time.
3. **Dynamic Logistics Optimization.** Where traditional routing software finds an efficient delivery path at one point in time, an Agentic AI can act as a real-time delivery coordinator. It reacts to new orders, weather delays, and traffic data, dynamically reprioritizing stops, reassigning drivers, and even coordinating drone or autonomous vehicle dispatch. Think of UPS's On-Road Integrated Optimization and Navigation (ORION) system evolving into a truly self-optimizing delivery network.
4. **Supplier Relationship Management Bots.** Instead of static dashboards, firms can deploy Agentic AI systems that track supplier performance, escalate compliance issues, and initiate corrective action workflows—all without waiting for procurement officers to manually intervene.

While the promise of Agentic AI is substantial, adoption comes with challenges. Delegating autonomous decision making is very risky and requires significant guardrails to prevent mistakes. Agentic systems must also interact across the entire software ecosystem, so they have to be integrated across the organization. Also, goals set for agents must align with organizational strategy and compliance norms to avoid unintended consequences. These consequences could be significant if autonomous systems are left unattended.

Yes, there are large risks with Agentic AI. However, they represent a step change in the evolution of AI, from predictive tools to autonomous collaborators. In SCM, Agentic AI enables dynamic, goal-driven optimization of complex, interdependent processes. Agentic AI is still an emerging capability, but early use cases show that it has the potential to become a strategic differentiator.

Agentic AI may be the next major leap forward—transforming not only how decisions are made but also who—or what—is empowered to make them.

1.3.4 AI-Driven Supply Chains

As the global economy becomes more uncertain, AI enables supply chains to be more agile, resilient, and customer centric. As a result, companies that lead in AI adoption are not merely optimizing costs, they are building competitive moats through faster delivery, smarter sourcing, and superior service. From predictive analytics and intelligent automation to dynamic decision-making and real-time visibility, AI is not just improving efficiency, it is redefining what modern SCM can achieve.

Let's look at some of these critical AI-driven supply chain capabilities:

1. **Predictive Demand Forecasting.** Traditionally, supply chains relied on historical data and static models to estimate demand. These methods often struggled with variability, seasonality, and rapid market shifts. AI has revolutionized forecasting by enabling predictive and even prescriptive analytics. Machine-learning algorithms can analyze vast amounts of data—including POS transactions, weather patterns, social media signals, and market trends—to produce highly accurate demand forecasts. For example, Walmart uses AI to integrate internal and external data in real time, adjusting forecasts based on shifting customer behavior and environmental events. This has helped reduce stockouts, minimize overstock, and enhance on-shelf availability across thousands of stores.

2. **Inventory Optimization.** AI enables dynamic inventory management through real-time monitoring and optimization algorithms. Rather than relying on static safety stock levels, AI models adjust inventory levels based on live demand signals, lead times, and risk assessments. Companies in fast fashion, such as Zara and H&M, leverage AI to analyze sales trends and adjust inventory replenishment daily. AI ensures that popular items are restocked while minimizing deadstock. This has not only improved customer satisfaction but also reduced waste and markdowns.
3. **Intelligent Transportation and Logistics.** AI has transformed logistics through route optimization, autonomous decision making, and predictive maintenance. Algorithms continuously evaluate traffic, weather, fuel prices, and carrier performance to select the most efficient shipping routes and modes. For instance, DHL employs AI to power smart routing in its delivery networks, using dynamic data to reassign deliveries in real time. Similarly, UPS's ORION system uses AI to analyze millions of route options daily, saving millions of miles driven and reducing carbon emissions.
4. **Supply Chain Visibility and Risk Management.** One of the most significant advances AI brings is real-time supply-chain visibility. Using AI-enabled platforms and IoT devices, companies can now track goods from supplier to shelf, anticipate disruptions, and proactively respond to risks. Maersk integrates AI with IoT sensors on shipping containers to provide real-time tracking, while tools like IBM's Sterling Supply Chain Suite use AI to simulate disruptions and recommend alternate sourcing or transportation options. This kind of visibility was critical during the COVID-19 pandemic, when companies had to quickly reconfigure supply routes.
5. **Warehouse Automation and Robotics.** AI-driven robotics and computer-vision technologies have revolutionized warehouse operations. Robots equipped with AI can navigate, pick, sort, and package items with remarkable accuracy and speed, reducing dependence on manual labor and improving fulfillment efficiency. For example, Amazon's fulfillment centers use AI-powered robots to move shelves to human packers, streamlining operations and increasing throughput. These intelligent systems adjust their workflows in real time based on incoming order volume and inventory location.
6. **Supplier Collaboration and Sourcing.** AI enables smarter sourcing by evaluating suppliers based on performance metrics, risk factors, ESG compliance, and geopolitical stability. Natural language processing and AI tools scan supplier contracts and news to identify potential disruptions before they occur. Companies like Unilever and Schneider Electric are using AI to improve supplier collaboration by sharing forecasts and inventory data in real time, aligning production schedules and improving resilience through joint planning.
7. **Human–AI Collaboration for Decision Making.** With all the AI capabilities, it is important to remember that AI does not replace human judgment—it enhances it. In sales and operations planning, for example, AI provides simulations and scenario planning, while human teams weigh trade-offs and make strategic decisions. The future of SCM will increasingly be a hybrid of human insight and machine intelligence, where data flows freely, decisions are made faster, and resilience is designed into every node of the network.

Let's look at an example of Procter & Gamble (P&G) that illustrates how AI connects purchasing, operations, logistics, and marketing into a seamless, data-driven supply-chain system.

COMPANY EXAMPLE: Procter & Gamble's AI-Integrated Supply Chain

Procter & Gamble (P&G), one of the world's largest consumer goods companies, has built a highly integrated AI-enabled supply chain. The company uses AI to bridge traditionally siloed functions—purchasing, operations, logistics, and marketing—into a coordinated, real-time, decision-making network.

Here are some examples of how AI insights from one function provide input to another:

1. **AI in Purchasing (Sourcing & Procurement).** P&G uses AI to predict demand trends and align procurement decisions with forecasted needs. For example, if AI detects a spike in demand for baby care products in Southeast Asia, it automatically recommends increased raw material orders from regional suppliers. Machine-learning algorithms assess supplier risk (e.g., geopolitical issues or factory delays) and adjust sourcing strategies proactively.

   🔗 Connection to Operations: These purchasing insights are directly fed into production schedules at factories, ensuring that operations are not disrupted due to shortages.
2. **AI in Operations (Manufacturing & Planning).** Inside P&G's digital factories, AI is used for production planning, quality control, and predictive maintenance. AI determines optimal production quantities based on real-time sales data and supply availability. It simulates different production scenarios to optimize cost, efficiency, and responsiveness.

   🔗 Connection to Transportation: Once products are produced, AI determines the optimal distribution centers to which ship the products, based on inventory levels and projected demand from downstream retail locations.
3. **AI in Transportation (Logistics & Distribution).** P&G leverages AI in logistics to optimize route planning, load consolidation, and shipment scheduling. AI dynamically adjusts routes in response to weather, traffic, and demand patterns, cutting delivery time and emissions. It can predict warehouse congestion and redirect shipments to avoid bottlenecks.

   🔗 Connection to Marketing: AI anticipates where promotions will drive spikes in regional demand and ensures that inventory is pre-positioned accordingly.
4. **AI in Marketing (Demand Shaping & Consumer Insights).** AI models process consumer behavior data from retailers, social media, and P&G's own e-commerce platforms to personalize marketing campaigns. If AI sees that eco-conscious consumers are engaging more with a certain detergent, it triggers localized digital campaigns. It also feeds demand signals back into the supply chain to influence how much of which product gets made and where it is distributed.

   🔗 Full Loop Integration: This consumer information feeds back into purchasing decisions (e.g., sourcing more sustainable materials), closing the loop from consumer demand to raw materials.

For P&G, AI acts as the connective tissue that links purchasing, operations, transportation, and marketing into a synchronized system. As a result, P&G has a top-performing AI-integrated supply chain. It has reduced out-of-stocks by more than 15 percent, lowered logistics costs by 8–10 percent, and shortened product launch cycles. It also improved marketing Return on Investment by double digits.

QUESTIONS

1. How does AI help break down traditional silos between supply chain functions like purchasing, operations, and marketing?
2. What organizational changes might be required to support this integration?
3. What are the potential risks or limitations of relying heavily on AI-driven decision making across the supply chain? How can companies mitigate these risks while still taking advantage of AI?
4. In what ways does consumer behavior data influence upstream supply-chain decisions in P&G's model? Can you think of other industries for which this feedback loop could be similarly powerful?
5. How does real-time AI-enabled decision making improve supply-chain resilience, especially during disruptions or demand spikes? Would this model have helped during the COVID-19 pandemic?

6. What are the ethical or strategic considerations when automating key decisions across purchasing, operations, and marketing using AI? Who should retain final decision authority, and in what situations?

1.4 Supply Chains: A Universal Flow of Value

When we strip down supply chains to their essence, we can see that supply chains are systems that move something of value from origin to consumption through a series of coordinated steps. Traditionally the "something" was a pallet of goods, but in 2025 the object in motion might just as easily be a playlist, a vaccine temperature reading, a cross-border payment, or a line of code. Wherever flows exist—materials, data, cash, energy, creative content, even human expertise—there is a supply chain. AI is rapidly becoming the orchestration layer that senses, predicts, and optimizes all these flows in real time.

We can say that supply chains are a universal flow of value. In turn, AI is the flow orchestrator of all supply chains. AI does not merely optimize supply chains; it enables the very idea of a chain—transforming it into an intelligent network of flows that span molecules, money, media, and moments of customer delight. This is true whether companies are delivering sneakers, streaming symphonies, or settling split-second payments. In a world where everything moves, AI is the orchestrator of flow systems.

AI excels at three tasks common to every flow system. First is sensing and providing visibility. AI can take in high-volume, high-velocity signals—through IoT readings, click-streams, telemetry, invoices—to create a live digital twin. Second is prediction and prescription. This involves forecasting demand, detecting anomalies, simulating scenarios, and recommending actions faster than humans can react. And the third is execution. This involves triggering decisions—such as rerouting, repricing, reallocating capacity—without waiting for managerial approval once guardrails are met. Because these tasks transcend industries, the AI toolkit (e.g., machine learning, knowledge graphs, generative optimization) becomes immediately portable, from a textile factory to a streaming platform.

Table 1.3 shows a cross-industry snapshot of flow systems and how AI adds value. The implications are significant. We need to think of supply chains in terms of flows, not functions. And this

Table 1.3 Cross-Industry Snapshot of Flow Systems & AI

Flow Systems	Product In Motion	How AI Adds Value	Example
Manufacturing & Logistics	Containers, spare parts, finished goods	Predictive maintenance, dynamic routine, digital twin simulations	Maersk and UPS use AI route platforms to cut vehicle downtime and carbon footprint by analyzing billions of daily data points.
Healthcare & Life Sciences	Vaccines and cold-chain integrity	Demand-surge prediction, cold-box temperature anomaly alerts, drone dispatch optimization	AI models anticipating African vaccine stockouts improve fulfillment lead times and reduce spoilage.
Financial Services	Payment instructions, KYC ("know your customer") identity documents, liquidity	AI-extracted invoicing data, fraud classification, real-time credit routing	JPMorgan automates supply-chain payments and is scaling Gen AI use cases across operations.
Transportation & Field Services	Technicians, vehicles, rides	Demand forecasting, workforce scheduling, route optimization	Mobility fleets use AI to align driver hours with predicted ride hotspots, cutting idle miles.
Entertainment & Media (Music)	Audio stems, licenses, royalties	Stem detection, rights tracing, dynamic royalty allocation	Startups like Deezer and Vermillio offer authentication AI solutions and protection of IP.
Streaming & Digital Content	Personalized artwork, trailers, recommendations	Contextual bandits, generative creatives, content delivery network pre-positioning	Netflix's AI continuously tests previews and pushes most-probable content to edge servers before viewers click.

flow transcends all industries and systems. This will require mapping every value stream—whether physical or digital—and treating it as a supply chain. This also means applying the lessons and principles of SCM, from strategic to tactical.

While AI is indeed the orchestrator of supply-chain flows—supply-chain principles apply to these flows regardless of industry. Certainly, there are issues specific to industry sectors. However, basic principles—such as bullwhip effect, scaling, bottlenecks, and others—apply to all supply chains. We will discuss these principles throughout this book and stress that everyone needs to understand the principles of SCM, regardless of the industry they work in.

Being a part of the global economy in the AI-driven post-pandemic era requires that everyone understand SCM, so we ensure that the mistakes and lessons of the pandemic are not repeated.

Let's look at an example of supply chains in the music industry and the role of AI.

COMPANY EXAMPLE: Taylor Swift's "Eras Tour"

Taylor Swift's "Eras Tour" ran from March 17, 2023, to December 8, 2024, spanning 149 shows in 51 cities across five continents. The tour was designed as a career-spanning retrospective, featuring more than 40 songs in 10 acts and lasting around 3 to 3.5 hours each night. More than 10 million fans were expected to attend, generating over $2 billion in revenue and having economic ripple effects for local economies—restaurants and hotels. At its conclusion in Vancouver, it became the highest-grossing concert tour in history, crossing the $2 billion mark with more than 10 million attendees.

However, Taylor Swift's "Eras Tour" was not just a musical event, it was a masterclass in modern SCM fused with cutting-edge AI technology. From ticketing to fashion, stage logistics, hydration, and digital systems, every aspect revealed huge successes and notable supply-chain challenges.

The supply-chain challenge was huge. Let's begin with the massive equipment rollout. More than 90 trucks carried stage sets, lighting, pyrotechnics, instruments, wardrobe, and crew gear across 146 global shows, mirroring a warehouse distribution network. The movement of what is basically a small city involved freight logistics. The international legs of the tour incorporated both sea and air freight: shipping containers abroad and domestic trucking ensured that equipment arrived precisely when needed. This meant that all bulky assets, from stage to stadium, required shipping by sea in containers. Then they were forwarded via trucks to be more cost-efficient and scalable abroad. The tour organizers also implemented JIT scheduling to ensure that the crew, instruments, and sets arrive exactly when needed—minimizing idle days and high costs.

A large production crew, plus 90 truck drivers, transported heavy equipment. Custom-built digital twins and simulation tools helped rehearse staging, crowd flow, and technical timing before each performance. Just consider that for a moment—a digital twin simulation running before each performance! In addition, there was an understanding that people are a critical part of the operation. As a result, the personnel involved received additional financial incentives for their work. The drivers and operators received premium compensation, which helped ensure dependable and punctual delivery, and underscores that this type of operation needs both technology and human commitment.

Technology, however, was the enabler that made all this happen. In fact, the description of the technology sounds more like an advanced manufacturing operation. For example, advanced logistics systems, TMS (transportation management systems) and WMS (warehouse management systems), were used to track multimodal shipments in real time, enabling proactive rerouting around customs delays or port disruptions. Engineering teams used 3D-rendered models and rehearsal software to simulate staging, block movements, and timing sequences before performances. And, as mentioned earlier, digital twins were used for rehearsal.

Then there was onstage technology and real-time response. The stage was made up of a T-shaped configuration with hydraulic platforms, moving blocks, LED wristbands, fireworks, lasers, and high-resolution projection systems. This setup required real-time orchestration managed by custom control systems. It also involved building in what were called mission-critical contingency plans. This meant that if anything went wrong there was a plan. So, for example, when Swift's mic failed mid-performance, crew members executed seamless mid-show tech switches, showcasing tight coordination and rehearsal of failure protocols. These were all rehearsed and planned for.

Let's not forget merchandise and the ripple effects on local supply vendors. Each concert sold roughly $3 million in merchandise, with long queues forming overnight. Local vendors also saw soaring demand for Swift-themed items, including friendship bracelets, sequins, cowboy boots, and rhinestones. This created a type of grass roots inventory expansion. Boutique retailers and crafts suppliers ramped up stocks—beads and sequins saw shortages as Swift-inspired fashion surged. Then there was tech tie-in. Some vendors integrated real-time e-commerce and inventory tools, tapping fan analytics and social media trends to predict and restock rapidly.

For fans, the outcome was an incredible tour with unforgettable performances. However, Taylor Swift's "Eras Tour" was not just a concert tour, it was a grand supply-chain feat. It wove together supply-chain precision, massive freight and production planning, digital fan infrastructure, crowd safety, technical artistry, and local economic stewardship. While it generated seismic excitement—it highlighted systemic supply-chain challenges that are involved in all supply chains. For industries from live events to retail to logistics, the takeaways are clear: 1. Scale systems for peak unpredictability. 2. Break down silos among digital, physical, and service logistics. 3. Plan for the unexpected, and rehearse for failure. 4. Leverage real-time data—from trucks to tweets—to adapt on the fly.

The "Eras Tour" has proved that modern supply chains, when elevated with technology and strategy, do not just move things, they also create seamless experiences and inspire global design.

QUESTIONS

1. What key supply-chain principles were tested during Taylor Swift's "Eras Tour," and how did technology help mitigate or amplify those challenges?
2. In what ways did the tour integrate JIT logistics, and what are the risks and rewards of this strategy in live event operations?
3. How did real-time data (e.g., from wearable devices, crowd metrics, and logistics systems) enhance the tour's responsiveness and performance quality?
4. What sustainability practices were embedded in the tour's supply chain, and how can large-scale events balance economic goals with environmental impact?
5. How did the merchandise supply chain—both official and unofficial—respond to demand spikes, and what lessons can be learned about forecasting and local sourcing?
6. In what ways did the "Eras Tour" blur the lines between a physical product supply chain and a digital experiential one? How might this apply to other industries like fashion, media, or retail?

Chapter Review

Key Points

1. The pandemic showed the world that supply chains are the backbone of the global economy. When supply chains stopped, so did the functioning of the world as we know it.
2. Disruptions and uncertainty have become part of the new normal. AI has become the necessary tool that helps make supply chains agile and resilient.
3. Supply chains are multi-flow dynamic systems of integrated functions that include purchasing, operations, logistics, and marketing. AI capabilities exist for each supply-chain function, but they must be integrated to optimize the supply-chain system.
4. AI is not a stand-alone tool but rather an umbrella of enabling technologies. They include decision tools (e.g., generative AI), processing (e.g., robotic systems), communication (e.g., IoT, blockchain) and integrative platforms (e.g., cloud AI). AI-driven supply chains rely on these tools.
5. A broad view is that supply chains are systems that move anything of value, from boxes and vaccines to a playlist and payment flows.
6. All these flows are AI-orchestrated and have universal supply-chain principles that apply.

Discussion Questions

1. What specific lessons did the COVID-19 pandemic teach governments, companies, and consumers about the importance of supply chains?
2. Can you name a product or service that was severely affected during the early stages of the pandemic?
3. How did global supply-chain disruptions during COVID-19 accelerate the adoption of AI technologies in business operations? What made AI more essential during times of uncertainty? Why is managing a global supply chain considered complex? What factors—such as geography, regulations, or dependencies—make coordination difficult?
4. What are the four core functions of SCM (buy, make, move, and sell), and how do they interact with one another? Why must these functions be viewed as a system rather than separate activities?
5. What are some specific examples of how AI is used in each of the four supply-chain functions (purchasing, operations, logistics, and marketing)? Which function do you think benefits most from AI, and why?
6. How has AI improved supply-chain visibility and responsiveness in real time? Can you think of a business that leveraged this to gain a competitive edge?
7. In what ways does AI help companies prepare for and recover from disruptions, such as natural disasters or pandemics? How does this differ from traditional risk management?
8. What are some of the challenges businesses face when trying to implement AI across a complex global supply chain? Is technology the biggest barrier, or are people and processes more difficult to change?
9. Can AI help reduce the trade-off between efficiency and resilience in supply chains? What does a "smart and resilient" supply chain look like in practice?
10. Looking ahead, do you think AI will make supply chains more localized, more global, or a mix of both? Why? What are the strategic implications of each path?

Case Studies

A Cautionary Tale of Amazon vs. Borders

Consider the fate of Borders. In 1971, the company opened its first store in Ann Arbor, Michigan, when the book industry was a different place. In 2011, 40 years later, the bookstore chain closed its doors.

So, what happened?

Borders fell behind the curve on embracing the digital world. Not understanding that the rules of the game had changed, Borders had outsourced its online bookselling to Amazon. So, any time you visited Borders.com, you were redirected to Amazon. Playing by the old rules made this seem like a smart decision. In the new world, however, there was a problem.

The competitive world Borders lived in was one where booksellers tracked which books sold and which did not. Loyalty programs could help tie purchases to individual customers. That was about it. Then shopping moved online. The ability to understand and track customers changed dramatically.

Online retailers could track every aspect of what customers bought. They could track what customers looked at, how they navigated through the site, how long they hovered over a site, and how they were influenced by promotions and page layouts. They were now able to develop microsegments of individual customers and groups based on endless characteristics. They could then create individually targeted promotions. Then algorithms were developed to predict what books individual customers would like to read next. These algorithms were self-teaching and performed better every time the customer responded to a recommendation. Traditional retailers like Borders simply could not access this kind of information. They could not compete in a timely manner.

And Amazon? In 2007, Amazon used Kindle to digitize books, turning them into a quantified format and capturing reading behavior to refine recommendations. By 2010 machine-learning models optimized everything from warehouse robotics to last-mile routing. Fast forward to 2025: Amazon begins using multimodal AI models to power an AI assistant that reads user-uploaded photos, suggests outfits, and links directly to shoppable items—closing the loop from demand sensing to fulfillment in seconds. Amazon understood that to achieve its goals requires using the latest in technology to manage its entire supply chain in a synchronized manner. Data and technology have been used to coordinate everything from customer orders to fulfillment, inventory management, labor, warehousing, transportation, and delivery.

Unlike Borders, Amazon was continuously evolving and adapting. That is the secret. A key takeaway is that technology is not static and staying current matters. Winning organizations treat technology as a moving target, investing in data infrastructure, experimenting with emerging AI, and reorganizing around rapid learning cycles. The cost of complacency is not just market share; it can be corporate extinction.

Questions

1. What strategic mistake did Borders make by outsourcing its online operations to Amazon? How did this decision limit its future competitiveness? Could this have been avoided, or was it inevitable given the business culture at the time?
2. How did Amazon's early investment in data and technology give it a long-term competitive advantage? Why is building a technology stack in-house often more powerful than outsourcing?

3. In what ways did Amazon's use of AI transform the traditional retail supply-chain model? How does real-time data affect inventory, pricing, and customer engagement?
4. Borders failed to adapt to the rise of e-books and digital retailing. What signals or trends should they have paid attention to earlier? What are some modern examples of companies ignoring similar warning signs today?
5. How does Amazon's use of generative AI (e.g., fashion assistants) demonstrate a shift from selling products to delivering personalized experiences? How should companies think about the evolving definition of customer value?
6. Why is continuous adaptation to technological change more important today than in the past? How does the pace of innovation change the life cycle of competitive advantage?
7. What lessons can today's business leaders learn from the Amazon versus Borders case about investing in AI and digital capabilities? What mind-set or structural changes are required to stay competitive in a rapidly evolving tech landscape?

Transcontinental Movement of Giant Pandas

In October 2024, two giant pandas—Bao Li and Qing Bao—were transported from the Dujiangyan Panda Base in Chengdu, China, to the Smithsonian's National Zoo in Washington, DC. The 19-hour, 8,250-mile journey exemplifies how AI is transforming supply-chain operations across every domain—even the safe, high-stakes movement of live animals.

This case challenges the traditional view of supply chains as pipelines for products, demonstrating instead that in the modern era, *anything*—from vaccines to endangered species—can move through global logistics networks.

And AI applies to all of it.

AI and Smart Logistics Integration

When the Smithsonian's National Zoo partnered with China to transport giant pandas across continents, the task called for more than just careful handling—it demanded precision logistics at a global scale. FedEx, renowned for its expertise in high-stakes deliveries, was selected for this unique mission. Leveraging its long history of transporting endangered species and its investment in smart logistics, FedEx orchestrated the journey using advanced AI technologies to ensure the pandas' safety and comfort, from departure in China to arrival in the United States.

The transcontinental panda transport was managed through FedEx's charter flight operations, guided by AI-enabled planning tools. These systems optimized flight routes, departure timing, and fuel stops by analyzing weather patterns, geopolitical constraints, and crew logistics. Even before embarking on the journey, FedEx used AI-driven simulations prepared for various contingencies. Then, en route, FedEx used AI real-time tracking for full visibility across air, ground, and quarantine checkpoints.

To ensure animal welfare throughout the journey, custom-built crates were outfitted with IoT sensors that continuously monitored temperature, oxygen levels, movement, and stress indicators. This data fed into AI-powered dashboards, allowing veterinarians and handlers to make dynamic adjustments mid-journey, demonstrating how smart logistics can extend beyond freight and parcels to protect precious, living cargo.

Human–AI Collaboration for Precision Care in Transit

Transporting live, endangered animals like giant pandas requires continuous and precise monitoring of their physical condition throughout the journey. The transcontinental relocation of the pandas used a combination of AI-enabled systems, IoT sensors, and veterinary expertise to ensure that the pandas' vitals were tracked in real time. This allowed the transport team to respond immediately to any signs of distress.

The pandas were first acclimated to their enclosures before travel using positive reinforcement techniques, all guided by AI-analyzed behavioral data. Then, during transport, veterinary teams used AI-supported health analytics to track changes in vital signs and anticipate stress-related complications—critical given the length and sensitivity of the 19-hour journey.

The custom-built travel crates—designed in collaboration with zoologists and logistics engineers—were embedded with environmental sensors to monitor:

- Temperature and humidity inside the enclosure (critical for thermoregulation)
- Oxygen and CO_2 levels, to ensure proper airflow during pressurized flight
- Movement patterns, detecting restlessness, pacing, or inactivity

These data points were continuously relayed to a dashboard monitored by the flight's veterinary team and logistics crew in real time. AI tools helped interpret subtle changes and distinguish between normal transport anxiety and critical health warnings.

A senior veterinarian from the Smithsonian National Zoo accompanied the pandas on the journey and used portable diagnostic equipment to check:

- Heart rate using a handheld monitor (during calm periods)
- Hydration via gum color and skin elasticity
- Intake of food and water, ensuring that pandas were eating and drinking normally

The AI-enabled tracking system provided alerts if environmental thresholds were crossed (e.g., a sudden drop in oxygen levels), allowing the vet team to intervene early. All monitoring data was logged and fed into the pandas' medical records, enabling continuity of care once they arrived in Washington, DC. It also informed the 30-day post-arrival quarantine plan, helping zookeepers track any delayed health effects of long-haul air travel.

This real-time, data-informed monitoring process represents the next frontier of supply chains, in which AI and IoT create a responsive care environment, even 35,000 feet in the air.

Questions:

1. How does the use of AI in the panda transport reflect broader trends in real-time supply-chain visibility and adaptive supply chains? How would AI-driven monitoring and routing systems be applied to other high-stakes cargo movements such as pharmaceuticals or high-luxury goods?
2. What are the ethical and operational considerations of using AI and IoT in the care and transport of live animals?
3. How should companies balance technological optimization with animal welfare—and welfare of precious cargo—with public perception?
4. How does AI-enabled planning enhance risk management and contingency planning for sensitive or diplomatic shipments? How might AI have mitigated risks in this case compared with traditional transport approaches?
5. What does this case reveal about the importance of cross-sector collaboration—between governments, conservation groups, and logistics providers—in executing complex global operations? What role does AI play in aligning the goals of and communication among such diverse stakeholders?

Chapter 2

Strategy, Competitiveness & AI "Fit"

In This Chapter You Will Learn

1. that AI is not a "one-size fits all" but should be selected to support and enable a company's unique strategic direction,
2. that business and supply-chain strategy and AI capabilities must be aligned,
3. about individual factors that drive AI technology selection,
4. about competitive priorities and how they are a critical factor in deciding where to make AI technology investments, and
5. how a company selects AI technologies that have the right fit with strategic objectives.

Every company has a unique way it competes in the marketplace.

Look at Walmart, focused on price competitiveness and mass scale, compared with Target, which has a slightly higher price point focused on shopping experience and curated style. In the entertainment space, Netflix focuses on volume-based, bingeable productions, while HBO (Max) is prestige focused with limited releases. In the airline industry, Southwest is low cost, no frills, and short hauls, while Emirates offers luxury experience, long hauls, and premium service.

Each company has their own strategic niche, and the company's AI needs to support this. Successful companies acquire AI capabilities to support their strategy and drive competitiveness—not copy what others are doing. Target cannot blindly copy Walmart, just as Emirates cannot copy Southwest.

When it comes to supply chains things are a little more complicated. This is because a supply chain is a system composed of multiple companies and functions, and to achieve success these multiple functions all need to be aligned. In Chapter 1, we learned that a supply chain has four functions: purchasing ("buy"), operations ("make"), logistics ("move"), and marketing ("sell"). A common mistake is to optimize AI for just one function—say, purchasing or marketing. Successful use of AI requires that the entire system of these functions is optimized, supporting the overall objectives of the business.

In this chapter we look at business and supply-chain strategies that drive decisions about AI to achieve competitiveness. We will also look at critical organizational factors that must be considered to develop a unique AI strategy for each company.

2.1 Turning Strategy into AI "Fit"

AI is not a one-size-fits-all solution. Rather, AI technologies should be selected to support and enable the company's unique strategic direction. Whether a business is focused on operational efficiency, customer experience, innovation, or market expansion, its use of AI must reflect and support those priorities.

Without alignment with company objectives, AI initiatives can become fragmented, underutilized, or even disruptive. AI is also expensive, and financial investments cannot be squandered. Selection of AI must be deliberate and targeted.

Let's see how this is done.

2.1.1 Business Strategy Drives Decisions

There are countless AI applications. *Which AI applications should a company invest in?*

A company cannot afford to implement all AI applications. In fact, a company doesn't need all of them. The key is to be strategic about AI selection. Selection of AI initiatives needs to be focused, manageable, and critically aligned with the company's broader strategy.

As we discussed in Chapter 1, the uncertain and disruptive global landscape is pushing businesses to increasingly embrace AI. However, many organizations fall into the trap of treating AI adoption as a stand-alone initiative rather than integrating it with core business strategy. Understanding the importance of alignment with business goals is critical for achieving meaningful impact.

The importance of aligning investments in AI with overall company goals cannot be overstated. Without it, companies risk wasting resources on AI solutions that fail to scale or solve real problems. However, when AI is embedded within the business strategy it brings efficiency, innovation, and competitive advantage. It also fosters cross-functional collaboration as technical teams work closely with business leaders to design AI solutions that deliver strategic value.

In summary, business strategy sets the direction, and then AI is selected to support these goals. Aligning the two ensures that AI is not just a tool but instead a strategic enabler, helping organizations not only do things better but also do better things.

Let's look at an example of this in practice.

COMPANY EXAMPLE: UPS

UPS's core business strategy is built around logistics efficiency, on-time delivery, and operational excellence. Its goal is to provide reliable package delivery at scale while maintaining cost leadership in a competitive, time-sensitive market. Strategic priorities include reducing delivery times, minimizing fuel costs, optimizing routes, and improving customer service.

UPS's AI strategy centers on using advanced data analytics and improving logistics performance. One of its most prominent initiatives is ORION (On-Road Integrated Optimization and Navigation), an AI-powered route optimization system. ORION uses AI to analyze millions of data points—traffic patterns, weather, delivery time windows, and road conditions—to optimize delivery routes for drivers.

The AI system is not just a tech project; it supports a core business goal: efficiency and cost reduction. By shortening routes, ORION saves UPS tens of millions of gallons of fuel annually and reduces carbon emissions, supporting both cost and sustainability goals.

How do they align?

UPS's business strategy demands efficiency, and the AI strategy delivers it by leveraging technology to reduce delivery miles, fuel costs, and idle time. AI is not used in isolation or for the sake of innovation; AI directly supports the company's long-standing strategic pillars of reliability, scalability, and operational excellence.

The alignment is evident. The business strategy goal for UPS is lower operating costs while improving delivery times. The AI strategy implementation, in turn, uses machine learning (ML) and real-time data to optimize delivery routes. The outcome has been a savings of more than 100 million miles driven annually, translating to hundreds of millions of dollars in cost savings

UPS exemplifies how targeted selection of AI technology, when intentionally designed to serve business objectives, becomes a strategic enabler. Rather than asking, "*What can AI do?*" the company asked, "*How can AI help us meet our business goals better?*" This alignment turns AI from a siloed experiment into a transformational asset.

QUESTIONS

1. How does UPS's ORION system illustrate the strategic value of aligning AI with core business goals rather than treating it as a stand-alone innovation? What risks might arise if AI is deployed without this alignment?
2. In what ways does UPS's AI strategy support both cost efficiency and sustainability simultaneously? Can companies pursue environmental and financial goals without trade-offs through AI?
3. How does real-time data (e.g., traffic, weather, road conditions) enhance the effectiveness of AI in UPS's delivery operations? What role does data infrastructure play in enabling this?
4. UPS uses AI primarily to improve logistics and operational efficiency. Should it also explore AI for customer experience, predictive demand, or other areas of the business? Where else might UPS gain competitive advantage from AI?
5. What lessons can other companies learn from UPS about scaling AI initiatives from pilot projects to enterprise-wide transformation? How important is cultural readiness and change management in that process?

2.1.2 Strategic Alignment

Every company must begin with a long-range business strategy to maintain a competitive position in the marketplace. A **business strategy** is an overarching plan for the company that clearly defines who they are, their long-term goals, how they plan to achieve these goals, and how the company will differentiate itself from its competitors. It defines the overarching goals, competitive positioning, and value proposition of an organization. It answers fundamental questions such as

- What markets should we compete in?
- How do we differentiate?
- Where should we allocate resources?

For example, the Hershey Company has a business strategy of making high-quality, innovative confectionary products in an efficient and sustainable manner with responsiveness to consumer needs.

A **supply-chain strategy** is a plan for the design and ongoing management of all supply-chain decisions that support the business strategy. Consider that the design of a supply chain should differ based on how the company intends to compete in the marketplace. To maintain competitiveness, companies must design their supply chains to be aligned with their business strategy, to satisfy the needs of customers, take advantage of the company's strengths, and remain adaptive.

Let's look at the Hershey Company again. To meet their business strategy, the company has created a supply chain focused on digitizing processes for efficiency, optimizing procurement, and integrating demand planning to reduce waste and improve inventory management. At the same time, they committed to responsible sourcing aiming for 100 percent sustainably sourced cocoa by end of 2025.

An **AI strategy**, on the other hand, is a plan that outlines how a company will use AI technology to achieve its business goals. It includes identifying where AI can create the most value, deciding which technologies and capabilities to invest in, building the right teams and infrastructure, managing risks and ethics, and ensuring that AI initiatives align with the company's mission, culture, and competitive positioning. It drives the adoption and use of AI technology that enables both the business and supply-chain strategies. A good AI strategy ensures that these objectives are achieved. Hershey, for example, uses AI to focus on efficiency and sustainability, such as identifying hidden capacity. AI helped the company identify $35 million worth of capacity in KitKat production.

Every decision made within the organization should support its overarching goals. This is the key to using AI for competitiveness and is shown in Figure 2.1. Without such alignment, individuals across the company or individual departments may pursue their own goals. This would ultimately cause the organization to lose focus, pursue scattered efforts, waste resources, and fall apart.

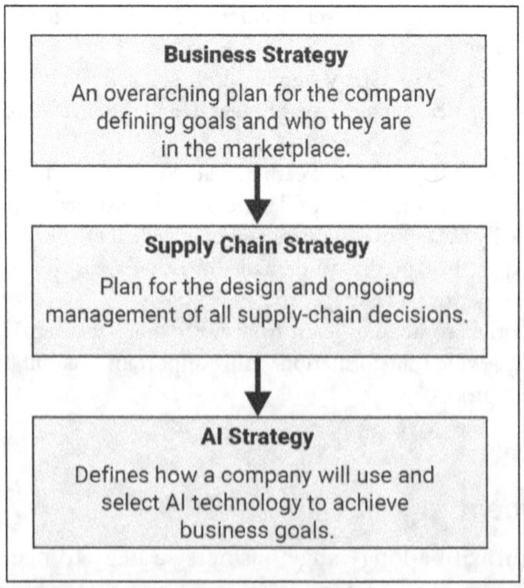

Figure 2.1 Strategic Alignment

Let's look at another example of alignment that works.

COMPANY EXAMPLE: ZARA

Zara is an excellent example of a company that leads in strategic alignment. Zara leads with its business strategy, which is to be a retailer that is "fast, focused, flexible." Zara's overriding goal is to democratize the runway: capture the newest looks, re-create them quickly, and sell them at accessible prices while demand is still peaking. That requires "time-based competition"—collections refresh every two to three weeks, and roughly two-thirds of annual styles drop in-season rather than being planned a year in advance. The brand also purposefully limits production runs to create scarcity and minimize markdowns, sustaining margins that are among the highest in global apparel.

Zara's supply-chain strategy is to use deep vertical integration to lead in "clock-speed," a high-performance tempo. To enable that tempo, Zara keeps the critical path in-house. Design, pattern cutting, and prototyping sit in Arteixo, Spain; 50-plus percent of finished goods are made in company-owned or tightly contracted plants in Iberia, Morocco, and Turkey. This is a short distance—a mere truck ride from its La Coruña and Zaragoza distribution hubs. Short lead times (as little as 15 days from sketch to shelf), twice-weekly store replenishment, and RFID-tagged units that track flow through distribution centers give Zara the agility to pull supply from real-time demand.

Zara's AI strategy is to invest in the AI capabilities that amplify speed and precision. Zara hasn't chased every shiny algorithm; it has cherry-picked AI capabilities that tighten the design-to-delivery loop. For example, Zara has invested heavily, yet strategically, in the following AI capabilities:

- **Trend sensing:** Zara uses natural language processing (NLP) and computer vision to follow more than half a million social posts daily, as well as runway imagery and search logs. As a result, it is able to identify micro-trends (e.g., oversized blazers) days ahead of rivals, directing designers to high-yield ideas.

- **Demand & allocation forecasting:** Zara invested in AI models that blend POS, weather, foot traffic, and event calendars. SKU-store predictions are updated every 48 hours, enabling dynamic purchase orders and has resulted in reduced stockouts.
- **Inventory automation:** Zara uses computer-vision cameras and RFID readers in distribution centers and stores. This has enabled accurate following of real-time stock levels.
- **Price & markdown optimization:** Zara uses AI engines to test price strategies against targets and determine the end-of-season markdown rate.

Being vertically integrated and owning the supply chain has helped Zara. This means the moment its AI tools flag an emerging look, in-house teams can sample, test in limited stores, and ramp up manufacturing within days—something an outsourced supply-chain network can't do. The same data streams through the supply chain, so stores receive the right sizes and colors mid-season, keeping shelves fresh and selling high. Crucially, Zara's tech team codes many models internally and deploys them on its edge-cloud, shortening iteration cycles from months to weeks.

This alignment is shown in Figure 2.2 and has resulted in high performance. Since its AI implementation the cycle time of "concept-to-rack" fell from 35 to 25 days over 2022–2024. Their gross margin held above 57 percent in 2024 despite inflationary headwinds. This is because of lower markdowns due to AI capability.

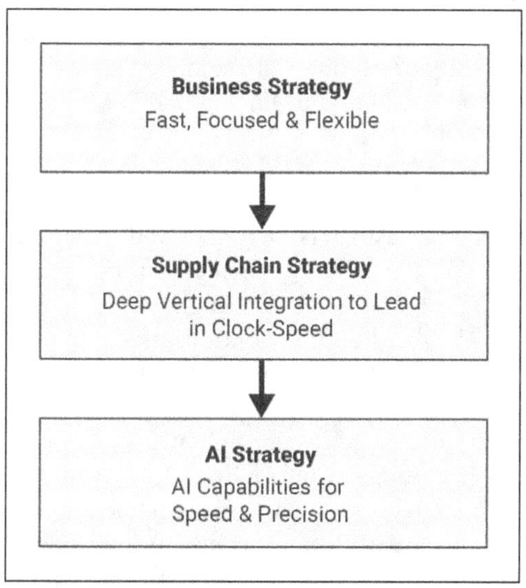

Figure 2.2 Zara's Strategic Alignment

Zara is now piloting generative AI design assistants and customer avatars for virtual try-on, but its litmus test remains the same: Does this algorithm compress lead time or raise inventory velocity? If the answer is "yes," Zara builds or buys; if "no," it waits—proof that technology is a servant to strategy, not the other way around.

Notice the alignment with strategy. That is the key.

QUESTIONS

1. How does Zara's vertically integrated supply chain enable its AI investments to create a competitive advantage that would be difficult for more traditional, outsourced retailers to replicate?
2. Zara uses AI to accelerate trend detection and reduce markdowns. How do these capabilities reinforce its business model of fast fashion, and what risks might arise from relying heavily on AI-driven trend prediction?
3. Compare Zara's selective approach to AI adoption with companies that pursue a more aggressive, broad AI strategy. What are the trade-offs between focus and breadth in digital transformation?

4. In what ways does Zara's AI-powered inventory management support sustainability goals, and how can supply-chain visibility enhance both profitability and environmental performance?
5. Zara tightly aligns its business, supply chain, and AI strategies around speed and responsiveness. How might this model need to evolve as consumer preferences shift toward sustainability, personalization, or slower fashion cycles?

As we can see in the example of Zara, there must be **strategic alignment** between the business strategy and supply-chain strategy. A company's supply-chain strategy should be developed to drive and support its business strategy. Then the investments are made in AI capabilities to support these strategies.

Consider an electronics company that has formulated a business strategy to compete on delivery excellence, such as order-fulfillment time. As a result, the supply-chain strategy may be designed for speed of delivery, although at higher cost. Now, imagine that the company decides to change its business strategy to compete on cost rather than delivery, considering current market competition and customer perceptions of value. If this change in business strategy is not communicated to other supply-chain functions, the supply chain will continue to focus on delivery rather than cost. The company will continue to excel at delivery, while incurring a higher cost and not meeting the new business goals.

Today's most successful companies, as illustrated by Zara, have achieved world-class status in large part due to a skillfully designed supply-chain strategy that is manifested in the design of its supply chain. Then they have invested in AI technologies that support these specific capabilities.

2.3 Factors Driving AI Selection

Every business has different goals, customers, strengths, and constraints. A company focused on high-touch customer service will use AI very differently from one prioritizing automation and cost reduction. For example, a luxury fashion brand might use AI to enhance personalized styling, while a logistics company might focus on route optimization and predictive maintenance. Even within the same industry, companies can have vastly different strategies, and their use of AI must reflect those differences. Copying another company's AI playbook without considering your own context can lead to poor results and missed opportunities.

Consider two retailers, such as Amazon and IKEA.

Amazon follows a business strategy centered on speed, scale, and customer personalization. Its AI initiatives support this strategy, with heavy investment in machine learning for personalized recommendations, supply-chain optimization, voice-enabled shopping through Alexa, and predictive analytics to forecast demand.

In contrast, IKEA has a business strategy focused on affordability, sustainability, and in-store experience. Its use of AI is more targeted: optimizing warehouse logistics, using computer vision

Table 2.1 Factors Driving AI Selection

Factors Driving AI Selection	Company Example
1. Competitive Priorities (cost, time, customer service, innovation)	Amazon (focus on cost, speed, and customer service)
2. Digital vs.Physical Products	Spotify (digital) vs. Ford (physical)
3. Service vs. Manufacturing	Airbnb (service) vs. Toyota (manufacturing)
4. Small vs. Large Firms	Local coffee shop (personalization) vs. Starbucks (scale)
5. Startups vs. Mature Firms	OpenAI (startup) vs. IBM (mature firm)
6. Degree of Regulation (highly regulated vs. unregulated)	Pfizer (highly regulated) vs. Meta (less regulated)
7. Degree of Uncertainty (volatile vs. stable industries)	Tesla (volatile industry) vs. Procter & Gamble (stable)

for inventory tracking, and offering AI-powered design tools to enhance customer planning experiences. While both companies leverage AI, they do so in ways that directly reinforce their distinct business strategies.

Companies need to consider numerous factors in developing their AI strategy. Ultimately these considerations will lead to something unique for each company—and that is the "secret sauce." These factors are shown in Table 2.1 as well company examples that illustrate the factor.

We discuss each of these factors next.

2.3.1 Competitive Priorities

Competitive priorities are the way a company competes in the marketplace. Business and supply-chain strategy and AI technology investments greatly depend on a company's competitive priorities. What a company competes on drives the technology it needs to invest in. A company that competes on cost, for example, will invest in technologies very different from one that competes on innovation.

All successful companies—Amazon, Walmart, Toyota, Nike, McDonald's, and IBM— follow this process. A big mistake is for a company to just mimic others in their industry and use AI that copies their competitors. This does not lead to a competitive advantage; it simply leads to being a follower in one's industry.

There are five primary competitive priorities listed in Table 2.2. A company cannot compete equally on all of these. Rather, a company must have a dominant competitive priority for which it is known in the marketplace. Also, every company has finite resources and needs to know exactly what priority to focus on. Then it can select their AI to support this.

Table 2.2 Competitive Priorities

Competitive Priority	Competition	Label	Example
Cost	Competing on price or cost	Low-Cost Leader	Price tag, discount retailer, dollar store signage
Time	Competing on speed of delivery or response time	Speed Advantage	Express delivery van, stopwatch, same-day shipping
Innovation	Competing on new products, services, or ideas	Creative Disruptor	Futuristic gadget, brainstorming session, tech expo
Quality	Competing on high quality	Excellence Focused	Craftsmanship close-up, quality seal or badge
Service	Competing on high customer service	Customer Champion	Friendly customer support, concierge interaction

Let's look at each of these competitive priorities in detail and see how they drive supply-chain design and AI selection.

2.3.2 Cost Competition

If the company's business strategy is to compete on cost, then both the supply chain and AI strategy must be designed to support this. Companies that compete on cost offer products at the lowest price possible. These companies are either maintaining market share in a commodity market, or they are offering low prices to attract cost-sensitive buyers. Once a company's business strategy has determined that the company will compete on cost, the other functional strategies—including the supply chain and AI strategy—should be designed to support this.

For example, competing on cost requires highly efficient, integrated operations that have **cut costs** out of the system. The supply chain plays a critical role in keeping both product and supply-chain costs down. It may also require going to the least expensive suppliers rather than focusing on high-quality components.

The low-cost supply chain focuses on meeting **efficiency-based metrics** such as asset utilization, inventory days of supply, product costs, and total supply-chain costs. The operation strategy is designed for **product and process standardization**, as are operations of the suppliers. Notice that the supply-chain design would be different had the chosen business strategy focused on something else, such as competing on customization. In that case, cost would be less of an issue, as would be the ability to provide a wide range of customized products.

Dell Computer Corporation has been a model of supply-chain excellence. However, Dell does not compete on cost. In fact, Dell does not claim to offer the least expensive computers but merely customized computers in record time. That is the major difference. Dell's computer prices are within the industry range. This is unlike Walmart, which promises the lowest prices but not special customized products. The important thing is that each company's supply chain is designed to support the chosen business strategy.

Although a company competes on efficiency and low cost, it cannot do so at the expense of service, innovation, or quality if one of these is an element of the business strategy. Focusing on price alone, without considering other aspects of the business strategy, may result in a poor supply-chain strategy.

An example of this is the apparel industry, where manufacturers have typically outsourced production to Southeast Asia. The manufacturers insist on fixed production schedules to keep their costs low. This, however, affects their flexibility and can hurt retailers when demands unexpectedly shift, such as when there is an unexpected surge of demand. This limits retailers in their ability to respond to demand. For many retailers, being out of stock on a regular basis can wreak havoc with customers and erode market share.

COMPANY EXAMPLE: Mattel

In August–September 2007, Mattel pulled about 21 million toys off shelves worldwide—among them Barbie accessories, Sarge die-cast cars, and dozens of Fisher-Price items—after US tests found lead-paint levels far above legal limits and hazardous magnets that could detach and be swallowed. The emergency cost the company an estimated US $110 million in returns, legal fees and extra testing and knocked the share price down by 6 percent in a single trading day. Beyond direct costs, Mattel endured congressional hearings, class-action lawsuits, and the passage of the 2008 Consumer Product Safety Improvement Act, which tightened testing rules for the entire industry.

Like many Western brands in the early 2000s, Mattel had chased the lowest unit cost by outsourcing more than half its production to contract factories in China. Minimal on-site auditing and weak data sharing meant design changes and paint-spec violations slipped through until products were already in consumers' homes. The crisis exposed the danger of treating offshore suppliers as black boxes instead of partners integrated into a real-time quality system.

The recovery meant moving from manual checkpoints to data-driven oversight. Mattel tripled third-party toy testing and opened in-house labs on three continents. They rolled out Global Manufacturing Principles scorecards for every plant. They also introduced barcode-level traceability down to paint lots and began shifting high-risk SKUs to plants in Mexico, Vietnam, and India. In fact, China's share of output is set to drop below 40 percent by the end of 2025, giving Mattel a more diversified footprint.

In June 2025, Mattel announced a multiyear partnership with OpenAI, giving its engineers access to ChatGPT Enterprise and model-building tools. The arrangement covers:

- Generative design—rapid iteration of toy concepts with compliance rules baked into prompts.
- Supplier-risk analytics—large-language-model agents that check inspection reports, tariff changes, and social-media whistle-blower posts to flag anomalies before shipment.

- Interactive quality manuals—voice bots for factory managers that answer in Mandarin or Spanish.

What is different this time?

Post-recall, Mattel put together supplier and shipment data into a single data lake; AI can now see processes in near real time. Mattel also added strict governance. Every AI tool now must pass the company's Product Safety Review Board, avoiding another race-to-the-cheapest scenario. Last, they prioritized resilience over pure cost.

QUESTIONS

1. Can AI truly solve quality problems, or does it simply make governance lapses harder to hide?
2. Describe the details of Mattel's move from reactive recalls to proactive anomaly detection.
3. How might Mattel's partnership with OpenAI reshape power dynamics between brand owner and contract factories? How would you add data transparency and worker upskilling in this partnership?
4. Diversification lowered Mattel's China exposure to less than 40 percent. Could deeper AI visibility have justified staying in a single low-cost hub instead?

2.3.3 Time Competition

Time is one of the most important ways companies compete today. Companies in all industries are competing to deliver high-quality products in as short a time as possible. Amazon, FedEx, LensCrafters, and UPS all compete on time. Customers today are increasingly demanding short lead times and are not willing to wait for products and services. Companies that can meet the needs of customers who want fast service are becoming leaders in their respective industries.

Making time a competitive priority means competing based on all time-related dimensions, such as rapid delivery and on-time delivery. Rapid delivery refers to how quickly an order is received; on-time delivery refers to the number of times deliveries are made on time. When time is a competitive priority, the job of the operations function is to critically analyze the system and combine or eliminate processes to save time. AI is a key technology that has helped speed up processes, including optimizing the use of a flexible workforce to meet peak demand periods and eliminating unnecessary steps in the production process.

COMPANY EXAMPLE: FedEx

"The World on Time" is not just a slogan for FedEx—it is a business strategy hardwired into every part of its global supply chain. Since its founding, FedEx has positioned itself as the premier choice for time-sensitive deliveries. While its 1990s slogan, "Absolutely, Positively Anytime," emphasized reliability, the 2024 version, "The World on Time," reflects the evolution of that promise into a high-speed, data-driven operation powered by AI, automation, and air dominance.

FedEx has time as a competitive priority. It has deliberately chosen speed and reliability as its main competitive differentiator. Unlike rivals such as UPS or regional carriers that emphasize cost efficiency or volume, FedEx prioritizes timeliness—even if it means higher operational costs. This is reflected in its global fleet of more than 680 aircraft, making it one of the largest airlines in the world by fleet size. The ability to control its own air network allows FedEx to operate on tighter delivery schedules, with fewer dependencies on third-party logistics.

To support this business model, FedEx has made AI and real-time data integration central to its logistics ecosystem:

- Sensor-based logistics: FedEx uses RFID, GPS, and barcodes to track more than 16 million packages daily, with real-time location updates accessible to both customers and operations teams.

- FedEx Surround: A launched collaboration with Microsoft, this platform uses AI to proactively detect risks such as weather disruptions, mechanical delays, or civil unrest and reroutes packages dynamically.
- Autonomous and robotic technology: FedEx continues to experiment with AI-powered delivery robots like Roxo and is investing in autonomous tugs and loaders for airport and hub operations, increasing speed and reducing human error.
- Predictive analytics: AI models forecast package volume by region, enabling faster staffing and vehicle deployment decisions, particularly during high-demand events like holidays or weather emergencies.

FedEx has deliberately made strategic trade-offs, prioritizing speed over cost. While FedEx's AI-driven infrastructure enhances delivery accuracy and speed, it comes at a cost. Maintaining its own fleet, investing in cutting-edge technology, and optimizing delivery performance at a global scale means that FedEx does not position itself as a low-cost provider. Its value proposition is clear: *pay more for guaranteed speed.*

Other operational functions—from warehouse layout to customer support—are tightly aligned to support this strategic priority. For instance, sorting hubs operate in near real time, using AI to prioritize urgent shipments. Route planning tools dynamically adapt to traffic and delays to shave minutes off delivery times. As of 2026, FedEx is pushing deeper into digital twins of aircraft and warehouse systems for proactive maintenance and resource allocation.

FedEx offers a compelling case study of how a company can build its entire supply chain, technology, and organizational infrastructure around a singular competitive priority: *time.* AI and data analytics aren't just tools—they are strategic enablers that help FedEx fulfill its brand promise to get the world's goods delivered faster than anyone else.

QUESTIONS

1. How has FedEx's investment in AI and autonomous technology helped maintain its strategic edge in time-based competition? Could it sustain this advantage without controlling its own fleet?
2. FedEx prioritizes delivery speed over cost competitiveness. What are the risks and limitations of this strategy in a world increasingly concerned with sustainability and affordability?
3. Platforms like FedEx Surround use predictive analytics to reroute shipments in real time. How might this approach be applied in other industries where time is critical (e.g., healthcare, food logistics)?
4. How can AI help FedEx strike a better balance between time, cost, and environmental impact as carbon regulations tighten? Should time always take precedence?
5. With growing competition from Amazon's in-house logistics and emerging drone services, how should FedEx evolve its AI and supply-chain strategy to stay competitive over the next decade?

2.3.4 Innovation Competition

Companies whose primary strategy is innovation focus on developing products that the customers perceive as must-haves and thereby pull the product through the supply chain with significant demand. Examples of such companies include Sony and Nike—companies that deliver innovative products customers want. Due to the must-have nature of these products, these companies can typically command a premium price, which is to their advantage because competing on innovation requires a sizable financial commitment. In addition to the supply-chain capability, the real ability to compete on this dimension lies in superior marketing and product development, which is directly related to the supply chain.

Companies that compete on innovation typically have a very short window of opportunity before imitators enter the market and begin to steal market share. They are aware that they must enter the market early with an innovative design. The supply chains of these companies typically focus on two features: *speed* and *product design*. This requires a carefully integrated supply chain that enables collaboration on product design between suppliers and manufacturers. Manufacturers with this type of supply chain have both internal integration between functions and integration externally with suppliers.

Another challenge for innovative supply chains is the ability to *quickly raise production volumes* should demand suddenly increase. An innovative product doesn't accomplish much if the company cannot quickly deliver a large volume of the product to the market to meet demand. Often this cannot be accomplished by one company's manufacturing process alone, and the ability to access production capacity when needed provides a significant competitive advantage. That is why most of these companies, such as Nike, are virtual companies. This means that marketing and product design are their strength, yet they outsource all the other aspects of their supply-chain and manufacturing processes, albeit through a tightly controlled system.

2.3.5 Quality Competition

Competing on quality means that a company's products and services are known for their premium nature. Two important elements of this competitive priority are *consistency* and *reliability*.

Examples of companies known for competing on quality are Mercedes, General Electric (GE), and Motorola. Many aspects of the supply chain are altered when companies compete on quality versus another dimension, such as cost. This includes sourcing of components, as well as the implementation of concepts such as total quality management and Six Sigma throughout the entire supply chain. This also means embedding quality in all aspects of transportation, delivery, and packaging. It is particularly challenging when items are perishable, fragile, or of high value, such as luxury goods.

As supply-chain management is a boundary-spanning activity, an important attribute of competing on quality is **product traceability**. This means that the supply chain can easily trace a product from point of origin in the supply chain, through to the customer, and back down the supply chain in the case of returns. This feature is becoming increasingly important with greater emphasis on security and sustainability.

Consider industries for which this might be of particular importance, such as the pharmaceutical industry, where safety is critical. Another industry is the area of luxury goods, where counterfeiting is a problem, and traceability is very important in the identification of goods.

In 2024, authorities uncovered one of the largest counterfeit luxury goods operations in New York City history, seizing more than $42 million worth of fake designer handbags in a coordinated raid across multiple storage facilities in Manhattan and Queens. The counterfeit items included knockoff versions of high-end brands such as Louis Vuitton, Gucci, Chanel, Hermès, and Prada, many of which were nearly indistinguishable from authentic products to the untrained eye. These goods were being sold through both physical markets and online platforms at a fraction of retail prices, targeting budget-conscious consumers and exploiting the growing demand for affordable luxury.

Unlike in previous years, AI technologies now play a critical role in detecting and preventing counterfeit activity. Tools such as computer-vision models can scan and authenticate products based on micro-pattern recognition, stitching irregularities, and material inconsistencies. Additionally, blockchain-based traceability systems—often enhanced by AI—are being adopted by luxury brands to track a product's origin and verify authenticity throughout the supply chain. As a result, the fight against counterfeiting has shifted from reactive enforcement to proactive, data-driven prevention.

2.3.6 Service Competition

Competing on service means that a company understands the dimensions that its target customers define as high service and has chosen to tailor their products to meet those specific needs. An important aspect of this strategy is that these companies typically build customer loyalty, which can often guarantee continued sales. These companies typically have exceptional order fulfillment systems, with fast invoicing that enables them to be consistent and reliable. These companies typically do not compete on cost and cannot offer the lowest-priced product. However, their target market is one that wants high-quality service and is willing to pay the extra cost.

From an operational viewpoint, companies with superior customer service often avoid unnecessary costs related to expediting orders or the costs of product returns, often faced by other companies. These companies also have a strong ability to segment their customers based on perceived value. This way they can offer services of one type to one market and a modified version to another. For these firms, the relationship between product cost and profitability is well defined. Consequently, highly customized services that are costly to deliver are offered only to customers who meet strict business criteria and tend to be in the high-value segment.

COMPANY EXAMPLE: Apple

Apple has built its global brand not just on innovative products but also on delivering a seamless, high-quality customer experience at every touchpoint. In 2026, this commitment remains a core pillar of its competitive strategy. Whether online, in-store, or through its devices, Apple invests heavily in creating a service ecosystem that reinforces trust, loyalty, and satisfaction.

Apple Stores are designed as experience hubs rather than simple retail outlets. Customers can test products, receive personalized advice, and book hands-on technical help through the Genius Bar. In recent years, Apple has enhanced in-store services with AI-powered appointment scheduling, real-time repair tracking, and in-app concierge guidance, offering a more tailored, efficient support process.

Across digital channels, Apple provides a wide array of customer support options, including 24/7 online chat, phone assistance, AI-driven diagnostics via the Apple Support app, and a robust self-service knowledge base. Its proactive customer care also includes automatic alerts for service needs, remote troubleshooting tools, and next-day device replacement programs through AppleCare.

Despite consistently premium pricing, Apple's focus on service quality—through ongoing software updates, product longevity, and a tightly integrated ecosystem—continues to set it apart. The seamless experience across devices (iPhone, Mac, iPad, Apple Watch) reinforces customer retention and satisfaction. In fact, as of 2026, Apple claims a customer loyalty rate exceeding 90 percent for its iPhone segment alone.

This strategic emphasis on exceptional service over low cost allows Apple to compete not just as a hardware company but also as a lifestyle and support brand, where customers are willing to pay more for reliability, ease of use, and peace of mind.

QUESTIONS

1. How does Apple's emphasis on customer service and experience help justify its premium pricing model? What psychological or practical value does this add for consumers?
2. In what ways has Apple integrated AI and digital tools to enhance its customer support ecosystem? How might these innovations improve service quality or operational efficiency?
3. Apple positions its retail stores as experience centers rather than traditional sales outlets. How does this strategy influence brand perception and customer loyalty?
4. Could Apple's service model be replicated successfully by lower-cost competitors, or is it uniquely tied to Apple's brand identity and ecosystem? What would it take for another brand to compete on service rather than price?
5. Apple has built a tightly integrated ecosystem of products and services. How does this integration affect customer support, and how might it contribute to customer lock-in or long-term loyalty?

2.3.7 Order Winners and Order Qualifiers

None of us can be all things to all people. The same is true for companies.

Successful companies understand that they cannot effectively compete on all dimensions. This is a very important and difficult point for most companies, as they typically want to be good at everything. Companies that succeed are those that understand which dimensions to excel on and then focus their energies on those dimensions.

This does not mean that a company will have poor performance on the other dimensions. In fact, companies must continually trade off one competitive dimension for another. However, this means that a company should have merely satisfactory performance, or stay within the norms of the industry, on those strategically less important dimensions.

Two important concepts that companies need to monitor are **order winners** and **order qualifiers**:

- **Order winners** are those characteristics that win the company orders in the marketplace.
- **Order qualifiers** are those characteristics that will qualify the company to be a participant in a particular market.

This means that a company just needs to stay within the industry standard on its order qualifiers to ensure that it is on par with competitors. It doesn't need to spend resources to excel on those dimensions. However, when it comes to order winners, that is where it should excel.

COMPANY EXAMPLE: Tesla

In the competitive and rapidly evolving electric vehicle (EV) market, companies must meet a set of fundamental expectations just to be considered viable players. These are known as order qualifiers—basic criteria that allow a brand to enter and remain in the market. For Tesla, meeting these expectations is essential. Consumers in the EV space demand vehicles that can travel more than 250 miles per charge, deliver high safety standards, and offer reliable, consistent performance. Tesla delivers on all fronts. Its models frequently outperform competitors in driving range, often exceeding 350 miles on a single charge, depending on the model. In addition, Tesla consistently earns top crash-test ratings and incorporates advanced safety features, including driver-assist systems.

These attributes—range, safety, and reliability—do not win Tesla customers outright, but they prevent the brand from being eliminated from consideration. If Tesla failed to meet these minimum thresholds, consumers would likely rule it out before ever evaluating its more innovative features. Thus, while these order qualifiers are necessary, they are not sufficient to differentiate the brand.

Where Tesla truly excels is in its **order winners**—the unique features and capabilities that give it a competitive edge and influence consumer preference. A standout example is Tesla's Autopilot and Full Self-Driving technology. While still evolving, these systems represent some of the most advanced autonomous driving capabilities on the market, providing Tesla with a clear technological advantage. Additionally, the Supercharger network—a proprietary, global fast-charging infrastructure—offers Tesla owners a level of convenience and speed that most competitors have yet to match.

Tesla's brand image also plays a critical role in winning orders. Positioned as futuristic, tech-forward, and environmentally conscious, the brand resonates with a loyal and passionate consumer base. Tesla has cultivated an identity that transcends cars, often seen as a lifestyle choice or technological statement, rather than just a transportation option. Further strengthening its position are over-the-air (OTA) software updates, which allow Tesla vehicles to receive improvements and new features long after purchase—an innovation that traditional automakers are only beginning to adopt.

These order-winning factors—autonomous capabilities, charging infrastructure, brand loyalty, and software-driven improvements—are what set Tesla apart in a crowded and increasingly sophisticated EV market. By building on strong order qualifiers and pushing boundaries with its order winners, Tesla not only competes, it leads.

QUESTIONS:

1. How do Tesla's order qualifiers enable it to compete in the EV market, and what would happen if the company failed to meet these basic expectations?
2. Which of Tesla's order winners do you think has the most long-term competitive advantage: Autopilot, the Supercharger network, brand image, or OTA updates? Why?

3. Can traditional automakers catch up to Tesla's order-winning features, or are these advantages too deeply embedded in Tesla's business model and technology stack?
4. How does Tesla's ability to push software updates to vehicles over time reshape customer expectations in the automotive industry?
5. What lessons can other companies—inside or outside the automotive sector—learn from how Tesla uses order qualifiers and order winners to shape its competitive strategy?

Recall that a supply chain must be designed to support the business strategy, so alignment between the two is a top priority. It should be remembered, however, that strategy is a dynamic rather than a static process. As such, strategies change over time, and supply chains must adapt accordingly.

2.4 Other Factors

In addition to competitive priorities, many other organizational factors must be considered in the selection of AI. These are listed in Table 2.1 (p. 32). Let's look at these in more detail here.

2.4.1 Digital versus Physical Products

A key factor that differentiates a firm's AI approach is whether its offerings are digital or physical products and services. A firm can structure its offerings in three ways: by providing only digital products and services, only physical products and services, or a combination of both. Each approach significantly influences how a company selects technologies to operate and deliver value. This shapes a company's overall strategy, influencing the management of resources, coordination of operations, and fulfillment of customer demands.

Companies that focus exclusively on digital offerings, like Netflix or Spotify, prioritize technological investments such as software development, data analytics, and cloud infrastructure to deliver seamless digital experiences. In contrast, firms that provide only physical products, like Toyota or IKEA, emphasize efficient supply-chain management, logistics optimization, global sourcing, and Lean manufacturing to streamline their operations.

For companies operating both digital and physical channels, such as Amazon or Walmart, the integration of these offerings enhances omnichannel supply-chain visibility. By connecting brick-and-mortar stores with digital platforms, companies can leverage real-time data to automate processes, improve demand forecasting, and optimize inventory management across their supply-chain networks. This integration enables better coordination between procurement, distribution, and customer service, ensuring a more agile and responsive supply chain.

The digital versus physical factor directly influences how companies allocate resources for technology. Companies excelling in both areas utilize advanced technologies to streamline operations and ensure smooth integration across their supply chain. By effectively managing the complexities of digital and physical offerings, companies can maintain efficiency while addressing diverse customer demands.

Recognizing the strategic value of this distinction, and investing in the appropriate technological infrastructure, allows companies to optimize operations and enhance agility in adapting to evolving market trends and customer needs.

2.4.2 Service versus Manufacturing

Service and manufacturing firms have fundamentally different business models, operations, and customer interactions, which means their needs for AI differ significantly. Service firms typically deal with intangible products, real-time human interaction, and high variability in customer experience. In contrast, manufacturing firms focus on producing tangible goods through repeatable, often standardized processes. These differences shape how each type of firm uses AI to optimize performance, reduce costs, and enhance value.

In service firms, AI is often used to enhance customer experience, personalize offerings, and automate knowledge work. For example, companies like Bank of America use AI-powered virtual assistants like Erica to help customers manage finances, answer questions, and flag suspicious transactions. In healthcare, Mayo Clinic uses AI to support diagnostic decision making and personalize patient treatment plans. These service applications rely heavily on NLP, predictive analytics, and recommendation systems. AI in this context must be adaptive, responsive, and capable of interpreting human nuance, often in real time.

On the other hand, manufacturing firms use AI to drive efficiency, consistency, and quality in physical production environments. AI in this space often powers predictive maintenance, robotic automation, supply-chain optimization, and quality control. For instance, GE uses AI to monitor equipment health and predict machinery failures before they occur, reducing downtime. Tesla uses computer vision in its factories for automated assembly and defect detection. These "use cases" require AI to integrate with physical systems (like IoT sensors), process vast amounts of structured data, and function in environments where precision and consistency are critical.

Ultimately, while both types of firms benefit from AI, the specific capabilities required, (human-centered adaptability versus process-centered automation) are shaped by the nature of the business. Understanding this distinction helps firms invest in the right kind of AI solutions to gain competitive advantage in their respective industries.

2.4.3 Small versus Large Firms

When developing a supply-chain strategy, it is important to understand the company's strengths and weaknesses to realistically determine what it can and cannot accomplish. The company must play to its strengths.

For many supply-chain companies, a large source of power comes from their sheer size. Companies like Amazon, Walmart, Toyota, and Home Depot are large firms with equally large market clout. These companies are sometimes called **supply-chain masters** and can strong-arm their suppliers into compliance, especially when adopting AI. However, not all companies have this advantage.

When designing its supply-chain strategy, a company needs to understand how much influence or power it has in the marketplace. Large companies have the advantage of being able to buy larger quantities of goods and command lower prices due to quantity discounts. Due to their size, these companies can impose their own processes and rules on suppliers and customers.

Large companies can also impose the supply-chain structure they want. They can impose their own process rules on suppliers and maintain a high degree of control over the entire supply chain. In the auto industry, for example, manufacturers stipulate that if a supplier's delay in delivery shuts down the production line, the supplier can be subject to a substantial penalty due to revenue lost while the line is down. Not every company can make this kind of demand.

Smaller companies have, however, developed their own supply-chain strategies. For example, when developing their supply-chain strategies, smaller companies should consider that size is relative. Few companies are large on a broad, global scale. When the scope narrows to a particular market segment or region, companies can find that they are really not that small. In that case, they can find select suppliers within that market segment with whom to work and develop strategies to compete within that market. When it comes to AI, these companies can also turn to many free options currently available. For example, the free version of OpenAI continues to be sufficient for many small firms.

2.4.4 Startups versus Mature Firms

Another factor that shapes a firm's AI technology approach is its classification as a startup or a mature company. Given their differing stages of growth, unique considerations must be considered.

Startups are newly established organizations that operate with smaller teams and less structured processes due to their early stages of development. These companies can take on more risk, often relying on limited capital or seeking multiple funding rounds to support growth. Startups can face challenges when building new relationships with suppliers and customers because they lack an established track record. However, their ability to move fast allows for quick pivots that improve innovation.

A startup's approach to innovation includes experimentation and learning from failures. They often use agile methodologies to iterate quickly on ideas and prototypes. Multiple iterations allow for a fast response to market feedback. By embracing new innovative technologies, they can gain a competitive advantage over mature companies in the market. This makes innovation a key part of their survival.

Mature firms, by contrast, are characterized by having a large workforce, well-established protocols, and corporate culture with extensive operations. These companies have had a long history of success, and they often dominate their industry, granting them greater access to resources and enabling them to have long-term relationships with customers and suppliers. However, due to their size, they likely have a more bureaucratic process, resulting in slower decision making.

When companies reach a mature stage, their growth typically slows down as their focus changes from rapid expansion to maintaining stable and profitable operations. As a result, mature firms are often less motivated to adopt new technologies and pursue innovation. Leaders in mature firms are often resistant to change and reluctant to invest money and resources in new technologies. Focusing on optimizing existing operations and altering current products is seen as less risky than pursuing innovative ideas, resulting in a slower adoption of new technology.

2.4.5 Degree of Regulation

The degree of regulation in an industry plays a critical role in shaping how businesses adopt and implement AI technologies. Highly regulated industries must balance innovation with strict compliance, data security, and ethical considerations, which can slow the pace of AI integration or require customized solutions that align with legal frameworks. In contrast, less regulated industries often have greater flexibility to experiment, iterate, and scale AI-driven initiatives quickly.

In highly regulated industries like healthcare, finance, and defense, AI adoption is often constrained by concerns about transparency, accountability, and risk. For example, UnitedHealth Group uses AI to assist with claims processing and patient risk prediction. However, it must also ensure that its algorithms are interpretable, unbiased, and compliant with HIPAA regulations. Similarly, JPMorgan Chase applies AI for fraud detection and algorithmic trading, but it operates under strict scrutiny from financial regulators to avoid systemic risk or consumer harm. In these industries, risks are high, and businesses must invest heavily in governance frameworks, explainable AI, and third-party audits.

In contrast, companies in less regulated sectors, such as retail or entertainment, face fewer barriers and can more rapidly deploy AI to personalize customer experiences, optimize operations, or drive engagement. For example, Netflix uses AI to recommend content based on viewer preferences without needing to navigate complex regulatory oversight. Amazon leverages AI across its supply chain, pricing algorithms, and voice assistants like Alexa, experimenting freely with large volumes of customer data to optimize results. These companies can prioritize innovation and customer-centric metrics without the legal constraints found in more tightly controlled sectors.

The regulatory environment acts as both a gatekeeper and a guide for AI use in business. Companies in regulated industries must take a more cautious, deliberate approach, often partnering with regulators to ensure compliance. Meanwhile, those in less regulated fields can serve as test beds for innovation, pushing the boundaries of what AI can do before similar technologies are adopted more cautiously in higher-stakes contexts.

2.4.6 Degree of Uncertainty

Degree of uncertainty is the extent to which future events, outcomes, or conditions affecting a business cannot be predicted or controlled. It is an important factor guiding technology investment. High uncertainty industries, such as the technology sector, are subject to fast changing environments with market trends or competition that can shift quickly. This environment requires the ability to be agile and to react and adapt to constant changes. High uncertainty industries may find it risky to be locked into a specific technology. They need to be flexible, and technology must be scalable in order to adapt to different scenarios in a quick manner.

Tech companies are an example of firms that invest in AI-driven systems or cloud-based infrastructure that allow for development in the long run, as well as adaptation as changes occur. Stable industries, in contrast, characterized by low uncertainty can have the ability to focus on longer term technological investments. Whereas stable industries can commit to longer term technologies, that may be more robust and focus on cost efficiency and reliability. For example, a utility company does not need constant evolutions in their technology; rather, they focus their resources on long-term contracts that prioritize consistency in their power grid to their customers.

COMPANY EXAMPLE: Moderna

Let's look at the pharmaceutical firm Moderna to illustrate the degree of uncertainty within an industry and how it affects a firm's technology approach. Although the pharmaceutical industry is normally viewed as stable, the biotech segment faces higher volatility due to long and uncertain research anddevelopment (R&D) cycles, shifting regulations, and competitive pressures to innovate rapidly. Moderna's rapid adaptation during the COVID-19 pandemic exemplifies how a biotechnology firm, operating in a volatile segment of a stable industry, leverages flexible technology.

Moderna focuses on mRNA technology, a revolutionary tool to develop new vaccines and treatments. Flexibility in their mRNA platform led the company to change with rapidly changing circumstances such as new viral variants and shifting health regulations across the globe during the COVID-19 pandemic. The company's agility translated into stock market success during the pandemic. This underlines the volatility and rapid growth that Moderna went through as opposed to that of traditional pharmaceutical firms.

Moderna's use of mRNA technology is inherently flexible and scalable, which is critical in an industry marked by uncertainty. While the more traditional development of vaccines usually takes a number of years, mRNA allows for rapid design and production once the genetic sequence of a virus has been ascertained. Moderna's mRNA platform allowed it to cut down development time, which is a key advantage in volatile market conditions.

Moderna's cloud-based architecture made it possible to scale R&D activities quickly. This has made it possible for Moderna to use AI to model, test, and develop vaccine candidates at previously unattainable speeds. Additionally, these technologies have made it easier to work in real time and on a global scale with manufacturers, research institutions, and regulators—a crucial illustration of how choosing technologies strategically can make all the difference in a volatile environment.

QUESTIONS

1. How did Moderna's mRNA platform and cloud-based architecture give it a strategic advantage in navigating the uncertainty of the COVID-19 pandemic? Could traditional pharmaceutical companies have responded as quickly without similar platforms?
2. In what ways does operating in a volatile segment of a traditionally stable industry (like biotech within pharmaceuticals) change a firm's approach to technology investment? Should companies in other stable industries plan for volatility in a similar way?

3. Moderna's use of AI and cloud technologies accelerated R&D and global coordination. How might this model be applied to other sectors that rely on collaboration and speed, such as automotive or consumer electronics?
4. How does the flexibility and scalability of mRNA technology challenge the traditional long-cycle drug development model? What risks might arise from such accelerated innovation?
5. What can Moderna's experience teach business leaders about selecting technologies not just for efficiency but also for resilience in unpredictable environments?

2.5 Selecting the Right AI

AI is not a single product but rather a portfolio of capabilities—vision, language, forecasting, optimization, generative creation—each with different data needs, deployment models, and organizational effects. The most successful companies treat AI selection as a strategy exercise: they start with **why** (competitive priorities) and **where** (industry context, firm attributes, regulatory landscape), then decide what kind of AI and **how** to deploy it.

We looked at multiple factors that drive AI selection. These had to do with context, essentially creating contextual filters that refine the choice of AI. These are shown in Table 2.3, which itemizes the core questions companies should ask and showcases AI examples. The table synthesizes the key factors and illustrates them with AI in practice.

Table 2.3 Factors Driving AI Selection

Competitive Factor	Core Question	AI Fit In Practice
Competitive Priorities	Which capability – *Cost, Time, Quality, Service* – creates advantage in our market?	Map each priority to an AI sweet spot (e.g., computer vision for zero-defect quality; generative design for innovation)
Cost Competition	Can AI lower unit cost or overhead cost faster than rivals?	Mattel is rolling out OpenAI tools both inside product-design studios and across back-office functions to offset rising supply-chain costs while keeping toy prices competitive.
Time Competition	How do we win on speed?	FedEx Surround combines SenseAware ID sensors with ML to predict disruptions and reroute in real time, protecting on-time-delivery promises for healthcare and e-commerce shippers.
Innovation Competition	Who invents the next big thing?	Generative-AI platforms let R&D teams prototype thousands of variants overnight, shrinking the idea-to-market cycle.
Quality Competition	Can AI drive defect rates toward zero?	Manufacturers pair edge-vision models with predictive-maintenance analytics to catch micro-defects and equipment drift long before they reach the customer.
Service Competition	Does service excellence create "stickiness"?	Apple is implementing a ChatGPT-style tool ("Ask") for AppleCare advisers and expanding on-device Apple Intelligence so support feels instantaneous and privacy preserving, turning service into a differentiator.
Order Winners & Qualifiers	Which AI features move a customer from "will consider" to "must buy"?	Tesla's first fully autonomous Model Y delivery demonstrates how bespoke AI chips and real-time perception are no longer nice-to-have qualifiers but headline order winners in EVs.

Putting it all together requires considering many factors simultaneously.

How can a company do that?

- *First*, start with clarity of strategy. It is important to identify the one or two competitive priorities that truly matter in the market segment. Not everything is equally important.
- *Second*, score AI options against the factors listed. Use the listed factors as a checklist, considering issues such as deployment environment, data availability, governance, talent, and risk appetite.

- *Third*, run pilot-portfolio experiments. For example, they can test multiple AI tools aligned to each priority. Then they can measure outcomes on cost, time-to-value, and adoption friction.
- *Fourth*, scale selectively. Look at pilot experiments that easily convert priorities into measurable business value and eliminate the experiments that don't. Pilots are very important in filtering the AI technologies that work.
- *Fifth*, continually recalibrate. Competitive priorities and regulatory landscapes evolve as do order winners and qualifiers. It is important to treat AI selection as an ongoing strategic capability, not a one-off purchase.

In summary, AI decisions need to be anchored in *what* wins in the market, *who* the company is, and *how* it operates. Then organizations avoid "AI for AI's sake" and invest instead in technologies that amplify their unique sources of competitive advantage.

Chapter Review

Key Points

1. AI is not a one-size-fits-all technology. AI should be selected to support and enable a company's unique strategic direction.
2. Business and supply-chain strategy must be aligned, and then AI technology should be selected to align with these strategies and support them.
3. A number of contextual factors need to be considered in AI technology selection. These factors will drive the financial investment in the technology and ensure that it is implemented across supply-chain functions in an integrated and aligned manner.
4. One of the most important set of factors in selecting AI technology are competitive priorities. This is how a company competes in the marketplace. Companies that compete on cost, for example, have different AI technology needs than those that compete on speed, customer service, or innovation. A company has to know what its competitive priorities are and design a supply chain that aligns with that.
5. Important factors to consider include digital versus physical products, service versus manufacturing, small versus large firms, startups versus mature firms, degree of regulation, and degree of uncertainty. There may be other factors that companies need to consider as well. Together this portfolio of factors should be mapped out to identify the key constraints in AI selection.

Discussion Questions

1. Why is it dangerous to treat AI as a one-size-fits-all solution, and what are the consequences of selecting AI without considering a company's strategic direction? Can you think of examples where a misaligned AI implementation caused problems?
2. How can businesses ensure that their AI investments reinforce, not undermine, their overall business and supply-chain strategies? What processes or teams should be involved in aligning these efforts?
3. What role do competitive priorities play in determining which type of AI a company should adopt? How would AI differ between a company competing on cost versus one competing on innovation?
4. How might AI implementation differ between a service-based company and a manufacturing-based company? What technologies or priorities might be unique to each?
5. Startups and mature firms often operate under very different constraints. How should this difference shape their approach to AI investment and integration?
6. To what extent should regulation influence the choice of AI tools in industries like healthcare, finance, or transportation? How can companies balance compliance with innovation?
7. How does the level of uncertainty in a company's market affect the kind of AI technologies it should adopt? Should more volatile industries prioritize flexibility over efficiency?
8. How can companies map and evaluate the full portfolio of factors (product type, size, regulation, etc.) to make a well-informed AI decision?
9. What are the risks of investing heavily in AI without fully integrating it across supply-chain functions? How can companies prevent AI from becoming a siloed or underused asset?
10. Looking ahead, how might emerging contextual factors—such as geopolitical tensions, climate risk, or consumer privacy concerns—shape future AI selection strategies? Should companies proactively adapt their AI strategies in anticipation of these factors?

Case Studies

Walmart

Walmart, the world's largest retailer, has long been at the forefront of leveraging technology to streamline operations and gain a competitive edge. With more than 240 million customer visits per week and more than 1.5 million transactions processed every hour, the company generates and manages vast quantities of operational and consumer data. Walmart's ability to harness this data through AI, ML, and advanced analytics has made its supply chain one of the most sophisticated and efficient in the world. Below are some of the key features of Walmart's AI-driven supply chain.

AI-Driven Supply-Chain Optimization

Walmart employs AI at virtually every stage of its supply chain—from sourcing and inventory management to last-mile delivery and customer personalization. The company's integrated technology platform ingests data in real time from stores, e-commerce channels, suppliers, warehouses, and external sources. This centralized system allows Walmart to

- forecast demand with high precision using AI-powered predictive models that account for seasonality, local trends, historical sales patterns, external variables such as weather conditions, regional events, and even competitor promotions;
- dynamically adjust pricing and inventory levels to optimize for margins and availability. (for instance, if weather forecasts predict a heatwave, Walmart can preemptively stock more water, fans, and summer apparel in affected regions while adjusting prices accordingly);
- simulate disruptions such as supplier delays, port congestion, or natural disasters using digital twins and scenario modeling (these simulations help Walmart create robust contingency plans and maintain supply-chain resilience).

Retail Link: A Strategic Collaboration Tool

A cornerstone of Walmart's supply-chain strength is its Retail Link platform—an internal, proprietary system that provides more than 100,000 suppliers with direct access to real-time sales, inventory, and performance data. This system, developed in the 1990s and continuously enhanced since, enables unprecedented transparency and collaboration between Walmart and its vendors.

Through Retail Link, suppliers can

- monitor how their products are performing in stores across various regions,
- track inventory turnover and detect stockouts or slow-moving items,
- analyze trends and adjust production or delivery schedules proactively, and
- benchmark performance against peers and receive AI-generated insights for improvement.

By democratizing data and empowering suppliers with actionable insights, Retail Link reduces inefficiencies and helps Walmart maintain in-stock levels while minimizing excess inventory.

Warehouse Automation and IoT Integration

At the operational core of Walmart's supply chain is its distribution centers and fulfillment hubs, where automation and IoT technologies play a critical role. AI-driven warehouse management systems (WMS) coordinate tasks such as picking, sorting, and packing.

The WMS technologies include

- IoT sensors that provide real-time visibility into inventory movement, temperature-sensitive goods, and equipment health;
- edge computing devices that process data locally at warehouses to enable low-latency decisions, such as rerouting a shipment or initiating predictive maintenance on malfunctioning equipment; and
- robotic automation for repetitive tasks, improving speed and accuracy while reducing human fatigue.

Advanced Logistics and Last-Mile Delivery

Walmart has significantly upgraded its logistics capabilities by integrating AI into its transportation management systems. These tools optimize delivery routes based on real-time traffic, weather, and order volume. They also allocate fleet resources dynamically to reduce fuel consumption and emissions. And they utilize autonomous delivery vehicles, drones, and smart lockers in select markets.

Customer-Centric Personalization

AI at Walmart isn't just about operations—it's deeply embedded in the customer experience. Walmart's personalization engines analyze browsing behavior, purchase history, location data, and event triggers to deliver tailored product recommendations on its app and website; adjust local assortments based on regional demand—for example, stocking certain snacks or apparel that perform well in specific states or communities; and predict surge demand patterns, such as the well-known spike in Strawberry Pop-Tarts before hurricanes, thus allowing for preemptive restocking at stores in affected regions.

Walmart's AI-driven supply chain exemplifies how digital transformation can deliver efficiency, agility, personalization, and sustainability at scale. Retail Link fosters supplier collaboration, IoT and edge computing enhance operational intelligence, and predictive models keep Walmart one step ahead of consumer demand and logistical challenges.

In 2025, Walmart began deploying generative AI to further enhance product search and discovery. By interpreting natural language queries and browsing patterns, these models can infer customer intent and offer more relevant, customized product bundles or suggestions. These capabilities not only keep Walmart competitive but also position it as a trailblazer in the future of intelligent retail.

QUESTIONS

1. What are the key advantages of Walmart's integrated data platform in managing its global supply chain?
2. How does Walmart's use of AI in demand forecasting improve both customer satisfaction and operational efficiency?
3. What role does AI play in Walmart's ability to personalize the customer experience? Can this be a competitive differentiator?
4. Walmart uses generative AI to better understand customer intent in searches. How might this technology reshape the future of online retail?
5. In what ways does Walmart's AI strategy reflect alignment with its broader business goals?
6. How does Walmart balance automation (e.g., Robotic process automation, AI-powered robots) with the human workforce in its supply chain operations?
7. What are the potential risks or downsides of relying heavily on AI and predictive analytics in supply-chain management?
8. How might Walmart's AI-enabled supply-chain strategy differ from that of a smaller retailer with fewer resources?
9. What ethical considerations arise from Walmart's extensive data collection and analysis practices, especially related to customer behavior?

Anduril: Startups versus Mature Firms

Founded in 2017 and headquartered in Costa Mesa, California, Anduril Industries is a defense technology startup redefining how military and security operations are conducted. The company develops a range of cutting-edge systems, including unmanned Aerial Systems (drones); autonomous vehicles for land, sea, and air; and autonomous mission-control software.

At the heart of these products is the Lattice OS, an AI-powered operating system that integrates data, recognizes threats, and autonomously coordinates system responses. Lattice enhances situational awareness while reducing human workload, making military decision making faster, smarter, and less labor intensive.

The Role of AI in Defense

Anduril's leadership embraces a bold vision: warfare must evolve through rapid iteration and intelligent automation. In a "Bloomberg Technology" interview, CEO Palmer Luckey emphasized that AI-powered systems can replace what once required large teams of electronic warfare experts and expensive, hardware-heavy deployments. This approach promises faster deployment, smarter operations, and lower overhead.

The company's value proposition is simple but powerful: leverage AI to deliver faster, more flexible, and more cost-effective defense solutions, giving the United States and its allies a crucial technological edge.

Winning in a Legacy-Dominated Market

Historically, the US defense industry has been dominated by a few entrenched players like Lockheed Martin, Boeing, Raytheon, and Northrop Grumman. These companies benefit from long-standing government relationships, deep regulatory know-how, and massive infrastructure. An Axios article notes that serious new entrants to the defense market have been virtually nonexistent for the past 50 years.

Anduril, however, has broken through these barriers. It exemplifies the possibilities of small startups with great vision and AI. With revenues growing from $236 million in 2022 to $420 million in 2023, much of the company's growth is attributed to rapidly secured government contracts. This achievement, sustained through both the Trump and Biden administrations, reflects Anduril's ability to align with bipartisan defense priorities by emphasizing technological innovation over political positioning.

Competing with Defense Giants: Raytheon versus Anduril

In 2025, Anduril pushed into new domains, including space and collaborative combat aircraft (CCA). One of its major initiatives is developing AI-powered drones that can work in tandem with piloted US Air Force aircraft, highlighting a future of human-machine teaming.

This places Anduril in direct competition with Raytheon, a legacy powerhouse known for missile defense systems, advanced radar technology, battle-tested infrastructure and a deep compliance with government regulations.

While Raytheon operates on tried-and-true systems, Anduril brings a tech-first approach. It has developed systems designed to adapt in real time, such as rapid data processing and integration. Minimal human intervention startups like Anduril are showing that speed and flexibility can outmaneuver scale and tradition.

The Pandemic Stress Test

The COVID-19 pandemic disrupted supply chains worldwide, and the defense sector was no exception. Raytheon weathered the storm better than most due to its extensive supplier networks, financial reserves, and experience operating in conflict zones. However, its reliance on global logistics and hardware-intensive projects still led to serious delays in critical programs such as missile systems and military aircraft. Anduril faced a very different reality, thanks to its software-first infrastructure, localized and modular supply needs, and AI-driven systems. The company not only avoided major delays but also thrived. Demand surged for unmanned systems and intelligent surveillance tools, precisely because these solutions didn't depend on human operators or complex physical infrastructure.

The Shift Toward AI and Autonomy

The contrast between Raytheon and Anduril highlights a fundamental shift in defense priorities:

- From hardware to software
- From scale to speed
- From manual control to autonomy

Generative AI and autonomous systems are becoming the new foundation of defense capability. These tools enhance operational speed, decision-making accuracy, and resilience, particularly in modern conflict environments where real-time adaptation is critical.

A New Paradigm in Defense

Traditional contractors like Raytheon are constrained by legacy systems and long procurement cycles. Meanwhile, Anduril—and companies like it—are building a new paradigm:

- digital first
- AI powered
- rapidly deployable

This transition is accelerating. From battlefield logistics to next-generation drones, agility and autonomy are becoming the price of entry for global defense leadership.

Conclusion: The New Guard Is Rising

Anduril exemplifies how startups are disrupting legacy industries through innovation, speed, and bold vision. While mature players like Raytheon continue to dominate key defense segments, they face growing pressure to modernize or risk obsolescence. Anduril's ascent shows that in defense, as in tech, the future belongs to those who move fast, think differently, and harness the power of AI.

Questions

1. What factors have enabled Anduril to successfully enter and scale within the highly regulated and traditionally closed US defense industry?
2. How does Anduril's use of AI differentiate it from legacy defense contractors like Raytheon, and what are the strategic implications of this difference?
3. To what extent does Anduril's success reflect a broader trend of startups disrupting traditionally government-dominated industries through tech innovation?
4. What risks and ethical concerns arise from the growing use of AI and autonomous systems in warfare, and how should companies like Anduril address them?
5. How might Anduril's agile, software-first approach influence the future procurement strategies of government defense agencies?
6. In what ways could Anduril's expansion into space and CCA reshape the competitive dynamics between traditional defense giants and emerging players?
7. Given Anduril's rapid growth in both the Trump and Biden administrations, how politically insulated is the company's business model, and what factors contribute to that resilience?
8. How did the pandemic expose structural weaknesses in traditional defense firms like Raytheon, and what can legacy companies learn from Anduril's experience?
9. What are the potential long-term advantages and limitations of Anduril's reliance on venture capital funding versus the traditional defense contractor model?

Chapter 3

Forecasting in the Age of AI

In This Chapter You Will Learn

1. that forecasting is about predicting the future,
2. that forecasts drive supply-chain decisions and identifying risks,
3. the basic rules of forecasting,
4. about the process of forecasting every organization should follow,
5. about different types of forecasting methods, including AI models, and
6. about collaborative forecasting across supply chains.

Forecasting is much more than predicting demand. It is about predicting anything.

Have you ever hosted a party and found yourself short on food,or visited a restaurant only to hear that the chef's specials have already sold out? Do you remember shortages of toilet paper and hand sanitizer during the COVID-19 pandemic? These situations reflect a simple truth: *When we fail to predict future needs accurately, we end up unprepared. This is the essence of poor forecasting.*

Consider the case of Ben & Jerry's, the iconic ice cream brand. To enhance their forecasting capabilities, they invested in AI business intelligence tools. Each pint of ice cream is now stamped with a unique tracking code and monitored through an Oracle database that records everything—from ingredient sourcing to final sale. This gives Ben & Jerry's end-to-end visibility and enables them to track the performance of each flavor, sales trends by region, and customer demand patterns over time. For instance, if Chocolate Chip Cookie Dough starts trending upward in Texas while Cherry Garcia holds steady in New York, Ben & Jerry's can adjust production and distribution accordingly. The result is a responsive and efficient supply chain, based on real-time forecasting data.

Ben & Jerry's is not alone. Companies like Procter & Gamble, General Electric, Lands' End, and Red Robin Gourmet Burgers also invest heavily in AI-driven forecasting tools. Their goal: to anticipate demand more accurately, reduce operational risk, and seize new business opportunities before competitors do.

In an increasingly complex and uncertain business environment, having AI-driven forecasting tools is not just helpful, it is a strategic necessity.

3.1 Forecasting Basics

3.1.1 What Is Forecasting?

Forecasting is the process of predicting the future. Any time we make a prediction or estimate a future event, we are forecasting. We are always making forecasts even though we may not think of it in a formal way.

In daily life, forecasting might look like

- predicting if it will rain this afternoon,
- estimating how long it will take to commute,

- anticipating grocery prices next week,
- guessing when the new iPhone will launch, or
- predicting if you will get a raise this year.

Whether it is predicting next week's weather or estimating next quarter's sales, forecasting is an essential activity in both our personal lives and professional decisions. In fact, we forecast more often than we realize—formally and informally. It is something we do instinctively in our personal lives. But in business and supply chains, the stakes are much higher. Every strategic and operational decision, from how much inventory to hold, to which new markets to enter, to how many employees to hire, relies on an underlying forecast. When that forecast is wrong, the consequences can ripple through the organization, leading to excess inventory, stockouts, missed sales opportunities, production delays, and wasted resources. We constantly make decisions based on what we expect to happen—and business operates the same way.

In the business world, forecasting extends beyond just predicting product demand. It is used to anticipate a wide range of critical variables that influence decision making:

- seasonal sales trends (e.g., demand for pumpkin pies at Halloween)
- supplier availability and reliability
- risks of shortages, like a dip in coffee supply due to drought
- legislative changes and their potential market effects
- labor needs and hiring decisions
- capital investments and resource planning

Forecasting is involved any time a business makes a decision about the future, whether formalized with data models or done through managerial intuition.

In supply chains especially, forecasting plays a foundational role. Accurate demand forecasting helps companies align production with customer needs, optimize logistics, reduce costs, and improve service levels. Poor forecasting, by contrast, introduces uncertainty and risk, often resulting in either shortages that frustrate customers or overstock that ties up capital and storage space. In today's fast-moving markets, where consumer preferences and global conditions shift rapidly, companies must leverage AI and data to forecast with greater precision.

COMPANY EXAMPLE: Walmart

Walmart has an AI-driven forecasting process that exemplifies how forecasting drives competitive advantage and economic efficiency. Walmart has thousands of stores globally and millions of Stock Keeping Units (SKUs), so managing demand fluctuations is a monumental task. It could not be done without AI.

The key to forecasting is the use of advanced predictive analytics tools that process vast amounts of data. This includes weather patterns, local events, historical sales, and real-time inventory levels, to forecast demand at the store level. For instance, before a hurricane, Walmart's systems anticipate increased demand for items like bottled water and flashlights and automatically increase stock levels in affected regions. This not only improves customer service but also reduces the financial risk of understocking or overstocking.

Walmart's forecasting accuracy—boosted by AI and machine learning—was particularly critical during the COVID-19 pandemic. When global supply chains were strained and demand patterns shifted unpredictably, Walmart used scenario-based forecasting models to adjust procurement and logistics in real time. This responsiveness helped avoid empty shelves and protected market share, while many competitors struggled with disruptions.

Walmart also integrates AI into its omnichannel operations, forecasting demand not just at physical stores but also across its e-commerce platforms, curbside pickup, and delivery services. This level of forecasting precision has allowed Walmart to maintain high service levels while keeping inventory costs under control.

QUESTIONS

1. How does Walmart's use of AI in forecasting create a competitive advantage, particularly during high-risk events like natural disasters or global pandemics? How does predictive accuracy translate into operational resilience and customer satisfaction?
2. What role does data variety—such as weather patterns, local events, and economic indicators—play in enhancing the accuracy of Walmart's forecasts? Why is it important to integrate diverse data sources, and how do they improve forecasting relevance?
3. How did scenario-based forecasting help Walmart respond more effectively than competitors during the COVID-19 pandemic? Explain how real-time adjustments and adaptive planning models can mitigate supply-chain disruptions.
4. What are the challenges and limitations of forecasting across Walmart's omnichannel operations, and how does AI help overcome them? Identify how forecasting needs differ between brick-and-mortar stores and digital channels.
5. In what ways could smaller retailers adopt forecasting strategies similar to Walmart's, even without the same scale or technological resources?

3.1.2 Forecasting versus Planning

Forecasts drive all business decisions, and this process is called planning. **Planning** is the process of deciding on specific actions as a response to what the forecast predicts will happen.

When we look at a weather forecast and anticipate rain, we may plan by carrying an umbrella. Similarly, a business may order more items to stock in anticipation of higher demand, such as higher ice cream sales in the summer. Have you ever gone to a restaurant and been told that they are sold out of their specials or gone to the bookstore and found that the books are on backorder? These are examples of poor planning as a result of poor forecasts. Planning for any event requires a forecast of the future. Whether in business or in our own lives, we make forecasts of future events. Based on those forecasts we make plans and take action.

In business, we often confuse planning with forecasting. Business managers often say something such as: "Our forecast is to produce 6,000 units in the first quarter." This statement blurs the two concepts. In this example, *the plan* is to produce 6,000 units, not the forecast. The plan is what we are going to do, whereas the forecast is a prediction of what will happen.

Is it possible to influence the forecast? Of course it is, especially through promotional campaigns. However, we need to be clear on what is the forecast of the future, independent of our actions. We need to separately understand the plan, which may in fact include our influence. Without separating these we cannot understand each of these variables and be able to understand our true performance in the marketplace.

Forecasts drive plans. Plans are then made in response to the prediction of the future made by the forecast. This is shown in Figure 3.1.

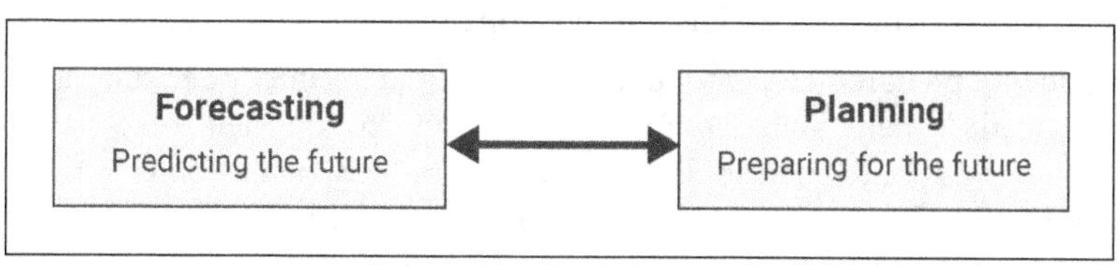

Figure 3. 1 Forecasts Drive Plans

Planning requires organizing resources in anticipation of the forecast. It is especially critical in supply chains as typically the entire supply-chain network shares a forecast and begins planning in anticipation.

Planning involves the following decisions:

1. **Scheduling existing resources**. For a business and its supply chain to be competitive it must use its current resources in the most efficient way possible. This includes the current production process, transportation, labor, facilities, and capital. An important aspect of planning is deciding how to best utilize existing resources. Consider an example such as the production of the well-known candies M&Ms, by Mars, Inc. M&Ms come in standard colors, with the company offering seasonal colors during holidays, such as Halloween or Christmas. The production facility operates at capacity. Producing candies in a new color means that other colors will be temporarily eliminated or made in smaller quantities. For M&Ms this is an important decision, as customers have color preferences so forecasts of desirability and demand are critical to allocate resources accordingly.
2. **Determining future resource needs.** In addition to efficiently utilizing current resources, organizations must determine what resources are going to be needed in the future. These decisions depend on forecasts of emerging market opportunities, new technology, new products, and competition. Let's look at another chocolate example, in this case the Hershey's corporation. The company has been investing in sustainable processes and has changed its packaging to be biodegradable. The company has forecast this to be an emerging opportunity and is changing processes accordingly. Its forecast is driving its plans.
3. **Acquiring new resources.** It takes time for companies to acquire new facilities, new technologies, new equipment, or expand to new locations. Plans must be made well in advance and procedures to acquire new resources and capabilities put in place well ahead of time. For example, in 2025, forecasts of global tariffs pushed companies to realign their entire supply chains and move to new locations.

3.1.3 Forecasts Drive Supply Chains

Forecasting is a cornerstone of supply-chain management, driving decisions in procurement, production, inventory, logistics, and customer service. Unlike forecasting in isolated business functions, supply-chain forecasting requires multitier visibility, real-time data integration, and cross-functional collaboration to align supply and demand efficiently.

Every major decision—what products to launch, how much inventory to carry, which markets to enter, or how to allocate capital—relies on assumptions about future events.

Poor forecasting can lead to

- missed revenue opportunities,
- overstocked or understocked inventory,
- inefficient labor and resource planning, and
- strategic missteps that can harm long-term competitiveness.

Forecasts influence procurement, production, logistics, and inventory decisions across the entire supply-chain network. When companies in a supply chain develop their forecasts independently, they often base their estimates on the demand of their immediate customer rather than on the final end consumer. This fragmented view leads to mismatches between actual demand and supply, as each firm optimizes against a different and potentially distorted signal.

Take the example of Apple and Taiwan Semiconductor Manufacturing Company (TSMC), which supplies Apple with advanced chipsets. Apple's forecast for iPhone or MacBook demand guides its orders for chip volumes, which TSMC then uses to plan its wafer production schedules—often months in advance due to the long lead times. If Apple and TSMC were to make independent forecasts without communication, discrepancies would likely arise, leading to shortages or excess inventory. However, Apple and TSMC engage in collaborative planning and real-time data sharing, ensuring that both companies align production and inventory levels with true consumer demand.

Historically, collaborative forecasting, planning, and real-time data sharing were pioneered by companies like Dell and Intel. Dell's build-to-order model required tight alignment with suppliers like Intel to ensure timely delivery of customized PCs. Today, many tech and retail companies have adopted advanced supply-chain visibility tools and integrated-planning systems that automate and enhance this type of collaboration.

Independent forecasting within the supply chain often gives rise to what is known as the **bullwhip effect**, a phenomenon in which small fluctuations in consumer demand result in progressively larger swings in inventory and orders as one moves up the supply chain. This happens when each company forecasts, stocks, and replenishes independently. The result is greater volatility in upstream orders, leading to excessive inventory buffers, more frequent stockouts, inefficient capacity utilization, and misallocated working capital.

To mitigate the bullwhip effect, many companies now employ collaborative planning, forecasting, and replenishment (CPFR) frameworks. Retailers like Walmart and suppliers such as Procter & Gamble share point-of-sale (POS) data, inventory levels, and promotional calendars, enabling synchronized forecasting and leaner operations. Similarly, Amazon uses real-time analytics and AI to dynamically adjust demand forecasts and communicate those signals throughout its supplier network.

Today, collaborative forecasting is supported by digital tools, AI, and shared data platforms and has become essential to ensure alignment with supply-chain partners. These tools are not limited to large companies and are equally essential for small firms.

COMPANY EXAMPLE: BrewCo—Local Coffee Roaster's Forecasting Challenge

BrewCo is a small, locally owned coffee roaster that supplies freshly roasted beans to regional cafés, farmers' markets, and a growing online customer base. Ahead of the winter holidays, BrewCo forecast a 40 percent increase in demand based on last year's holiday sales, social media engagement, and early preorders. Acting on this forecast, the company ramped up production, ordered higher volumes of specialty beans from its Colombian supplier, and increased packaging inventory.

However, a combination of factors—unseasonably warm weather, delayed promotional campaigns, and unexpected competition from a national brand—resulted in demand rising only 10 percent As a result, BrewCo was left with excess inventory of roasted beans that had a limited shelf life. The overstock led to spoilage, write-offs, and cash flow strain from unsold stock and prematurely paid supplier invoices.

Worse, the warehouse space became clogged, delaying the arrival of new green bean shipments for January blends. This caused ripple effects in their production schedule and damaged their ability to fulfill new wholesale contracts promptly. The forecast error not only created immediate financial losses but also hurt BrewCo's reputation with some key partners, who began seeking more reliable suppliers.

This example illustrates how a relatively small forecasting mistake—exacerbated by slow market signals—can disrupt the entire supply chain for a small business, from sourcing and production to distribution and customer satisfaction.

QUESTIONS

1. How could BrewCo have improved its demand-forecasting process to better anticipate the seasonal surge in sales? Consider what types of data BrewCo might use (e.g., past sales, weather trends, social media activity) and whether they should adopt qualitative, quantitative, or AI-based forecasting models.
2. What are the risks of over-ordering or under-ordering ingredients for a small company like BrewCo, and how can those risks be mitigated? Think about perishability, working capital constraints, and supplier lead times. What supply-chain strategies (e.g., safety stock, vendor-managed inventory) could help?
3. In what ways could collaboration with upstream suppliers have helped BrewCo align supply with expected demand more effectively? Explore the role of supplier relationships, shared data, and joint planning in reducing mismatches between forecast and supply.

4. How does the bullwhip effect apply to BrewCo's situation, and what could be done to prevent it from escalating in future supply-chain cycles? Discuss how small errors in demand planning can ripple upstream and the mechanisms that can either amplify or dampen those effects.
5. Should BrewCo invest in digital forecasting tools or continue relying on manual spreadsheets and intuition? What are the trade-offs in terms of cost, accuracy, and scalability?

3.1.4 Forecasting in Turbulent Times

The current global supply-chain environment has become increasingly turbulent, characterized by persistent uncertainty, volatility, and complex interdependencies. This instability is the result of a confluence of geopolitical tensions, shifting trade policies, labor shortages, climate-related disruptions, pandemics, and rapid technological changes. In such an environment, the ability to forecast is critical.

Given this volatility, businesses can no longer rely on static, historical data to drive decision making. Instead, dynamic and adaptive forecasting has emerged as a vital tool for managing risk and maintaining resilience across supply chains. This is where AI and machine-learning (ML) forecasting tools become essential.

Why is the global supply-chain environment so uncertain?

- **Geopolitical Tensions and Trade Disruptions:** Escalating trade wars, tariffs, sanctions, and regulatory shifts, such as those seen between the United States and China, have disrupted long-standing supply routes and sourcing strategies. For example, the US CHIPS Act, aimed at reducing dependence on Asian semiconductor manufacturing, has created ripple effects across the electronics supply chain globally.
- **Post-Pandemic Recovery Challenges:** While the COVID-19 pandemic has subsided, its impact continues to reverberate. Port congestion, elevated freight rates, and uneven recovery timelines across regions have created persistent bottlenecks. Global inventories are still misaligned with demand, and consumer behavior has shifted unpredictably in its wake.
- **Climate Change and Environmental Disruptions:** Increasing frequency of extreme weather events—such as floods in Germany, wildfires in Canada, or typhoons in Southeast Asia—has made certain supply routes unreliable. Climate-driven regulatory shifts, such as carbon border-adjustment mechanisms in the EU, further complicate sourcing decisions and cost structures.
- **Labor and Capacity Constraints:** Labor shortages, both skilled and unskilled, in logistics, warehousing, and manufacturing have strained operational capacity. Additionally, the shortage of containers and lack of diversification in freight options have exposed the fragility of global logistics networks.
- **Technological Shifts and Digitization Gaps:** While digital transformation presents opportunities, the uneven adoption of technologies like AI, IoT, and blockchain means that some players are better equipped to adapt than others. This digital divide introduces asymmetry in visibility and responsiveness across the chain.

Why is forecasting critical for risk management in SCM?

In this volatile context, forecasting is a core element of strategic risk management. Accurate and adaptive forecasting enables firms to

- **Anticipate Disruptions:** Companies can monitor signals across geopolitical, economic, and environmental indicators to forecast potential supply-chain shocks.
- **Manage Inventory Strategically:** Forecasting helps optimize safety stock levels, avoiding both stockouts and costly overstock situations.
- **Improve Supplier Coordination:** Collaborative forecasting across suppliers enables proactive communication and alignment, especially during shortages or unexpected demand surges.

- **Enable Scenario Planning:** By creating multiple forecast scenarios (e.g., best case, worst case), firms can prepare contingency strategies and allocate resources accordingly.
- **Support Financial Planning:** Forecasting informs cost projections, pricing decisions, and investment in capacity or new sourcing strategies.

A striking example of forecasting for risk management was Toyota's approach during the global semiconductor shortage. Unlike many automakers that faced severe production halts, Toyota was better prepared due to its investment in scenario planning and forecasting systems. After experiencing supply-chain shocks during the 2011 Fukushima earthquake, Toyota restructured its forecasting and risk management practices, maintaining a strategic stockpile of semiconductors and closely collaborating with tier-1 and tier-2 suppliers.

In 2021, when global demand for consumer electronics surged during the pandemic, diverting chip supplies away from automakers, Toyota managed to maintain production for a longer period than many competitors. While it eventually faced constraints, its initial resilience was a testament to robust forecasting and risk planning.

3.2 Principles of Forecasting

Forecasting models differ in their complexity, data requirements, and methodology, ranging from simple trend analysis to advanced AI-powered predictive analytics. However, there are several foundational principles that apply to virtually all forecasting approaches. Understanding these principles helps businesses set realistic expectations and develop more resilient supply-chain and operational strategies.

1. Forecasts Are Never Perfect.

Forecasting the future is inherently uncertain. Even the most sophisticated models cannot account for every unexpected variable, such as sudden geopolitical conflicts, pandemics, or natural disasters. Therefore, forecasts should be viewed as probabilistic estimates, not exact predictions. The primary goal is to reduce error over time, continually improve forecast accuracy, and use the forecast as a guide for better decision making, not as an absolute certainty.

Example: In 2020, many auto manufacturers faced forecast errors due to the sudden drop in car sales during COVID-19 lockdowns, followed by a sharp rebound in demand. Companies like Ford and Toyota had to rapidly revise production plans. Those with more flexible supply chains weathered the shifts better than those relying on fixed forecasts.

2. Aggregate Forecasts Are More Accurate Than Individual Forecasts.

Forecasts tend to be more accurate when applied to groups or categories of items rather than individual SKUs. This is due to the principle of error cancellation; fluctuations in demand for individual items often balance out when viewed in aggregate. This is known as the law of large numbers. For example, forecasting total sales for the entire men's apparel category is generally more accurate than forecasting sales for a specific size and color of a T-shirt.

Example: Target forecasts overall monthly demand for its grocery segment more accurately than it does for individual SKUs like a specific brand of pasta sauce. This helps them plan space and replenishment cycles more effectively, even if individual items vary in popularity week-to-week.

3. Forecasts Are More Reliable Over Shorter Time Horizons.

Short-term forecasts tend to be more accurate than long-term forecasts. The further into the future a forecast extends, the more variables and uncertainties come into play. This is because short-term data is less likely to be disrupted by major market changes or behavioral shifts. In contrast, long-term forecasts must account for higher levels of uncertainty due to evolving consumer preferences, economic cycles, and disruptive technologies. For instance, predicting demand for the next month is typically more accurate than projecting demand two years into the future.

Example: A fast-fashion retailer like Zara forecasts demand for its seasonal collections just a few weeks ahead based on recent sales and social media trends. This short-term, data-driven strategy enables rapid inventory turnover and minimizes markdowns, in contrast to long-term forecasting models used by traditional apparel brands.

4. Forecast Accuracy Depends on Data Quality and Relevance.

Forecasts are only as good as the data that informs them. Inaccurate, outdated, or irrelevant data leads to distorted outputs—what's known as "garbage in, garbage out." Accurate forecasts require timely, clean, and structured data sources that reflect real-world demand drivers. This is especially important in AI/ML systems, where data biases can lead to significant errors at scale.

Example: Amazon relies heavily on real-time and historical data for demand forecasting, including web traffic, regional sales trends, and delivery times. If the data were inaccurate—say, due to a glitch in inventory records—it could trigger false stockouts or over-ordering. To prevent this, Amazon continuously audits and refines its data inputs.

COMPANY EXAMPLE: Nike

Nike has significantly advanced its demand-sensing and forecasting capabilities by integrating AI and ML technologies. These innovations have enhanced forecast accuracy, optimized inventory management, and bolstered supply-chain resilience.

First, Nike uses an AI-driven demand-forecasting process. Building upon its data-driven forecasting foundation, Nike now employs AI to analyze a multitude of data sources, including

- historical sales data,
- real-time POS information,
- digital engagement metrics,
- weather patterns,
- market trends and signals, and
- social-media sentiment.

This approach is comprehensive and enables Nike to predict demand with greater precision. This, in turn, allows for dynamic adjustments in production and distribution strategies to meet consumer needs effectively.

Second, the integration of AI has led to notable improvements in forecast accuracy. Nike has achieved a forecast accuracy rate of approximately 75 percent, contributing to a reduction in inventory holding costs of an estimated $1 billion annually. This heightened accuracy minimizes the risks of overstocking or stockouts, ensuring that products are available where and when consumers demand them.

Third, Nike's AI-powered demand-sensing capabilities facilitate real-time analysis of consumer behavior and market conditions. This agility allows the company to respond swiftly to demand fluctuations, optimizing inventory levels across various regions and channels. By leveraging technologies such as RFID and IoT, Nike maintains visibility into its supply chain, enabling proactive decision making and efficient resource allocation.

Fourth, Nike has achieved strategic integration between its short- and long-term forecasts.

Nike's forecasting strategy emphasizes the integration of short-term demand sensing with long-term planning. By aggregating data across product families and ensuring high-quality inputs from both digital and physical channels, Nike aligns its operations with actual market demand.

Nike's investment in AI for demand forecasting exemplifies a forward-thinking approach to supply-chain management. The company's ability to accurately predict consumer demand and adjust its operations accordingly positions it as a leader in leveraging technology for operational excellence and customer satisfaction.

QUESTIONS

1. How has Nike's use of AI and ML improved its demand-forecasting accuracy? What implications does this have for inventory management and customer satisfaction?
2. What types of data inputs are most valuable in Nike's AI-driven forecasting model? How do these inputs enhance responsiveness to shifting market conditions?
3. Nike integrates both short-term demand sensing and long-term planning. What are the risks and benefits of this dual-forecasting approach, especially in highly volatile markets?
4. In what ways can Nike's AI-based forecasting contribute to sustainability goals, and how might reducing overproduction and waste create a competitive advantage?
5. What challenges might Nike face in maintaining or scaling its AI forecasting systems across global markets, especially in regions with less digital infrastructure or data availability?

3.3 The Forecasting Process

3.3.1 Steps in the Forecasting Process

Why do we need a forecasting process when we just want to forecast?

The reason is simple. It has been shown that following a process that has been well established will ensure forecasts with better accuracy. Otherwise, we will just be generating meaningless numbers. Understanding this process and the underlying principles will generate better forecasts.

The key to generating good forecasts is to follow a well-established forecasting process. In the eagerness to quickly generate forecasts, there is a temptation to rely too much on technology, automated systems and software, and to cut corners when it comes to the basics. The basics can seem tedious and boring. However, the basic principles of forecasting have been well established over many decades. This is analogous to a person trying to get in physical shape through drastic calorie cuts versus following a sensible plan of eating right and exercising. The latter will always produce better and longer-lasting results. The same is true with forecasting.

To get good forecast results we must follow a documented process that has been shown to lead to success. This is a process we will follow regardless of what is being forecast, whether predicting the weather or forecasting the sale of avocados at the grocery store.

This process is the foundation of forecasting with seven steps, as shown in Figure 3.2. Let's look at each one of the steps in detail.

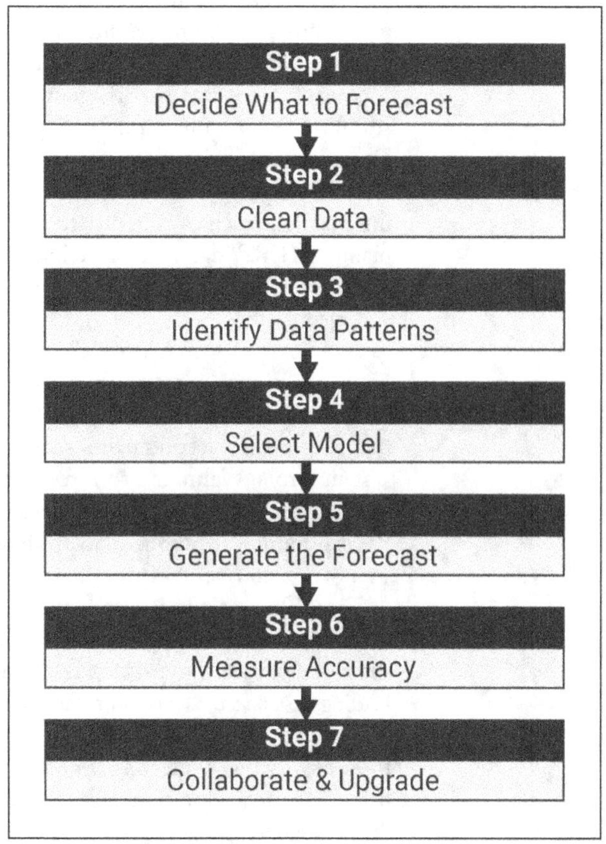

Figure 3.2 Steps in the Forecasting Process

3.3.2 Deciding What to Forecast

The first step is deciding what to forecast. This may seem overly simple, but it actually requires thought. There are multiple issues to consider. AI can serve as a guide, but you need to be the expert here.

Remember that we are generating a forecast to answer a question. One simple example in business may be asking whether we want to forecast sales or total demand. These may seem the same, but they are not. Company sales is likely only one part of total demand. The data and process we use would be different, as well as the insights these forecasts would provide, and different entities in the organization may need the different forecasts.

A company the author worked with offers an example of the challenge of identifying what to forecast. The company was struggling with correct inventory levels and believed that they needed to better forecast demand. However, upon further analysis, they realized that the actual problem was variation in supplier lead times. So rather than forecasting demand, the company began forecasting lead times, which led to large improvements.

This is a reminder that forecasts are generated to help plan for the future. To do so, we have to decide what forecasts are actually needed to guide the plan.

Another issue is to consider the time period to be covered by the forecast, such as whether the observation is continuous or at an instant of time. For example, a delivery of goods from a supplier is an instant of time—say, when a shipment arrives. The opposite is true of demand for goods. In retail, for example, sales are spread out more uniformly over a period of time. This will have an effect on the time period over which you want to make your forecast.

Yet another important decision is deciding on the units of the forecast. For example, units may be dollars versus boxes versus pallets. Remember that units will be used for comparison of performance over time, so this decision is important.

3.3.3 Clean Data

Most companies have access to an abundance of data. However, we have to remember that a forecasting model is based on the quality of the data, and the results will only be as good as that data. The adage "garbage in, garbage out" applies here. This means that we must ensure that the data used is credible and represents the future. With today's access to data, this is more important than ever. Data must be cleaned and checked for accuracy.

Clean data—data that is accurate, complete, consistent, and timely—is an essential prerequisite for effective forecasting. When data is dirty—has errors, missing values, inconsistencies, or biases—it can lead to costly errors in forecasts that affect the entire supply chain.

This is especially complicated, since data is gathered from a wide range of sources. This includes internal data, such as sales orders, inventory records, supplier performance, and enterprise resource planning (ERP) data. It also includes external data, such as POS data, market trends, weather patterns, and macroeconomic indicators. Data cleaning ensures the removal of errors, missing values, and inconsistencies that could distort the forecast.

For example, Walmart combines POS data from thousands of stores with supplier lead times and promotional calendars to forecast product needs by region. To ensure the accuracy of these forecasts, Walmart invests heavily in data cleaning processes. This includes automated systems that flag and correct anomalies such as duplicate transactions, missing values, or outliers caused by system errors. Walmart also integrates ML tools to identify inconsistent data patterns and align inputs from disparate systems, such as supplier feeds and weather data, into a standardized format. By maintaining rigorous data hygiene, Walmart improves forecast reliability and ensures that supply decisions are based on high-quality, real-time information.

Bias. One major concern is bias in data collection or interpretation. Bias in data occurs when the information used to train forecasting models is not representative of the real-world conditions the model is meant to predict. In AI-driven forecasting, biased data can severely distort outcomes, leading to inaccurate or misleading predictions. For example, if past sales data overrepresents certain regions, products, or customer segments, the model may learn to overestimate demand in those areas while neglecting others. This leads to stock imbalances, missed opportunities, and inefficient resource allocation. Because AI models learn patterns from historical data, any embedded biases—whether due to human error, systemic exclusion, or sampling flaws—can be amplified in the forecasting process, compounding the risk of poor business decisions.

Bias can creep in through faulty sensors, nonrepresentative sampling, or human errors in manual data entry. For instance, if a company relies heavily on historical sales data from a limited number of regions, it may overestimate demand in newer markets where consumer behavior differs. Algorithmic bias can also arise when AI forecasting models are trained on skewed data sets, perpetuating systemic inaccuracies that favor certain suppliers, customer segments, or geographies over others.

Missing Observations. Another issue with dirty data is missing observations. It is typical that data does not get recorded or is lost for some reason. It is important to replace this data with a more typical value. Otherwise, computations may assume that the value is zero, which will distort the forecast. Remember that zero has a different meaning from just a missing data point. Gaps in sales data, customer orders, shipment histories, or supplier performance metrics can distort patterns and trends. For example, if a retailer's holiday sales data is incomplete, a demand-forecasting model might underestimate peak season needs, leading to stockouts and lost revenue. In manufacturing, incomplete lead-time data can disrupt procurement schedules, delaying production and increasing costs.

Inconsistent and duplicated data are further culprits of forecasting failure. When multiple systems store different versions of product codes, pricing, or inventory levels, it becomes difficult to generate a unified and reliable forecast. These discrepancies often result in conflicting demand signals that confuse planners and delay timely decisions.

Trading Day Adjustments. There are differences in the number of trading days or working days, from year to year. The reason is that weekends and holidays do not fall on the same date every year. The result is that the same month of different years may have more or less working days, which distorts the activity. For example, the number of working days in the 2005 calendar does not match up until 2011, 2022, and 2033. This means that one month of one year may have a higher value in activity, such as sales, than that of another year. However, that may just have to do with the fact that there were more sales days. Therefore, an adjustment needs to be made to compare the same month across years.

Outliers (or Special Event) Adjustments. Outliers are data points that significantly deviate from the typical pattern of a data set—either much higher or lower than expected. In business contexts, outliers often arise due to unique or nonrecurring events such as holiday shopping spikes, promotional campaigns, product launches, or unforeseen disruptions like natural disasters or supply-chain delays. While these events may be valid and real, they do not reflect normal operations and can distort analytical models if left unadjusted.

When forecasting future demand, production needs, or inventory levels, including outliers, without proper treatment can lead to inflated projections, inaccurate safety stock levels, or under-preparation for typical demand periods. Therefore, it's crucial to identify, analyze, and—when necessary—adjust or remove these anomalies to maintain the integrity of the data set and improve forecast accuracy.

Consider a retailer like Target. During Black Friday and Cyber Monday, sales of electronics and household items may spike by 300 to 500 percent compared to average daily sales. If this data is used as-is in a demand forecasting model, it may lead to overstocking inventory in December and January, months when sales typically taper off. To mitigate this, Target uses outlier detection algorithms to flag and isolate such peaks, treating them separately in seasonal or promotional forecasts rather than letting them skew baseline demand planning.

Constant versus Current Price Data. When forecasting using monetary values—such as revenue, costs, or trade volumes measured in currencies like US Dollars or Japanese Yen—it's crucial to distinguish between current and constant prices. Current price data reflects nominal values at the time of transaction and includes the effects of inflation or deflation. Constant price data, on the other hand, adjusts for changes in the purchasing power of money over time, allowing for more accurate comparisons across different time periods.

Failing to account for these differences can lead to flawed conclusions. For instance, an increase in sales revenue over several years might appear as growth when, in reality, it is merely the result of inflation. To correct for this, analysts use a deflator index such as the Consumer Price Index (CPI) or Producer Price Index (PPI) to convert current prices into constant prices, effectively removing the effects of inflation and isolating real growth.

Imagine a logistics company analyzing transportation costs in Japan from 2015 to 2025. If it uses current Yen values, the data might show that shipping costs have increased by 25 percent. However, after adjusting for inflation using a deflator index, the real increase in cost might only be 5 percent The rest of the apparent increase is due to the declining value of the Yen over that period. Relying on unadjusted figures could lead to overestimating cost pressures and making misguided pricing or investment decisions.

AI Can Help. The consequences of relying on dirty data in forecasting are significant. Poor forecasts lead to either excess inventory, which increases holding costs and risk of obsolescence, or insufficient stock, which damages customer-service levels and brand trust. Moreover, inaccurate forecasts undermine supplier relationships and reduce the ability to respond quickly to disruptions or market shifts. Clean data is not just a technical concern—it is a strategic asset in supply-chain management. Organizations that invest in data governance, real-time data integration, and validation protocols are better positioned to generate reliable forecasts and gain competitive advantage. Clean data reduces uncertainty, enhances agility, and enables the proactive planning necessary in today's volatile supply-chain environments.

AI offers a range of powerful tools and methods for cleaning data to improve forecasting accuracy in supply chains. Technologies called automated data cleaning platforms use ML algorithms to detect and correct anomalies, outliers, and missing values by learning patterns from historical data. Natural language processing (NLP) can standardize unstructured data like supplier notes or customer feedback. Additionally, cloud-based master data management (MDM) systems can ensure consistent and authoritative data across enterprise functions by maintaining a single source of truth. These innovations collectively reduce manual errors, improve data transparency, and make it possible to scale data cleaning processes across supply chains.

3.3.4 Identify Data Patterns

Analyzing data for underlying patterns is a critical step in the forecasting process. While not every data set contains every type of pattern, recognizing the ones that are present helps determine the most suitable forecasting model. Also, multiple patterns are often present simultaneously. Effective forecasting focuses on modeling the identifiable patterns—level, trend, seasonality, and cycles—while accounting for the unavoidable effect of random variation. Let's look at these:

Level (or Horizontal). This pattern appears when data fluctuate around a constant average, showing no clear upward or downward movement. It is the simplest pattern and generally the most stable to forecast.

Example: A company that produces table salt may observe steady weekly sales with little fluctuation, allowing it to use a simple moving average model to predict future demand.

Trend. A trend exists when data consistently increase or decrease over time. Trends may be linear or nonlinear and reflect long-term shifts in consumer behavior, technology, or market conditions.

Example: A solar panel manufacturer might observe a rising trend in sales over the past decade as consumers shift toward renewable energy, prompting the use of a trend-adjusted forecasting model.

Seasonality. Seasonality refers to patterns that repeat at regular intervals, such as daily, weekly, monthly, or annually. These are driven by recurring events or behaviors.

Example: A toy retailer typically sees a significant spike in sales during November and December due to holiday shopping, requiring seasonal adjustment in forecasting models.

Cycles. Cycles are irregular patterns influenced by broader economic or business factors such as recessions, booms, or product life cycles. Unlike seasonality, cycles are not consistent in timing or intensity, making them harder to forecast.

Example: An automotive parts supplier may face unpredictable swings in demand during economic downturns, requiring scenario-based models or economic indicators to aid in forecasting.

Random Variation. All data include some level of randomness—unpredictable fluctuations caused by unforeseen events or anomalies. While these cannot be forecasted, they must be acknowledged.

Example: A clothing retailer may face an unexpected dip in sales due to a sudden weather change or a viral social media trend affecting consumer choices.

AI Can Help. The better a company understands the patterns of its data, the more accurate and actionable its forecasts will be. AI is highly effective at identifying patterns in data—often far beyond what humans or traditional statistical models can detect.

How can AI help?

First, AI can detect complex and hidden relationships in data. AI, especially through ML, can uncover non-obvious patterns in large and complex data sets. For example, neural networks can detect nonlinear relationships and interactions between multiple variables that influence demand. An example may be an online retailer like Amazon that uses AI to find subtle patterns in user behavior (like search history, time of day, and weather) that predict purchase likelihood—even when no clear seasonal or trend signal is present.

Second, AI can identify multiple pattern types simultaneously. AI models can simultaneously identify and model trends, seasonality, cycles, and even abrupt changes, combining them into a single prediction. Traditional models often need these patterns to be separated and manually adjusted. An example may be a grocery chain that can use AI to forecast banana sales by accounting for long-term growth (trend), weekly peaks (seasonality), and irregular disruptions (like supply-chain delays), all at once.

Third, AI learns and adapts over time. AI models improve with more data. Unlike static statistical methods, AI can continuously retrain on new information, refining its pattern recognition and adapting to emerging trends or market shifts. An example is a fashion retailer using AI to detect that sales patterns for certain items have changed due to a viral TikTok trend and quickly adjust its forecasts and inventory planning.

Fourth, AI works well with noisy and incomplete data. AI can recognize patterns even when data is messy or partially missing, using techniques like data imputation or anomaly detection to fill gaps or remove outliers. For example, a logistics firm can use AI to identify delivery delays caused by weather, even when sensor data from trucks is intermittent or inconsistent.

3.3.5 Select Model

Once data patterns have been identified—whether level, trend, seasonal, or cyclical—the next step is to select the most appropriate forecasting model. While there are many forecasting techniques available, the key to success lies in choosing a model that aligns with the specific characteristics of the data. The selection process often begins by narrowing the options to a few candidate models and testing their performance on historical data to determine which yields the most accurate forecasts.

We discuss different categories of models in the next section. However, there are many available forecasting models, and selection can seem overwhelming. Fortunately, there are guiding principles that simplify model selection. Not every method is appropriate for every context, and several critical factors should be considered.

What are the key factors in selecting a forecasting model?

Type and Quantity of Data. Different forecasting methods require different data inputs. Traditional statistical models often need structured historical data, whereas AI and ML models thrive on large, diverse data sets—including real-time and unstructured data. When historical data is limited or unavailable—as in the case of new product launches—judgmental or analog-based models may be more appropriate.

Forecast Accuracy versus Cost. The level of accuracy needed must justify the complexity and cost of the model. For high-stakes decisions—like pharmaceutical production or aerospace inventory—advanced ML models may be worth the investment. Conversely, for low-cost items, such as paper clips, simpler models like moving averages or exponential smoothing are sufficient.

Forecast Horizon. The length of the forecast period is another key determinant. Short-term forecasts (e.g., next week's retail sales) may benefit from high-frequency data and real-time AI-driven methods, while long-term forecasts (e.g., energy demand over the next decade) require models capable of capturing structural trends and macroeconomic cycles.

Recognized Data Patterns. It is essential to match the model to the dominant data patterns. Failing to do so often results in poor predictions. For instance, applying a level model to data with a strong upward trend will lead to consistent under-forecasting. AI systems are increasingly capable of automatically identifying complex patterns, such as multiple overlapping seasonal patterns—and selecting or adapting models accordingly.

Flexibility and Adaptability. Today's dynamic business environment favors adaptive models. AI-powered forecasting systems, such as those used by Amazon or Walmart, adjust in real time based on new inputs, anomalies, or changing demand signals. These models can outperform static statistical approaches by learning from incoming data and continuously improving over time.

AI Can Help. AI is transforming the model selection process itself. Modern AI platforms not only help identify data patterns but can also automate the selection, tuning, and validation of forecasting models based on those patterns. These systems use techniques like automated machine learning (AutoML) and ensemble modeling, which combine the outputs of multiple models to generate more robust and accurate forecasts.

For example, AI tools can detect whether a time series contains trend and seasonal components and recommend hybrid models, such as recurrent neural networks (RNNs), based on the specific structure and behavior of the data.

Remember that model selection is both a technical and strategic decision that depends on data characteristics, business needs, and forecast objectives. As AI continues to advance, it not only expands the toolkit available to forecasters but also reduces the guesswork in choosing the right method, allowing businesses to forecast with greater speed, precision, and agility.

3.3.6. Generate the Forecast

Once a forecasting model has been selected—based on data patterns, business needs, and available data—the next step is to generate the forecast. This is often seen as the most straightforward stage in the forecasting process, largely due to the availability of user-friendly forecasting software and AI-enabled platforms. In many cases, the actual generation of the forecast may appear as simple as pressing a button. However, this apparent simplicity belies the importance of the underlying assumptions, model structure, and data inputs.

At its core, generating the forecast involves applying the chosen model to the cleaned and structured data. Forecasting software—such as SAP IBP, Oracle Demand Management, or AI-powered platforms—automates much of the mathematical computation and outputs the resulting numerical projections. These may include product-level demand by region, forecasted revenue, or labor requirements across different time horizons.

In recent years, AI has transformed this step from a static to a dynamic process. Rather than relying solely on historical data, AI-driven forecasting engines continuously learn from new data streams, adjust their parameters in real time, and even generate scenario-based forecasts that consider uncertainties such as supply-chain disruptions or sudden demand shifts. These systems are particularly valuable for complex, volatile environments in which human forecasters may struggle to adapt quickly.

Still, while the software handles the heavy lifting, it is essential for business analysts and decision makers to understand what the system is doing. This includes knowing what data was used, how it was processed, and which modeling assumptions were applied. A forecast is only as good as the logic and inputs behind it. Misinterpreting the output—or accepting it uncritically—can lead to poor decisions in procurement, inventory management, or production planning.

Procter & Gamble employs AI-enhanced demand-forecasting tools that integrate internal sales data with external signals such as weather patterns, market trends, and promotional calendars. Once a suitable model is selected for a product category, such as paper towels or baby diapers, the system automatically generates weekly and monthly forecasts at a granular level. These forecasts are not static; they are continuously updated as new data flows in from stores, distributors, and online platforms. The ability to quickly and accurately generate forecasts has helped P&G improve shelf availability, reduce stockouts, and better synchronize production schedules with consumer demand.

Ultimately, generating the forecast is where the analytical groundwork translates into actionable insights. But the usefulness of this step hinges on the quality of prior steps—model selection, data cleaning, and pattern recognition—and on the expertise applied when interpreting and applying the forecast results.

3.3.7 Measure Accuracy

After the forecast is generated and actual events happen, we need to evaluate forecast performance. This is done by measuring forecast error. Forecast error is simply the difference between what we forecast and what actually happened and is easily computed. For example, we may have forecasted sales for January to be 300 units, but we actually sold 320. Therefore, the forecast error for January would be 320–300, or 20.

Notice, however, that the number by itself doesn't tell us much. Computing error for one time period does not provide enough information. We need to measure forecast accuracy over time. If we track performance over time, such as measuring the forecast for every month of the year, we will be able to track forecast accuracy and have a good idea as to how we are doing. This information should be used to improve the forecasting process. Remember that forecasting is an ongoing process. It is always changing as new information and data become available.

When we are forecasting, we are predicting the future based on past data. This is shown in Figure 3.3. Recall that one of the principles of forecasting is that forecasts are rarely perfect. The question you are probably asking is: *Why should we even bother with forecasts?* The reason is simple. Without a forecast we have no way of anticipating the future or preparing for it. Even if our forecast is not perfect, a forecast provides us with an idea as to what to expect. Also, if we follow the right forecasting process we are assured of a solid average forecast performance over time. The best that we can do in forecasting is to generate forecasts with a good average performance over time. However, what constitutes a good average performance will vary and be different based on the characteristics of what is being forecast.

Some things are difficult to forecast. They have what is called low forecastability, which means how easy or difficult something is to forecast. A good example would be forecasting sales of a new product such as the very first iPhone. With no historical data this would be difficult to forecast. Forecast errors would likely be high until historical data becomes available. Other things may be easy to forecast. They have a high forecastability. Just consider, for example, sales of a mature product like table salt at a grocery store or ice cream sales in July. In this case there is plenty of historical data and there aren't any large variations, making it highly predictable.

How do we know if we are doing well with our forecasts?

Evaluating forecast performance is achieved by measuring forecast accuracy, or benchmarking performance. This should be done on a routine basis as part of the forecasting process. In fact, measuring forecast accuracy is a critical aspect of forecasting. This tells us how our forecasting methods are performing and enables us to improve forecast performance over time.

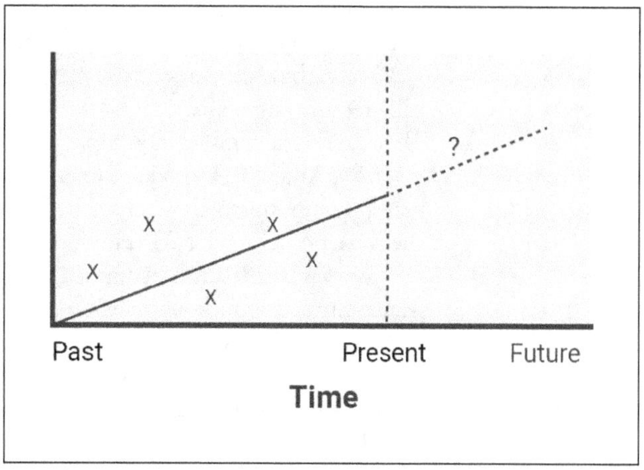

Figure 3.3 Forecasts predict the future based on past data

Data can change over time, and models that once provided good results may no longer be adequate. The model's accuracy can be assessed only if forecast performance is measured over time.

There are many metrics that measure forecast error, and each provides different insights. At the simplest level, forecast accuracy is the difference between an actual and forecasted value taken as an average across the measured time periods, called Average Error (AE). Here both positive and negative errors are averaged and, although they cancel each other out the net effect provides the direction of the forecast or the bias. Another metric is the Mean Absolute Deviation (MAD), computed as the average of the sum of absolute errors, rather than algebraic errors, providing a measure of AE regardless of sign or direction. There are many other metrics, and forecasters should use multiple metrics for accuracy.

3.3.8 Collaborate & Update

Forecast results need to be shared across the supply chain and planning functions—demand planning, procurement, manufacturing, logistics—to align decision making. Collaborative forecasting (CPFR) with suppliers and distributors improves synchronization and reduces the bullwhip effect discussed earlier. Procter & Gamble, for example, works with retail partners to share forecasts and codevelop replenishment schedules, ensuring availability without overstocking. Forecasts are continuously updated as new data arrives or conditions change (e.g., price fluctuations, disruptions). Monitoring tools and dashboards allow planners to adjust inputs and rerun models frequently, especially in volatile or seasonal environments. For example, during a supply-chain disruption (e.g., port delays), a pharmaceutical company adjusts its forecast in real time to reroute inventory and manage demand.

3.4 Categories of Forecasting Models

Forecasting is a critical activity in business, economics, supply-chain management, and many other domains. It involves predicting future outcomes based on historical data, patterns, and analytics. Forecasting models can be classified into several categories based on methodology, input data, and usage. Traditionally, forecasting models have been grouped into two main categories:

1. Qualitative models (based on expert opinion and judgmental)
2. Quantitative models (based on statistics and mathematics)

Although ML and AI models originate within quantitative forecasting many experts and academic sources now consider them a distinct third category for several key reasons.

First, AI and ML models have a distinct methodology. Traditional quantitative models, like AutoRegressive Integrated Moving Average (ARIMA) and regression rely on explicit statistical assumptions and parametric modeling. However, AI/ML models (e.g., neural networks, ensemble trees) use non-parametric learning, often without assuming specific data distributions.

Second, they have unique data requirements and complexity. For example, AI/ML models typically require larger and richer data sets, including unstructured data (e.g., images, text, IoT signals). They can learn complex nonlinear relationships and detect patterns that traditional models miss.

Third, AI and ML models are built on different foundations (optimization, pattern recognition, neural computing). They rely heavily on high-performance computing and automation, making them part of a broader AI transformation rather than just statistical evolution.

Fourth is a difference in application scope and autonomy. AI/ML systems are often embedded in real time, autonomous decision-making environments (e.g., dynamic pricing, personalized recommendations, fraud detection). These systems adapt and learn over time, distinguishing them from static statistical models.

As a result, many academic programs, textbooks, and consulting frameworks now distinguish AI/ML as a third class of forecasting tools. For instance, consulting firms such as Gartner and McKinsey, as well as academic journals in operations and analytics, increasingly separate AI/ML

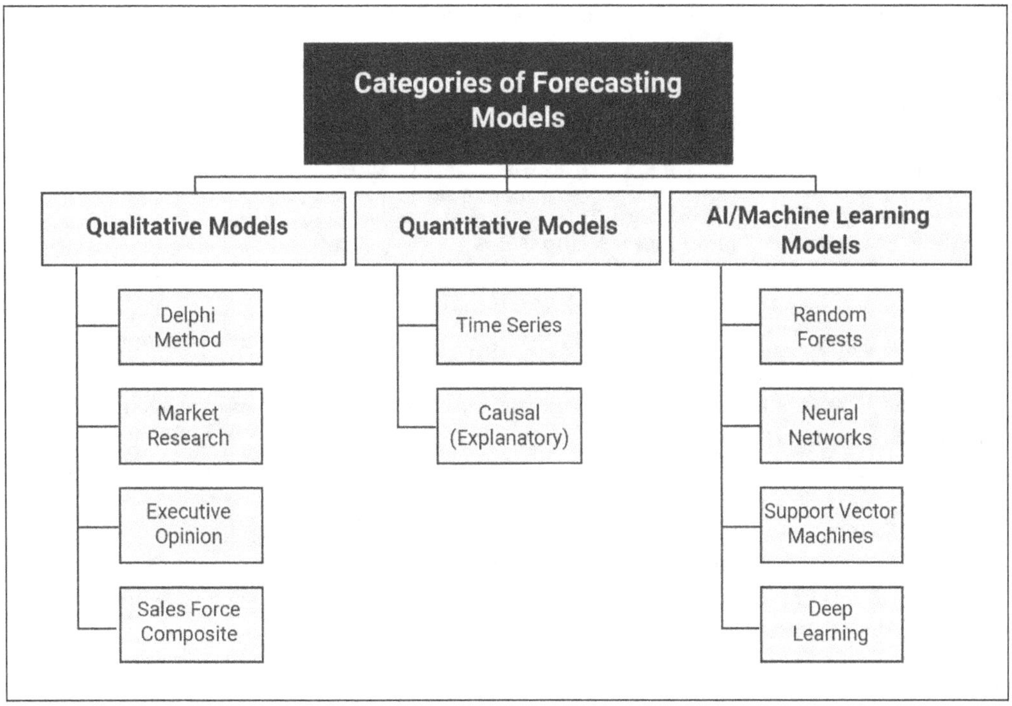

Figure 3.4 Categories of Forecasting Models

from traditional time series and causal models due to their unique operational features.

Today AI and ML models are recognized as a third category of forecasting models, distinct from both qualitative and traditional quantitative methods. They represent not just a set of tools but also a paradigm shift in how businesses forecast in complex, dynamic environments. This is shown in Figure 3.4. Let's now look at each category in a bit of detail.

3.4.1 Qualitative Forecasting Models

Qualitative forecasting models—often called judgmental models, are methods based on subjective opinions and judgment of individuals—such as managers, sales staff, or customers. Asking customers whether they would buy a particular product—called intention surveys—are a type of qualitative forecasting method. Another one is called sales force composite and occurs when the sales staff make a group forecast of upcoming sales. These models rely on expert judgment, intuition, and experience rather than numerical data. They are particularly useful when historical data is limited or unavailable, such as during the launch of a new product or in emerging markets.

Qualitative or judgmental forecasts are made by people who often have the latest information. These forecasts are subjective and based on opinions. As a result, they are subject to numerous human biases, such as optimism, wishful thinking, and political manipulation. Just think if your annual budget was set based on a forecast. You would naturally be inclined to use a model that gave you the highest value to justify a higher budget. On the other hand, think about if your performance was going to be compared against your annual forecast. The opposite would be true, and you would likely select the model with the lowest possible forecast value for comparison. These behaviors are natural and part of being human. They do, however, bias the forecast.

Strengths

- highly responsive to the latest changes in environment
- can include "inside" and "soft" information difficult to quantify

- can compensate for one-time or unusual events
- provide forecaster with sense of ownership

Weaknesses

- lack of consistency
- optimism
- wishful thinking
- difficulty in understanding causal relationships

Key Types

- Delphi Method: A structured process that gathers expert opinions through multiple rounds of anonymous surveys to converge on a consensus.
- Market Research: Uses surveys, focus groups, and customer interviews to gauge future behavior or preferences.
- Executive Opinion: Involves gathering input from high-level managers or specialists to make educated forecasts.
- Sales Force Composite: Relies on the knowledge of the company's sales team to estimate future demand based on customer interaction.

Best Use: New product development, strategic planning, and long-range forecasting in uncertain or innovative contexts.

3.4.2 Quantitative Forecasting Models

Quantitative forecasting methods are forecasting methods based on mathematics and statistics. These methods are objective and consistent and are capable of handling large amounts of data and uncovering complex relationships. These models depend on historical numerical data and mathematical computations to make projections. Provided that good data are available, these methods are far more accurate than qualitative methods.

These methods are not subject to the numerous biases of human forecasting. They are also consistent and will provide the same numerical value on the same data set. They can also process huge numbers of data and variables. However, these models are only as good as the data upon which they are based. In fast-moving environments, some of the latest changes may not be incorporated in the data set. Judgmental correction, or judgmental override, may have to be made. This may involve a snowstorm delaying shipment or a pandemic that has just occurred.

It is true that we can model almost everything. However, it may be costly. When we are forecasting thousands of items it may not be possible to model every recent event. Also, such levels of modeling are possible but require deep technical understanding, which takes skill, time, and cost. This added requirement may simply not be realistic.

Strengths

- objective
- consistent
- can process large amounts of data
- can consider many variables and complex relationships

Weaknesses

- slow to react to changing environments
- only as good as the model formulation and the data it is based upon
- can be costly and time consuming to model soft information
- requires technical understanding

There are two broad categories of quantitative models: time-series models and causal models.

Time-Series Models focus on patterns over time and assume that the future will follow historical trends.

Key Types

- **Naive** Model Assumes that the next period's forecast is equal to the most recent actual value. Simple but useful for stable processes.
- **Moving Average** Calculates the average of a fixed number of past periods to smooth fluctuations.
- **Exponential Smoothing** Weighs past observations with exponentially decreasing weights. Variants include Single, Double, and Triple Exponential Smoothing (e.g., Holt-Winters).
- **ARIMA** Combines autoregression, differencing, and moving averages. It is suitable for complex and nonstationary time-series data.

Best Use: Demand planning, sales forecasting, inventory control.

Causal Models (Explanatory Models) identify and quantify relationships between the variable to be forecast and other influencing factors.

Key Types

- Linear Regression forecasts outcomes based on one or more independent variables.
- Econometric Models incorporate economic theory and multiple equations to understand interdependencies.
- Input-Output Models often used in macroeconomic planning; these trace inter-industry relationships.
- Leading Indicator Models use variables (e.g., housing starts, interest rates) that tend to change before the target variable.

Best Use: Economic forecasting, marketing analytics, and policy planning where cause-effect relationships exist.

3.4.3 AI/ML Forecasting Models

AI and ML forecasting models represent the latest evolution in forecasting and predictive analytics capability. These methods use self-learning algorithms that automatically detect patterns in large and often unstructured data sets. Unlike traditional quantitative models, which rely on explicit mathematical formulas, AI/ML models can dynamically adjust as new data becomes available. As a result, they are particularly useful in complex, fast-moving, and high-variability environments such as retail, e-commerce, logistics, and energy demand forecasting.

A core strength of AI/ML forecasting is adaptability. These models are not static: they improve over time as they are exposed to more data. They do not require the forecaster to specify all relationships between variables in advance. Instead, the algorithms learn these patterns on their own. For example, neural networks can uncover nonlinear and hidden interactions that traditional models may miss. This makes them extremely powerful when working with real-world data sets that contain irregularities, seasonal shifts, or external factors like weather or social trends.

Just like quantitative models, AI/ML models are objective and free from human biases. However, they take data-processing capabilities to an entirely new level. AI models can integrate data from a wide range of sources, including social media, satellite images, web traffic, and IoT sensor feeds. This allows companies to respond to market signals in real time. Imagine trying to manually analyze millions of daily transactions, customer reviews, and shipment updates; AI can do this effortlessly.

AI models are not without challenges. First, they require large volumes of clean, labeled data to perform well. Poor data can mislead the model, and unlike simpler models, AI can be a black box, making it difficult to interpret how a forecast was produced. Furthermore, developing and maintaining AI models often requires specialized technical skills in data science and engineering. In many companies, judgmental or statistical models are still used in parallel with AI to validate or override certain results.

Key Types

- Decision Trees & Random Forests are excellent for forecasting where multiple decision rules are involved—like product returns, promotions, or customer churn. Trees are interpretable, and forests increase accuracy through ensemble learning.
- Gradient Boosting Machines (e.g., XGBoost and LightGBM are useful for structured, tabular data with high dimensionality. They are often applied in demand forecasting for retail or financial time series.
- Neural Networks (including Deep Learning & Long Short-Term Memory [LSTM] are best suited for forecasting complex, nonlinear time series, such as demand patterns in fashion or traffic flow. LSTM networks are especially effective for long-range dependencies in time-series forecasting.
- Reinforcement Learning is used in dynamic pricing or supply-chain routing, where the model learns optimal strategies based on outcomes over time.

Strengths

- learn automatically from large, dynamic data sets
- handle nonlinear, complex relationships
- integrate structured and unstructured data
- improve accuracy over time with continuous learning
- capable of near real-time forecasting

Weaknesses

- require vast amounts of clean data
- often opaque or hard to interpret (black box issue)
- high computational and implementation costs
- require technical expertise in AI/ML
- may overfit if not tuned or validated properly

Best Use: Real-time forecasting, dynamic pricing, fraud detection, and personalized recommendation systems.

The choice of forecasting model depends on the context, data availability, forecast horizon, and required accuracy. While traditional methods such as time series and regression continue to be widely used for their interpretability and simplicity, ML is revolutionizing forecasting with its capacity to handle vast data sets and capture nonlinear, nonobvious patterns. A hybrid approach, combining human judgment and machine intelligence, often provides the most resilient and accurate forecasts in today's complex business environments.

3.5 Combining Forecasts

3.5.1 Why Combine?

Much evidence exits to support the combining of two or more independent forecasts. The overall conclusion from these studies is that gains from combining are greatest when the methods combined differ substantially and are based on different sources of information. Also, the constituent forecasts should be absent of bias or have biases that cancel each other out. Last, combining is most effective when the correlation between forecast errors of constituent forecasts is low, meaning that each forecast brings different information to the integration process.

While forecasting plays a central role in business decision making, no single forecasting method is perfect. Each has its own assumptions, strengths, and limitations depending on the context, data quality, and planning horizon. To improve accuracy and build resilience into planning processes, many organizations now use a combination of forecasting methods. This approach, often referred to as forecast combination or ensemble forecasting, leverages the strengths of multiple techniques to mitigate the weaknesses of individual ones. Combining forecasts not only enhances accuracy but also provides a more balanced view of uncertainty, especially in volatile business environments.

Following are three main benefits of combining:

1. **Improved Accuracy Through Model Diversification.** Just as diversifying investments reduces financial risk, combining forecasts from different models reduces the risk of relying on a single, potentially flawed prediction. For example, a time-series model might be good at detecting historical patterns, while an ML model might excel at recognizing nonlinear relationships in real-time data. By combining both, the resulting forecast can capture a broader spectrum of insights.
2. **Improved Forecast Stability.** Markets and supply chains often experience shifts that make a previously effective model less reliable. Combining methods provides robustness to structural changes in data, such as seasonality breakdowns, sudden trend reversals, or external shocks (e.g., a pandemic or supply-chain disruption).
3. **Reduction of Bias and Noise.** Individual models often introduce specific types of biases. Judgmental forecasts, for example, may be optimistic, while statistical models may lag during turning points. A blended approach can cancel out extreme values or model-specific errors, leading to a more stable and less biased final forecast.

3.5.2 Methods of Combining

Combining forecasting methods is not just a technical enhancement, it is a strategic capability. By integrating statistical models, ML tools, and human expertise, organizations can generate more accurate, adaptable, and insightful forecasts. This is especially critical in today's dynamic business environment, where market volatility and supply-chain disruptions are common. As AI continues to evolve, the interplay between human judgment and algorithmic intelligence will define the next frontier of demand planning and business resilience.

There are established techniques for combining forecasts:

Simple Average: The most basic method where forecasts from different models are averaged. It is surprisingly effective when models are reasonably well calibrated.

Weighted Average: More sophisticated than a simple average, this method assigns weights to different forecasts based on historical performance or model confidence. For example, if a neural network historically outperforms a linear regression during promotional periods, it may receive a higher weight during similar future periods.

Bayesian Model Averaging: A probabilistic approach that incorporates uncertainty about model selection. It combines forecasts based on the posterior probabilities that each model is the best one, given the observed data.

Ensemble Learning Techniques: Used in ML methods to combine the predictions of multiple models to create a meta-model that often outperforms any individual input model.

Combining Human Judgment with AI Models: Some of the most effective forecasts emerge when AI is combined with human insight. AI and ML models are powerful at processing vast, complex data sets and detecting patterns that are imperceptible to humans. However, these models may miss context—such as emerging geopolitical tensions, regulatory shifts, or sudden consumer sentiment changes—that are apparent to human experts. A common strategy is where experts provide individual forecasts, review AI outputs, and converge toward a revised final forecast through iterative discussion. Businesses also use override thresholds, wherein AI forecasts can be adjusted manually if certain conditions or anomalies are detected. These are all effective methods to have human experts add their knowledge to the data-processing superiority of AI.

COMPANY EXAMPLE: The Retail Industry

The retail industry offers numerous examples of combining forecasts. One such example is Target Corporation, a company that exemplifies effective blending of forecasting methods. The retail giant uses a suite of AI and ML models to predict demand across thousands of SKUs, factoring in promotions, seasonality, and regional demographics. These models are particularly good at crunching data from online searches, weather forecasts, and in-store behavior.

However, during atypical events—such as a new product launch or an unexpected social trend—Target incorporates input from category managers and marketing teams. For instance, during the COVID-19 pandemic, Target's AI models initially struggled to predict surges in demand for cleaning supplies and home fitness products. Human judgment, informed by news trends and supply-chain constraints, was used to adjust forecasts and inform procurement decisions.

Target also uses forecast blending by combining time-series models for historical data, ML for real-time demand sensing, and human insights for strategic overrides. This three-pronged approach helped the company avoid widespread stockouts during peak pandemic periods and enabled more precise replenishment as consumer behavior stabilized.

Another example is Procter & Gamble, a global leader in consumer goods. P&G uses forecast combination as part of its advanced demand planning strategy. With more than 65 brands sold in more than 180 countries, P&G deals with high variability in consumer demand across regions, products, and channels.

To manage this complexity, P&G integrates forecasts from different models, including time-series models for historical sales trends, ML models incorporating promotional calendars, retail inventory, and weather data, and then judgmental inputs from sales and marketing teams during product launches or campaign periods

These forecasts are then combined, initially through weighted averages and increasingly through AI-driven ensemble methods, to improve accuracy. This hybrid approach has enabled P&G to reduce stockouts, improve service levels, and optimize production schedules globally. In fact, the company reports improved forecast accuracy of up to 30 percent in certain product categories, leading to significant savings in working capital and logistics.

Combining forecasts is a powerful strategy to improve accuracy, reduce uncertainty, and build resilience into the decision-making process. Whether through simple averaging or sophisticated ML ensembles, the integration of multiple models offers a hedge against the limitations of any single method. As data environments become richer and more dynamic, businesses that embrace forecast combination will be better equipped to anticipate demand, manage risks, and serve customers efficiently.

3.5.3 Scenario Planning

In an increasingly volatile business environment, traditional forecasting methods—often based on historical trends and linear assumptions—can fall short in anticipating disruptive change. Scenario planning offers a powerful alternative. Rather than predicting a single outcome, scenario planning explores multiple plausible futures. It is a qualitative forecasting tool that equips organizations to better understand uncertainty, assess risk, and develop flexible strategies in the face of change.

What is Scenario Planning?

Scenario planning is a strategic forecasting method that involves creating a set of diverse, yet plausible future scenarios based on key uncertainties and driving forces. Each scenario represents a different combination of potential developments, such as economic conditions, technological innovation, regulatory changes, or geopolitical shifts. These narratives help organizations think through how their operations, markets, or supply chains might be affected under different conditions.

Unlike point forecasts that produce single-number outcomes, scenario planning asks "What could happen?" rather than "What will happen?" This shift in perspective is particularly valuable in environments marked by complexity, ambiguity, or systemic risk.

How Scenario Planning Is Used in Forecasting

Scenario planning complements traditional forecasting by embedding qualitative foresight into strategic decision making. It is especially useful in long-term planning, risk management, and capital-intensive industries where future uncertainty can significantly affect outcomes.

The process typically involves

1. Identifying key drivers of change, such as climate change, consumer behavior shifts, or regulatory policies
2. Determining critical uncertainties, factors with high impact but unpredictable outcomes
3. Constructing a set of scenarios, usually two to four, each reflecting different futures
4. Assessing implications for supply chains, market strategies, product development, or investment
5. Developing contingency plans to enhance organizational agility and resilience

By exploring multiple futures, organizations can test their strategies under different assumptions, identify early warning signals, and build robust plans that work across a range of possibilities.

AI Can Help. While scenario planning is rooted in qualitative thinking, AI can enhance its accuracy and strategic value. Here are some ways AI can help:

- **AI-Enhanced Environmental Scanning.** AI tools can process vast unstructured data (e.g., news, social media, policy documents) to identify emerging trends and weak signals, helping teams spot disruptions early.
- **Predictive Analytics.** AI can generate probabilistic forecasts within scenarios, simulate economic or climate models, and stress-test financial outcomes.
- **Natural Language Processing (NLP).** NLP allows AI systems to analyze scenario narratives, compare them against real-time data, and detect which scenario is becoming more likely.
- **Risk Quantification.** AI helps quantify the probability and effect of risks within each scenario, transforming qualitative insights into actionable risk profiles.

This convergence of AI and scenario planning represents a major leap in how organizations approach strategic forecasting. It brings together the strengths of human imagination and machine intelligence.

The company Royal Dutch Shell is widely recognized for pioneering scenario planning. For decades, Shell has used scenarios to navigate oil price fluctuations, geopolitical risks, and shifts in energy policy. In its "Sky" and "Mountains" scenarios, Shell explored divergent futures—one in which rapid decarbonization reshapes energy markets, and another in which national interests delay climate action.

These scenarios guided investment decisions in renewable energy, helped Shell anticipate the rise of electric vehicles, and informed its transition strategy. Today, Shell incorporates AI-driven market models and carbon risk simulations into its scenario planning process—blending strategic foresight with cutting-edge analytics.

Scenario planning is more than a forecasting tool; it is a mind-set for resilience. By considering multiple futures and preparing for uncertainty, organizations can turn risk into opportunity. The integration of AI into scenario planning strengthens its predictive power, enabling businesses to model complexity, detect signals of change, and make smarter decisions. As demonstrated by leaders like Shell, scenario planning is an essential capability for navigating an uncertain world, especially when the future is anything but linear.

3.6 Collaboration Across Supply Chains

Today no company operates in isolation. Every organization is part of a broader supply-chain network, and the accuracy of demand forecasts affects not only internal operations but also the performance of suppliers, distributors, and retailers. **Collaborative forecasting**, a process in which supply-chain partners share data and jointly create forecasts, has emerged as a best practice to improve accuracy, responsiveness, and efficiency across the entire value chain.

Collaborative forecasting transforms fragmented supply-chain planning into a unified, strategic effort. By sharing data, aligning forecasts, and using AI to harmonize planning efforts, businesses can mitigate risk, enhance responsiveness, and build stronger supply-chain ecosystems. As volatility and uncertainty become the norm, collaboration is no longer a competitive advantage; it is a necessity.

Why Collaborative Forecasting Matters

1. **Reduces the Bullwhip Effect.** When companies forecast demand independently, small changes in customer demand can become exaggerated as they move up the supply chain. This is the bullwhip effect we discussed earlier, and it leads to overproduction, excess inventory, or stockouts. Collaborative forecasting is the primary way to address the bullwhip effect and helps align expectations, reducing inefficiencies.
2. **Improves Forecast Accuracy.** Collaborative forecasting draws on multiple perspectives and richer data sets, such as retail sales, marketing promotions, and supplier capacity. It is shared intelligence that results in more accurate, realistic forecasts.
3. **Enhances Supply-Chain Agility.** Real-time sharing of demand and inventory data enables faster response to disruptions, shifting customer behavior, or sudden supply shortages. This agility is critical in environments like fashion retail, electronics, or perishable goods.
4. **Strengthens Relationships and Trust.** Joint forecasting encourages transparency and alignment of goals. By working together, partners build trust, reduce conflict, and foster long-term strategic collaboration rather than transactional relationships.

How Collaborative Forecasting Works

Collaborative forecasting typically involves several key steps:

1. **Data Sharing.** Partners share relevant data such as sales history, point-of-sale (POS) data, promotional calendars, inventory levels, and market trends. Cloud-based platforms and APIs make real-time data exchange more seamless than ever before.
2. **Joint Forecast Development.** Using shared data, partners develop forecasts collaboratively. This may involve negotiation, reconciliation of assumptions, or use of joint planning software. AI tools increasingly support this process by suggesting consensus forecasts based on historical collaboration patterns.
3. **Consensus Forecast Approval.** Once a forecast is developed, all partners agree upon a single plan that guides procurement, production, and distribution activities across the chain.
4. **Continuous Monitoring and Updates.** Collaborative forecasting is not a one-time activity. Regular updates and feedback loops are essential. AI-powered platforms now offer rolling forecasts and scenario analysis to support real-time adjustments.

AI Can Help. AI significantly enhances collaborative forecasting in supply chains by improving data integration, real-time responsiveness, and decision accuracy among supply-chain partners.

First, AI allows for real-time data sharing and integration with partners. AI enables seamless data exchange across supply-chain partners—manufacturers, suppliers, distributors, and retailers—by automating the integration of large, heterogeneous data sets. These data sources may include POS systems, enterprise resource planning (ERP) platforms, inventory levels, promotions, and even weather or social trends. AI harmonizes and preprocesses this data for shared visibility, eliminating silos and inconsistencies. For example, in the consumer electronics industry, AI integrates real-time sales data from retailers with supplier capacity data to produce a shared demand forecast, reducing both overstock and stockouts.

Second, AI enables superior demand-pattern recognition across partners. Current AI/ML models can detect demand patterns that individual partners may not see on their own. ML algorithms identify sales spikes, seasonal shifts, and promotional effects across the network. These insights help all parties align on the true end-consumer demand rather than relying on their forecasts alone.

An example of this is when a retailer's upcoming promotion causes demand to surge. AI can alert upstream manufacturers and logistics providers to ramp up in advance, based on predicted promotional lift.

Third, AI allows for automated consensus forecasting. AI aggregates inputs from multiple partners and can recommend a unified forecast based on weighted accuracy, historical biases, or priority levels. This reduces the friction of manual negotiations and helps align all parties faster.

Example: In automotive manufacturing, where parts arrive from global suppliers, AI helps the original equipment manufacturers (OEMs) and suppliers arrive at consensus demand forecasts for engines, chips, and assemblies.

Fourth, AI allows automated monitoring of forecast performance across partners. It can identify which partner's forecast contributed most to errors and adjusts weights or models accordingly for the next forecast cycle. This continuous learning loop improves trust and accountability in collaborative forecasting.

Example: If a supplier consistently overestimates demand, AI adjusts its contribution to the final forecast until performance improves.

COMPANY EXAMPLE: Dell

Dell provides a leading example of collaborative forecasting in action. Known for its build-to-order model, Dell relies heavily on close coordination with suppliers like Intel, Seagate, and other component manufacturers to ensure timely production and delivery.

To avoid mismatches in supply and demand, Dell shares its sales forecasts and order data directly with its key suppliers. In return, these suppliers share information about production capacity, lead times, and component availability. Dell's collaborative forecasting model includes

- weekly forecast updates to reflect the latest demand signals,
- vendor-managed inventory (VMI) arrangements where suppliers manage stock levels at Dell's facilities, and
- use of AI tools that simulate various demand scenarios and recommend optimal procurement schedules based on shared constraints.

This is an integrated approach to collaborative forecasting that has helped Dell reduce inventory holding costs, shorten lead times, and maintain service levels, even during periods of supply-chain disruption, such as the 2020 semiconductor shortage.

Chapter Review

Key Points

1. Forecasting is about predicting the future. Forecasting involves making informed predictions based on historical data, current trends, and expected future events.
   It is not about being perfectly accurate but instead about being directionally correct and reducing surprise. Forecasts guide organizations in aligning resources with future needs.
2. Forecasts drive supply-chain decisions and identify risks. Forecasts directly influence decisions related to sourcing, production, inventory, logistics, and staffing. Forecasting helps organizations anticipate and mitigate risks such as demand surges, supply disruptions, and geopolitical events. Scenario forecasting and sensitivity analysis are tools for managing risk and uncertainty.
3. Basic rules of forecasting include that forecasts are rarely perfect, so we need to measure and manage forecast error. Short-term forecasts are usually more accurate than long-term ones. Aggregate forecasts (e.g., company level) are more accurate than disaggregated ones (e.g., per SKU).
4. Forecasting should follow a formal organizational process.
5. Different types of forecasting models exist —including AI models.
6. Collaborative forecasting is important in supply-chain management.

Discussion Questions

1. Why is forecasting important beyond just predicting sales or demand? Can you think of other areas in business or life where forecasting plays a key role?
2. How do forecasts drive supply chains, and what could go wrong in a supply chain if forecasts are inaccurate or missing? Provide an example of real-world consequences such as stockouts, overstocks, and bullwhip effects.
3. What are some of the basic rules of forecasting that can help improve its accuracy and usefulness? Why do you think these rules are important?
4. How might the forecasting process differ between a small business and a large multinational organization? How do scalability and complexity affect forecasting?
5. Can you identify a situation where judgmental forecasting might be more appropriate than quantitative forecasting? Why?
6. What are the strengths and weaknesses of different forecasting methods (e.g., time series, causal models, AI/ML-based models)?
7. What are the benefits of collaborative forecasting across a supply chain, and what challenges might arise in trying to implement it?
8. How might new technologies like AI or ML enhance forecasting accuracy, and what risks do they introduce?

Case Study

Royal Dutch Shell—Inventing and Mastering Scenario Planning

Royal Dutch Shell is not only one of the largest energy companies in the world—it is also recognized as the corporate pioneer of scenario planning. Since the early 1970s, Shell has integrated scenario thinking into its strategic decision making to manage deep uncertainty in energy markets, geopolitics, and environmental regulation. The company's approach to scenario planning has since become a global benchmark for long-term foresight and risk management.

Shell's influence on the business world extends far beyond the energy sector. By transforming scenario planning from a theoretical exercise into a core corporate discipline, Shell demonstrated how companies can thrive amid complex and unpredictable global forces.

The Origin

Scenario planning emerged at Shell in the late 1960s and early 1970s, during a time of increasing geopolitical uncertainty and growing volatility in oil markets. The technique was adapted from military and academic uses—particularly from the RAND Corporation—but Shell was the first major company to institutionalize it as a strategic business tool.

The turning point came in 1973, when the Organization of the Petroleum Exporting Countries (OPEC) declared an oil embargo, triggering a massive spike in oil prices and a global energy crisis. While most oil companies were blindsided, Shell was comparatively well prepared.

Why? Because Shell had developed a scenario that envisioned a world in which OPEC exerted stronger control over oil production and prices. Though no one could predict exactly when such a crisis would hit, Shell had mentally and operationally rehearsed for it. As a result, the company responded more nimbly than its competitors, gaining market share and solidifying the internal credibility of scenario planning.

How Shell Uses Scenario Planning Today

Shell has refined its approach over five decades, evolving scenario planning into a disciplined, multilayered process that is embedded in both corporate strategy and risk assessment.

- **Strategic Narrative Development.** Shell typically develops two to four scenarios, each representing plausible futures based on key uncertainties such as global energy demand, technological shifts, climate policy, and geopolitical realignment. Scenarios **are not predictions**; they are rich narratives describing how the future could unfold, often with emotionally resonant names like Sky, Mountains, and Waves.
- **Executive Engagement.** Shell's executive teams actively engage with the scenarios through workshops and strategic dialogues. The aim is not to pick a most likely future but to test the robustness of current strategies against all scenarios and develop early-warning indicators.
- **Quantitative Integration.** Shell augments qualitative scenarios with quantitative models, including energy demand curves, carbon pricing forecasts, and regional supply-demand simulations. Scenarios inform everything from capital investment decisions to long-term R&D, including transitions to renewables and hydrogen infrastructure.
- **Public Transparency.** Shell publicly shares its scenario reports, making them a valuable resource for policymakers, academics, and industry leaders. This transparency has helped position Shell as a thought leader on the future of energy.

The Sky Scenario

In 2018, Shell released its Sky scenario, a bold vision of a world that achieves net-zero carbon emissions by 2070. It was not presented as Shell's strategy but instead as one of several possible pathways the world might follow if society, governments, and companies worked aggressively toward climate goals.

Key assumptions in Sky included

- rapid electrification of transport,
- a global price on carbon by the early 2030s,
- a dramatic scale-up of renewable energy and carbon capture technologies, and
- a shift in consumer behavior favoring low-emission lifestyles.

The Sky scenario helped Shell identify where its legacy oil and gas portfolio could become stranded assets in a net-zero world. It informed Shell's decision to increase investments in renewables, hydrogen, and Electric Vehicle (EV) infrastructure. Shell used insights from Sky to communicate with investors about long-term climate risk and energy-transition planning.

By contrast, Shell also developed more conservative scenarios (like Mountains), in which global coordination on climate falters. Comparing strategies across these divergent scenarios helps Shell manage uncertainty and hedge risks across its vast portfolio.

Integration with Ai

In recent years, Shell has begun using AI and ML to enhance its scenario-planning capabilities. It applies NLP to global news, policy documents, and social media to detect emerging weak signals that may influence scenario probabilities. AI-based energy models simulate thousands of policy, technology, and demand permutations to quantify the effect of each scenario. Then digital twins and operational simulations are used to allow Shell to test how its assets would perform under extreme future conditions (e.g., high carbon price, regulatory shocks).

These innovations are making scenario planning more dynamic, enabling Shell to update its foresight more frequently and respond faster to global developments.

Lessons

Shell's leadership in scenario planning offers a compelling example of how a company can proactively prepare for uncertainty rather than react to it. By combining narrative foresight, executive engagement, quantitative modeling, and now AI, Shell has transformed scenario planning from a theoretical framework into a living part of strategic decision making.

The core lesson is not to predict the future but to be ready for many possible futures. For organizations navigating climate change, supply-chain disruptions, or global regulatory shifts, Shell's example shows the power of strategic imagination coupled with analytical rigor.

Questions

1. Why was Shell better prepared than its competitors during the 1973 oil crisis, and what does this reveal about the limitations of traditional forecasting? How did qualitative insight give Shell an advantage in a time of deep uncertainty?
2. What distinguishes scenario planning from other forecasting methods, and why is this distinction important for long-term strategic decision-making in volatile industries like energy?
3. Shell's Sky scenario outlines a pathway to net-zero carbon emissions by 2070. How should a company balance ambitious long-term scenarios with more conservative or realistic short-term business constraints?
4. In what ways can scenario planning influence specific corporate decisions, such as capital investment, product development, or market entry? Can you think of an industry where scenario planning might be especially critical today?
5. How does the integration of AI and advanced analytics enhance Shell's scenario planning process? Are there risks in relying too heavily on AI-generated signals in long-term strategy development?
6. Shell's scenario-planning process involves publishing scenarios publicly. What are the advantages and potential disadvantages of this kind of transparency? How might it affect stakeholder trust, investor perception, or regulatory relationships?
7. If you were leading strategic planning for a different type of company (e.g., in retail, healthcare, or technology), how would you adapt Shell's scenario planning approach to your own industry's uncertainties? What drivers of change would you focus on?

Chapter 4

Impact on Marketing, or "Sell"

In This Chapter You Will Learn

1. how AI drives supply chains through marketing or the Sell function;
2. howAI applications affects the Four Ps of marketing: product, price, place & promotion;
3. how personalization at scale has been achieved with AI-enhanced tools, such as micro-segmentation, cross-selling, in-store behavior analysis, and sentiment analysis;
4. how price optimization, merchandizing, and location-based marketing all evolved with AI capabilities; and
5. how customer-centric marketing is created by linking the marketing mix for optimal outcomes.

People think marketing is just advertising—like putting up a flashy sign in front of a store.

In reality, it is much more. It is not just about having a good product; it is about understanding who the customer is, what message will resonate, and what channels will reach them and when. Get it wrong, and you have wasted more than just money; you have lost trust, relevance, and market momentum.

Now consider Glow & Root, a growing natural skincare brand with a loyal online following and four physical locations in Colorado. The founder, Maya, built her brand through word-of-mouth and Instagram buzz. But as the business scaled, traditional marketing methods stopped delivering; social media algorithms shifted, email open rates dropped, and customer preferences fragmented. Maya found herself running expensive campaigns that did not deliver.

That is when Glow & Root adopted an AI-powered marketing platform designed for mid-sized consumer brands. The system analyzed customer behavior across email, web, and social media touchpoints and began delivering real-time recommendations on customers. Within eight months, the brand saw a 30 percent lift in campaign engagement and a 20 percent boost in repeat purchases. Personalized promotions replaced generic blasts. AI transformed marketing from a guessing game into a precision engine.

This is the power of AI in marketing.

4.1 Marketing "Sell" Drives the Supply Chain

Marketing is the business function that identifies, anticipates, and satisfies customer needs. In business jargon it is often referred to as **Sell**. However, marketing goes well beyond communicating with customers—it is a strategic engine that drives demand and shapes modern supply chains.

Marketing is also the function responsible for linking the organization to its customers and is concerned with the downstream part of the supply chain. The marketing task is to identify what customers need and want, create demand for a company's current and new products, and identify market opportunities. Marketing plays a critical role, as providing value to customers drives all actions of the organization and the supply chain.

We have all heard of AI applications being used by companies to gain insights into customer behavior—from customer loyalty programs to analyzing online clickstream data. But AI enables so many more capabilities for Sell, including creating demand.

Company Example: The Barkery

Consider the Barkery, a gourmet dog bakery with the marketing campaign "Treat Yourself Like Your Dog Would." The Barkery launched a campaign flipping the script on how pampered pets are. They leaned into the idea that people spoil their dogs more than themselves. So, the company created a pop-up experience where humans could try dog-inspired gourmet treats, like peanut butter bacon cupcakes, sweet potato scones, and "puppuccino" mocktails.

The key tactics the company uses are driven by AI technology, such as a social media challenge where people post side-by-side pics of how their dog gets pampered versus how they treat themselves (e.g., pup on a plush bed vs. human working hunched over a laptop). The company also posts Influencer Dogs where they have local dog influencers review the treats and events (yes, actual dogs with Instagram accounts). They even sell limited merchandise such as sassy t-shirts, mugs, and hats with phrases like "My Dog Gets Me" and "Sniff First, Ask Questions Later." The campaign went viral locally, increased foot traffic by 40 percent, and got picked up by Buzzfeed in a "Most Extra Dog Events" roundup.

QUESTIONS

1. How did The Barkery leverage emotional connection and humor in its marketing campaign, and why is this approach effective for both pet owners and viral potential?
2. What role did AI play in shaping the Barkery's social media strategy, and how does this differ from traditional marketing approaches?
3. How does user-generated content (like the photo challenge) contribute to the success of a campaign, and what are the risks and rewards of relying on it?
4. In what ways did the Barkery use cross-promotion and influencer marketing differently by featuring "dog influencers?" How might this approach affect brand credibility and engagement?
5. How can companies measure the return on inveswtment (ROI) of experiential marketing campaigns like the Barkery's pop-up event? What metrics would you prioritize in evaluating success?

"Sell" links the organization—and all the other business functions—to its customers (Figure 4.1). Unlike purchasing (Buy) that looks upstream and interfaces toward suppliers, marketing (Sell) is the engine that gets the downstream information that drives the rest of the supply chain. It signals exactly what goods and services the supply chain should supply, to which specific customers, and at which specific time.

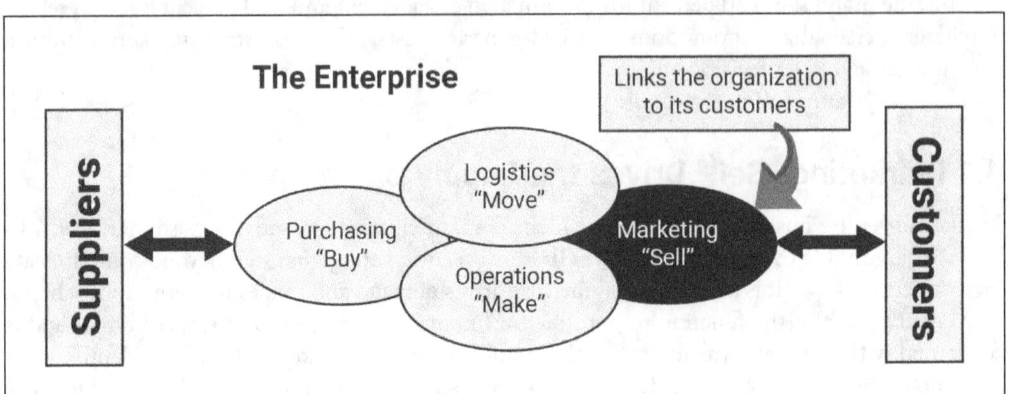

Figure 4.1 The Marketing Function

For an organization and its supply chain to be competitive, it must be better than competitors at meeting customer needs. Sell is responsible for identifying what these needs are and determining how to create value for customers and building strong customer relationships. AI tools enable this capability by capturing data that analyzes customer demand, creates micro-segmentation, and predicts consumer behavior.

Marketing provides answers to questions such as

- What do customers really want? (not just what they say)
- How do we create value for each market segment?
- How much are customers willing to pay? (truthfully, not just what they say)
- What product features result in actual sales, and which features do not matter?
- How much money should we invest in advertising, and how much lift are we getting in return?

Company Example: Spotify and Stitch Fix

Modern AI on the Sell side is superb at answering the above questions. Spotify uses AI to identify the precise music customers want by analyzing their users' listening habits, search queries, playlist choices, and even the time of day they listen. By analyzing all this data, Spotify's AI-driven recommendation engine predicts and serves up the music that customizes the playlist to match individual tastes. This personalized experience keeps users engaged, increases listening time, and boosts customer satisfaction, directly contributing to Spotify's growth and retention strategy.

Then there is Stitch Fix, a company in the fashion industry that uses AI to personalize clothing recommendations for customers. Their algorithms analyze style preferences, size data, feedback from previous orders, and even trends in the fashion world to predict what each customer is likely to love. Human stylists then use these AI-driven insights to create highly curated personalized clothing boxes, blending data with human judgment. This approach helps reduce returns, increases customer satisfaction, and drives repeat purchases.

QUESTIONS

1. How does AI-driven personalization at companies like Spotify and Stitch Fix enhance customer engagement? Why is this a key driver of competitive advantage?
2. What types of data are most valuable in building effective AI recommendation engines? How do these differ between industries like music and fashion?
3. In the case of Stitch Fix, how does the collaboration between AI and human stylists strengthen the overall customer experience? Could AI fully replace the human role?
4. What potential ethical or privacy concerns arise from companies collecting and analyzing personal data for AI-driven personalization? How should companies address this issue?
5. How might AI on the Sell side evolve in the next five years? What other industries could benefit from a similar approach to predictive, personalized selling?

4.1.1 What Drives Customers

Managing the Sell function is far more complex than it appears. The reason is that customers are driven by much more than their basic needs. They also often do not understand what drives their own behaviors. Steve Jobs was famous for saying that customers do not know what they want. He referred to a story about Henry Ford and the Model T. Had Henry Ford asked customers what they wanted, the response would have been "A faster horse!"

The key is to understand and develop the combination of goods and services that precisely meet the expectations for a cleverly targeted set of customers. Remember that resources are finite—so we do not want to spend them on efforts that do not have a big impact.

Consider this: If most customers of athletic shoes only care for three different color choices, offering the product in five colors makes little sense. There would be little benefit, and the supply-chain costs—extra sourcing, manufacturing, distribution, inventory—would be significant. It is critical to understand exactly what satisfies customers and then precisely match product and service offerings. In a world with rapidly shifting markets, companies must anticipate future needs of their customers. Utilizing AI, businesses are becoming adept at this predictive approach.

A prime example is **Nvidia**, which uses AI to effectively predict surging demands for its chips in various tech sectors, enabling precise adjustments in production and supply-chain processes to meet market demands efficiently.

Marketing is further complicated by the fact that there is no single market for any given product or service. All markets can be broken down into multiple segments, each of which has somewhat different requirements. A critical challenge is **market segmentation**—dividing broad markets into specific groups in a meaningful and effective way. Segments that are too broad do not permit the company to target a narrow enough group. On the other hand, segments that are too narrow may not be cost effective.

Target marketing focuses on a specific group of customers rather than the entire market, recognizes the diversity of customers, and does not try to please all customers with the same product offering. The power of target marketing is clear, as the average annual growth rate for total US retail sales is approximately 4.1 percent, while online retail sales are growing at a significantly faster pace—around 8.9 percent. This underscores how data-driven, personalized strategies are fueling online retail's continued surge.

The challenge is to identify the most effective market segments, and their needs, and coordinate with the rest of the supply chain. The goal is to make sure that production and delivery are feasible and to ensure that this can be done in a cost-effective manner. After all, marketing can identify an endless array of customer needs. Meeting them, however, may not be financially feasible. It is for this reason that the link to the supply chain is critical.

Think how marketing strategies have evolved with AI. There was a time, not very long ago, when consumers had to visit multiple stores in search of a shirt that matched their pants, often leaving empty-handed. Today, AI systems have radically transformed this scenario. Now you might merely mention a desired item during a casual conversation at home, and soon after, when you check your phone, targeted advertisements for that very item appear. This is today's target marketing, which not only anticipates but also reacts to consumer behavior in real time.

COMPANY EXAMPLE: Napa Valley Winery

A case in point is a small, family-run winery nestled in the heart of Napa Valley, an area renowned for its premium vineyards, scenic beauty, and boutique winemaking. A few years ago, this winery's marketing team proposed an ambitious idea—rebranding their signature wine with a unique, smoke-colored glass bottle. Their aim was to stand out on crowded retail shelves and evoke a sense of elegance and mystique that they believed would resonate with high-end consumers.

Excited by the concept, the team envisioned how the smoky hue would complement the winery's branding, rooted in the rugged sophistication of Napa's volcanic soil and misty mornings. However, when the idea was passed along to the procurement department, challenges quickly emerged. Smoke-colored bottles were not readily available through the winery's existing suppliers. To source them, the team would need to import from overseas, at significant cost, with added risks related to long lead times and potential supply-chain disruptions.

Still, the idea persisted. Before committing, the winery decided to apply a more analytical lens to the decision, a growing trend in the industry. They gathered customer feedback through surveys, blind tastings, and A/B testing with different packaging options. The results were eye opening. Customers overwhelmingly focused on the label design and wine quality, not the color of the bottle. In fact, many did not even notice the difference. What marketing had believed to be a potential differentiator turned out to be irrelevant to the consumer experience.

This revelation marked a turning point for the winery. It underscored the importance of data-driven decision making over gut instinct, especially in a business where margins can be thin, and marketing investments must be carefully measured. Had they moved forward with the smoke-colored bottles, they would have incurred unnecessary expenses for no tangible return.

Today, the winery uses AI-powered tools to track customer sentiment, test new concepts virtually, and more closely align marketing initiatives with actual buyer behavior. It is a powerful reminder that in the modern supply chain, smart marketing must be informed by real-world data, not just good storytelling.

QUESTIONS

1. What does this case reveal about the risks of relying on intuition or aesthetics alone in marketing decisions, particularly in industries with tight margins like winemaking?
2. How did customer feedback and data challenge the assumptions made by the marketing team, and what does this suggest about the importance of consumer research in product development?
3. What role could AI play in helping small businesses like this winery make smarter, more cost-effective marketing and procurement decisions in the future?
4. How can companies balance creative branding ideas with operational realities in the supply chain? What structures should be in place to evaluate feasibility early in the process?
5. What lessons does this case offer about the importance of cross-functional collaboration between marketing, procurement, and data teams in a modern business environment?

4.1.2 The Impact of AI

As consumers, we are all familiar with the use of AI applications designed for tracking and influencing our behavior. Companies along the Sell function are using every aspect of AI to drive their supply chains, such as location-based marketing, sentiment analysis, and in-store behavior analysis. Retailers are also using AI for merchandising, particularly price and assortment optimization. AI applications have provided more accurate signaling for improved supply-chain performance. Companies rely on point-of-sale (POS) data to create the demand signal, then use the information to inform suppliers and drive the entire supply chain (Figure 4.2).

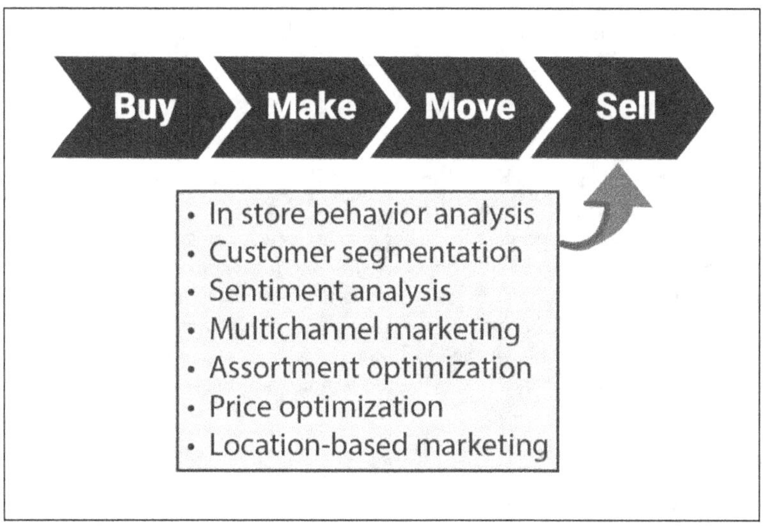

Figure 4.2 AI Applications in Marketing

4.2 The Marketing Mix

Marketing provides intelligence to guide the actions of all other functions within the firm, as they modify operations to accommodate new customer preferences. For marketing to be successful, it

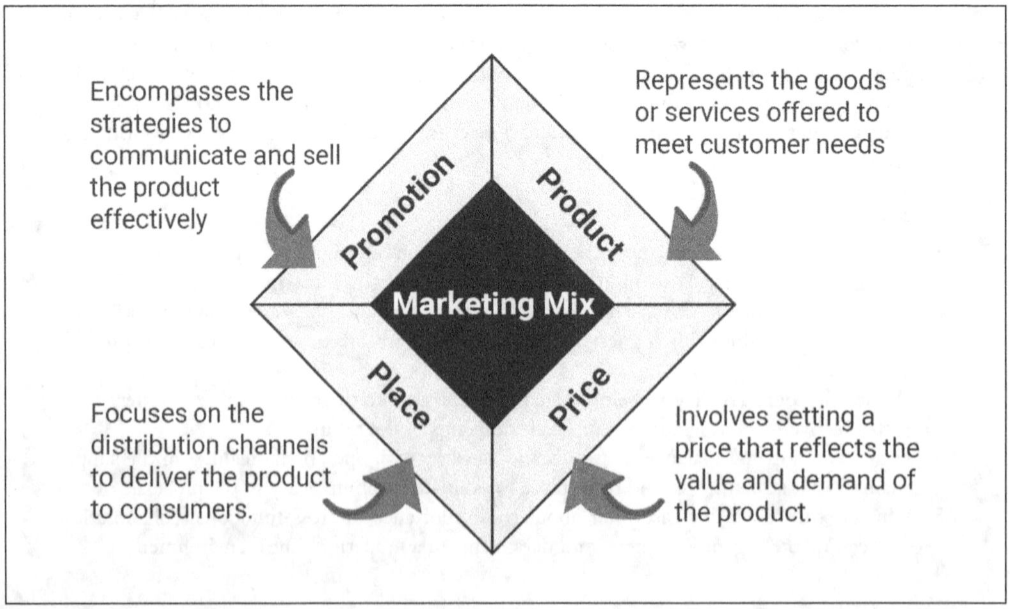

Figure 4.3 The Four Ps

must be supported by the rest of the organization, as the products customers want must be available. Remember that customer demand is worthless if there are no products to sell.

Just as the Sell lever drives the decisions of the organization to produce and deliver exactly what customers want, it also drives supply-chain decisions. Marketing decisions generally fall into four distinct categories known as the **marketing mix**, or the Four Ps (Figure 4.3). These decisions are **Product, Price, Place,** and **Promotion.** Together, these help marketing identify the right product, sold at the right price, in the right place, using the most suitable promotion. These decisions directly drive supplychain management. Unlike in the past when these decisions have been made speculatively, today most have all been optimized using AI.

COMPANY EXAMPLE: Walmart

Walmart exemplifies this approach with its use of AI to align inventory levels and minimize waste. Analyzing historical data, seasonal patterns, and weather forecasts, Walmart anticipates demand fluctuations across its product lines. For instance, recognizing a spike in strawberry Pop-Tart sales preceding hurricanes in specific regions, Walmart preemptively stocks these items to boost sales and improve customer satisfaction during these events. Walmart's predictive analytics system processes more than 2.5 petabytes of data every hour (equivalent to 167 times the information contained in all the books in the US Library of Congress), contributing to a 10 percent reduction in out-of-stock items and bolstering the company's annual revenue, which has recently surpassed $600 billion.

QUESTIONS

1. How does Walmart's use of AI to forecast demand illustrate the value of predictive analytics in supply-chain management, and what are the key data inputs driving its accuracy?
2. What does the example of Pop-Tart sales before hurricanes reveal about the importance of recognizing localized consumer behavior in large-scale retail operations?
3. In what ways does minimizing out-of-stock items through AI-driven forecasting affect customer satisfaction, operational efficiency, and revenue growth for a company like Walmart?
4. How might smaller retailers replicate Walmart's success with predictive analytics, even if they lack the same volume of data or technological resources?

5. What are the risks or limitations of relying heavily on AI for inventory and demand forecasting, and how can companies build resilience into their systems?

4.2.1 Product

Product involves decisions that encompass the bundle of product characteristics that satisfy customer needs. They relate to brand name and functionality. They also address decisions of quality and packaging, which are directly supported by supplier standards and logistics packaging decisions.

Advancements in AI are now unlocking the path for companies to enhance their product offerings by identifying customer preferences and creating highly tailored bundles. Notably, retailers with an online presence have excelled in utilizing consumer data to suggest personalized product combinations. These capabilities are being expanded by other companies, which not only identify related products but also assist customers in selecting items for specific occasions or individuals. This effectively lets the customer know what they want before they even ask. For example, Nike and other clothing brands have implemented size guide quizzes to analyze customer inputs and past purchasing data to recommend the best-fit sizes. This improves customer satisfaction and reduces returns due to poor fit. The deployment of analytics in these scenarios ensures that customers enjoy a more personalized and accurate shopping experience, transforming the way they interact with products online.

COMPANY EXAMPLE: Sony

Cross-selling, another technique to better utilize the Sell lever, enables correlating customer preferences and understanding the bundles customers buy. Online retailers such as Amazon, Barnes & Noble, and Overstock.com have been identifying such customer product preferences for years. Other companies are taking this capability to the next level by not only identifying **correlated products** but also helping customers select products for specific occasions or select individuals. Sony is using a new technology labeled "online discovery." The system identifies products that may surprise or delight customers. It bases its recommendations on past user behavior and the search terms users have employed. Further, Sony's website is integrated to take consumers immediately to the local dealer's website, where they can experience a highly trained and knowledgeable sales force.

QUESTIONS

1. How does cross-selling enhance the overall customer experience and increase sales? What role does AI play in making these recommendations more precise?
2. What are the key differences between traditional cross-selling and Sony's "online discovery" approach? How might these differences affect consumer engagement?
3. How can analyzing customer behavior and search patterns help retailers anticipate unspoken needs or uncover new selling opportunities?
4. What are the advantages of integrating online recommendation systems with off-line dealer networks, as seen in Sony's model? How does this affect the sales funnel?
5. What are some potential risks or limitations of relying on AI-driven cross-selling strategies, particularly when it comes to customer privacy, trust, or recommendation fatigue?

4.2.2 Price

Price refers to the pricing strategy developed for the product. This includes volume and sales prices, seasonal pricing, and price flexibility. Also, pricing strategies and price flexibility are directly tied to supply-chain functionality and costs. We must also remember that this is the only item of the marketing mix that generates revenue. The others are all costs, so we have to make sure to get pricing right.

AI capability now enables price optimization for specific target markets with real-time changes, called **dynamic pricing**. Canadian apparel retailer Northern Group Retail reports that price optimization software has helped it increase gross margins while reducing inventories. Similarly, department store retailer Kohl's credits price optimization with an increase in gross margin.

COMPANY EXAMPLE: Starbucks, Kroger, and Overstock.com

Starbucks leverages its loyalty program data to implement a dynamic pricing strategy, adjusting prices based on customer demand, time of day, and local economic conditions. With more than 100 million transactions per week globally, and nearly 31 million active members, there is a lot of data to choose from. This approach allows Starbucks to maximize revenue during peak times and boost sales during off-peak hours by attracting more customers with lower prices. By analyzing customer behavior and preferences, Starbucks can also offer personalized promotions, enhancing customer satisfaction and loyalty. This data-driven pricing strategy not only improves customer engagement but also optimizes real-time pricing decisions, driving higher revenue and growth for the company.

The effectiveness of different pricing strategies using different marketing mediums for different target groups can now be tested. For example, Kroger uses analytically targeted coupons on which it gets a huge redemption rate, well above the industry average. In addition, Kroger believes the targeted price promotions have increased overall sales. Similarly, the online retailer Overstock.com used an analytics-based gift recommendation system available on its website. It found that customers who used it bought more than double those who did not.

QUESTIONS:

1. How does Starbucks' use of loyalty program data enhance its dynamic pricing strategy, and what are the potential advantages and challenges of adjusting prices in real time?
2. In what ways can personalized promotions, such as those used by Kroger and Starbucks, improve both customer satisfaction and company profitability?
3. What ethical considerations should companies take into account when implementing variable pricing based on customer behavior and local economic conditions?
4. How can companies test the effectiveness of different pricing strategies across marketing channels and customer segments? What metrics should be used to evaluate success?
5. What does Overstock.com's success with its analytics-based gift recommendation system suggest about the power of AI in influencing consumer behavior and increasing basket size?

4.2.3 Place

Place deals with having the product *where* and *when* it is needed. These decisions include the selection of distribution channels and market coverage. They also include traditional logistics, sourcing, and operations decisions, such as inventory management, order processing, transportation, and reverse logistics. The place component of the marketing mix has become especially complicated as multichannel order fulfillment becomes a requirement for almost every retailer. Direct-to-consumer is currently the highest growth and highest potential distribution channel. Tracking and managing inventories across multiple channels—and ensuring fulfillment—is one of the biggest challenges.

Adding to the complication is a large variability across **demand channels**, avenues by which companies reach different markets. Consider that in the United States shipping volumes for brick-and-mortar stores peak in the fall, while direct-to-consumer demand typically peaks from Thanksgiving through the days before Christmas. Not only are there different channels' needs, but those needs are also shifting and evolving. Ensuring fulfillment while keeping costs down is a huge challenge. Estimates are that roughly one-fifth of the cost of a product goes into getting the product to the customer. AI assists in many ways, including analyzing when delivery constraints can be relaxed without negative effects on customer satisfaction.

COMPANY EXAMPLE: Pine & Peak

In a trendy urban neighborhood of Portland, Oregon, a boutique outdoor apparel brand called Pine & Peak has created a seamless omnichannel experience that blurs the line between physical and digital retail. Known for its eco-friendly hiking gear and minimalist aesthetic, the brand has cultivated a loyal following both online and in store. But what truly sets Pine & Peak apart is how it integrates data, design, and customer behavior, across every touchpoint.

When a customer walks into one of Pine & Peak's stores, the space is more than just a showroom—it's a curated experience. RFID tags on clothing trigger interactive displays that tell the story of where each garment was sourced and manufactured. A customer can scan a QR code to see available sizes, read reviews from other hikers, and even watch trail footage of people using the gear in the wild. If the item is not in stock, an associate can place an order on the spot and have it shipped directly to the customer's home, often arriving the next day.

At the same time, the online store mirrors the in-store experience. If a shopper browses a jacket online but doesn't purchase, they might receive an invitation to an exclusive in-store event featuring that same product line. Once in store, personalized recommendations based on their browsing history pop up via an in-app assistant. The online and off-line experiences are not just connected; they inform and enrich each other.

Brick-and-mortar, in this model, becomes a place for brand immersion and discovery, not just inventory. Meanwhile, the online store acts as an extension of the physical space, offering convenience, deeper content, and personalized follow-ups. This dual-channel strength allows Pine & Peak to respond to trends quickly, balance inventory more effectively, and keep customers engaged, whether they are hiking a trail, browsing on their phone, or stepping into a store on a Saturday afternoon.

QUESTIONS

1. How does Pine & Peak's integration of digital and physical retail channels create a more personalized and immersive customer experience? Why is this important in modern retail?
2. What technologies (e.g., RFID, QR codes, in-app assistants) are most critical to Pine & Peak's omnichannel strategy, and how do they enhance operational efficiency and customer satisfaction?
3. In what ways does Pine & Peak's approach to inventory management and fulfillment—such as on-the-spot ordering and next-day delivery—differentiate it from traditional retail models?
4. How can data, collected across online and in-store interactions, be used to refine marketing strategies, improve product design, and drive customer loyalty?
5. What lessons can larger or more traditional retailers learn from Pine & Peak's omnichannel model? What challenges might they face when trying to replicate it at scale?

4.2.4 Promotion

Promotion is an advertising technique to increase product visibility and desirability. Decisions involve promotional strategies and advertising, sales promotions, and public relations. The cost associated with promotion, or advertising goods and services, often represents a sizable proportion of the overall cost of producing an item. However, successful promotion increases sales, so that advertising and other costs are spread over a larger output.

COMPANY EXAMPLE: Overstock.com

AI has taken promotion to another level, transforming the traditional one-size-fits-all approach into *personalized, predictive*, and *highly dynamic marketing*. A standout example of this evolution is Overstock.com, the online discount retailer known for offering home goods and furniture at value prices. Over the past several years, Overstock has quietly built a powerful AI-driven marketing engine that redefines how promotions are timed, personalized, and delivered.

One of Overstock's most effective tools is its targeted and event-driven email marketing system, powered by AI. Rather than sending blanket discounts or weekly sale updates, Overstock tailors its emails to individual customer behaviors and contexts. For instance, if a customer has been browsing mid-century modern sofas, they might receive a personalized email featuring complementary products—like rugs or lighting—priced dynamically based on their browsing history, price sensitivity, and even regional trends.

Even more compelling is how the company responds to event-driven triggers. These could include abandoned carts, seasonal buying patterns, or life events inferred from customer behavior (like moving to a new home or getting married). AI models process real-time data streams to identify the right moment to send an offer, whether that is a limited-time discount, a bundled promotion, or even a reminder that an item they browsed is now on sale. This contextual timing creates a sense of urgency and relevance that drives significantly higher conversion rates.

Overstock has reported that targeted email promotions using AI have led to a measurable lift in both revenue and average order size. Customers are more likely to engage with offers that feel tailored to their needs, and they tend to spend more when those offers align with their preferences and timing. Additionally, AI helps Overstock reduce waste in promotions by limiting discounting to situations where it truly adds value, preserving margins while boosting effectiveness.

What is most notable is that AI does not just automate the process; it learns and adapts. Overstock continually refines its promotional algorithms based on customer responses, A/B tests, and emerging trends. The system then evolves with each campaign, getting smarter about what motivates different customer segments.

QUESTIONS

1. How does Overstock's use of AI in targeted, event-driven promotions differ from traditional email marketing, and why is this shift more effective in today's digital landscape?
2. What types of customer data are most valuable in creating personalized promotions, and how does Overstock balance personalization with concerns about privacy and data ethics?
3. In what ways do event-driven triggers (like abandoned carts or inferred life changes) enhance the relevance and timing of promotional offers?
4. How does Overstock's AI approach help maintain profitability by reducing unnecessary discounts, and what does this reveal about the role of AI in strategic pricing?
5. What are the long-term implications of continuously learning AI systems for marketing teams, and how might roles, skills, or decision-making processes need to evolve?

4.2.5 No "Average" Customer

In the past, companies primarily focused on meeting the demands of the average customer. This meant supplying products to customers without trying to determine or satisfy the unique needs of each individual customer. Today, consumers are knowledgeable; they demand what they want at the highest quality, lowest price, and delivered at record speed. Also, today's customers expect a world-class organization to provide high levels of customer service. To compete in this new economy, companies are putting forth a great deal of effort and money to precisely understand their customers and provide unprecedented levels of customer service and one-on-one customization. This has resulted in large changes in the supply-chain management of companies and the relationships between supply-chain partners who aim to understand and deliver what consumers want.

COMPANY EXAMPLE: Netflix

A compelling example of personalized customer experiences, enabled by advanced data analytics and AI, is Netflix, the global streaming giant. Netflix continues to push the boundaries of personalization by leveraging big data and machine-learning algorithms. They analyze not only user viewing habits and preference, but also nuanced behavioral signals such as watch time, pause frequency, scrolling patterns, and content discovery paths.

As of 2025, approximately 80 percent of the content watched on Netflix is still driven by these hyper-personalized recommendations, reinforcing how critical real-time data insights are to customer engagement. But what is new is the integration of generative AI—used to create dynamic trailers, personalized preview reels, and even customized cover art based on each viewer's taste profile—into the recommendation engine. This deep personalization is designed to not only surface content that users are likely to enjoy but also to present it in a way that immediately captures their attention.

These innovations have helped Netflix maintain a strong competitive edge in an increasingly saturated market. The platform's ability to anticipate what users want next—and deliver it seamlessly—has been a key driver in reducing churn, increasing average watch time, and maintaining high user satisfaction. Netflix has also used analytics to inform its original content production, deciding what to greenlight based on audience insights derived from global and regional trends.

Netflix's success is a prime illustration of how data-driven personalization can revolutionize not only user experience but also broader operational decisions, from content investment to marketing spend. Companies that can mirror this level of insight-driven strategy across their value chain—from product development to customer engagement—are better positioned to thrive in an economy in which personalization is no longer a luxury but an expectation.

This example also shows that the links between marketing, product delivery, and supply chain are increasingly blurred. Understanding individual customer preferences at scale enables organizations to align their entire operation around the end-user, driving growth, loyalty, and long-term sustainability.

QUESTIONS

1. How has Netflix evolved its use of AI from recommending content to personalizing how that content is presented? Why is this level of personalization so impactful for user engagement?
2. What are the benefits and potential risks of using granular behavioral data (e.g., pause frequency, scrolling patterns) to shape customer experiences?
3. How does generative AI expand the possibilities for personalization? What other industries might benefit from similar applications beyond entertainment?
4. In what ways does Netflix's data-driven personalization strategy influence decisions beyond marketing, such as original content production and global expansion?
5. What does Netflix's example reveal about the importance of integrating marketing, product, and operations around the customer? How can other companies replicate this approach to stay competitive?

4.3 Personalization at Scale

AI has brought tremendous advancement to marketing methods and tools. Let's expand on these, specifically micro-segmentation, cross-selling, in-store behavior analysis, and sentiment analysis.

4.3.1 Micro-Segmentation

Market segmentation has long been at the core of the Sell lever, and marketers have been segmenting customers for years.

What is different with AI?

The amount of data available for segmentation has exploded, and the increasing sophistication in analytical tools has enabled the division of customers into ever more granular segments called **micro-segmentation**. New channels of reaching customers and gathering data, such as mobile devices, are constantly emerging. Data on historical purchases can be combined with behavioral data of individual customers—including online clickstream data—enabling companies to track, segment, and personalize offerings. This is true customer-driven marketing now driving the entire supply chain.

Certainly, customers are going to be more likely to respond to offers and promotions that are specific to their needs. Companies know this and have been striving to personalize products and target promotions for years. In the past, offers were customized primarily based on demographic segments or clusters—age, gender, income, geographic location—mostly based on syndicated data.

Now companies are focusing on actual customer behavior as a predictor of future buying behaviors. One example is in-store buying behavioral data to target offers. Regardless of the source and type of data, targeting is much more successful with the identification of unique customers at the individual or household level.

Loyalty Programs. A key reason for loyalty programs and proprietary credit cards is to identify and track customers, creating micro-segments and personalizing offerings. These offerings may be based on simple extrapolations of existing customer behavioral patterns—buying and return patterns—or more complex propensity models. Offers may include products with a high likelihood of customer interest. This can easily be computed with correlations between products that are typically purchased in a bundle. Offers can also include targeted promotions and customized online product displays.

COMPANY EXAMPLE: Starbucks

Starbucks continues to lead in customer engagement by harnessing advanced data analytics and AI across its global operations. With more than 90 million and growing active loyalty members, Starbucks leverages its mobile app as a powerful data engine, tracking preferences, order patterns, and location-based behavior in real time. As of 2025, mobile orders accounted for more than 45 percent of US transactions, highlighting the central role of digital touchpoints in shaping the Starbucks experience.

Through this vast data ecosystem, Starbucks does not just track purchases; it uses machine learning to anticipate customer needs, deliver personalized offers, and even recommend seasonal items based on weather, time of day, and individual taste profiles. Promotions are dynamically generated and tested in real time, improving conversion rates while fostering stronger brand loyalty.

Beyond marketing, Starbucks uses these insights to optimize inventory management at the store level, ensuring that popular items are in stock while reducing waste, especially crucial in managing perishable goods. Its AI-driven forecasting tools also support staffing models and supply-chain planning, adapting quickly to fluctuations in demand across regions.

This data-first strategy has become a core driver of Starbucks' operational agility, customer satisfaction, and sustained revenue growth. By weaving AI into every layer of its business, from barista training to app design, Starbucks exemplifies how digital transformation is not just about tech but also about delivering smarter, more human-centric experiences at scale.

QUESTIONS

1. How does Starbucks use its mobile app and loyalty program data to personalize customer experiences, and why is this important for driving engagement and repeat business?
2. What role does real-time machine learning play in shaping Starbucks' promotional strategies, and how does this affect conversion rates and customer loyalty?
3. In what ways does Starbucks integrate AI beyond marketing—to inventory management, staffing, and supply-chain planning—and what are the business benefits of this approach?
4. How does Starbucks balance the use of advanced data analytics with the need to maintain a consistent and human-centered in-store experience?
5. What lessons can other businesses, especially those in food service or retail, learn from Starbucks' approach to embedding AI and data into daily operations and customer touchpoints?

COMPANY EXAMPLE: Other Retailers

Similarly, **Sephora**'s Beauty Insider loyalty program exemplifies a high-impact, data-driven approach to micro-segmentation. With millions of members worldwide, Sephora uses data analytics to track customer behavior, including browsing patterns, purchase history, and product preferences. AI algorithms further segment customers and offer personalized beauty recommendations and promotions tailored to individual tastes. Through targeted marketing and custom beauty advice delivered via their mobile app and online platform, Sephora has successfully increased customer engagement, conversion rates, and loyalty among its beauty-focused clientele.

Tesco, a UK-based retailer, claims to track almost all sales through its ClubCard program. Drugstore chain **CVS** tracks customer purchases through its ExtraCare card. Data on customer purchase patterns is then used to customize offers provided on register receipts, in kiosks, and in emails. Similarly, **Hallmark** uses its Gold Crown card to segment customers, customize direct mail offers, and assess customer reactions to promotions. These loyalty programs have proved invaluable in gathering customer data and creating personalized marketing.

Success of loyalty programs, however, is in part due to the ability of retailers to offer substantial premiums to loyalty customers, especially high-end retailers such as **Nieman Marcus** and **Nordstrom.** These companies have sufficiently large revenues per customer to be able to offer premiums to their most loyal customers. Nieman Marcus, for example, claims that the top customers in its 20-tiered loyalty program account for almost half of its revenues. Top customers can win free fur coats and even a Lexus luxury car. Lower-margin retailers cannot offer these types of premiums. However, they can still take advantage of the benefits of loyalty programs. such as targeted coupons and gathering customer data. Differentiation of loyalty programs will become increasingly important as customers reach a saturation point in their memberships.

Collaboration. A few major problems exist for companies here. First is the saturation of loyalty programs. Studies suggest that the high proliferation of loyalty programs has resulted in saturation. Most consumers typically carry multiple loyalty cards. In fact, the average household already belongs to more than a dozen loyalty programs. Related to this is the inability for many companies to offer a substantial premium to loyal customers. With so many loyalty programs, consumers are selecting those with the highest rewards. Retailers with high margins are much more able to offer such premiums compared to low-margin retailers.

COMPANY EXAMPLE: Best Buy

The consumer electronics retailer Best Buy faces this issue with its Best Buy Rewards program, which has more than 60 million active members across the United States. The consumer electronics market is highly competitive, with numerous loyalty programs vying for customer attention. On average, consumers enroll in roughly 14 loyalty programs. To stand out, Best Buy Rewards offers valuable perks, including purchase points, exclusive member-only deals, and free shipping on select items, which helps drive engagement and customer retention. By leveraging customer data collected through its loyalty program and online interactions, Best Buy personalizes product recommendations and promotions for its members, enhancing customer engagement despite the competition.

QUESTIONS

1. In a saturated loyalty program landscape, what makes Best Buy's Rewards program effective at driving customer engagement and retention?
2. How can data collected through loyalty programs be used to personalize product recommendations and promotions, and what effect does this have on customer behavior?
3. With consumers enrolled in an average of 14 loyalty programs, what strategies can Best Buy use to remain top of mind and foster deeper brand loyalty?
4. What are the potential risks or limitations of relying heavily on loyalty programs for customer retention, and how can these be mitigated?
5. How does Best Buy's use of loyalty data compare with other industries (e.g., grocery, airlines, streaming), and what cross-industry practices might Best Buy adopt to improve its program further?

The second challenge is to get reliable customer data. Firms relying on credit card–based identification can typically uniquely identify no more than half of customers. Further complicating matters is the Gramm-Leach-Bliley Act in the United States, which places additional restrictions on uniquely identifying customers based on credit card data, and affects data collection strategies for companies like Best Buy. To navigate this, Best Buy must balance data privacy compliance with personalization efforts, continuing to refine its loyalty program to better engage and retain customers within these regulatory boundaries.

A good way to address both problems is through collaboration among retailers and supply-chain partners. Collaboration can enable combining loyalty programs and can provide sharing of data. For example, Air Miles is the name of a separately operated loyalty program in Canada, the Netherlands, and the Middle East. Points are earned on purchases at participating merchants and can be redeemed against flights with specific airlines. Air Miles tracks spending across a variety of retail and service sponsors; members can redeem the accumulated miles for free flights and other rewards.

Collaboration helps in the sharing of costs. CVS, for example, gathers data on its customers through the ExtraCare program, but the manufacturers of products pay for the coupons. Collaboration can also help in the capturing and sharing of multichannel data, particularly for new and less-traditional channels where data may be difficult to capture.

4.3.2 Cross-Selling

Cross-selling is using analytical approaches to recommend other product offerings for specific customers. The online recommendations we see are typically based on correlations between the item purchased and other items purchased by the same customers. This is known as **collaborative filtering**. Recommendations may also be based on specific product attributes. These may provide more accurate recommendations. However, a database of product attributes is more difficult and costly to develop if it does not already exist within an industry. Delivery of product recommendations is usually online but could also be through salespeople. The idea is to increase the purchase bundle, and it is a very effective tool.

State-of-the-art cross-selling uses all the data that can be known about a customer: demographics, purchase history, preferences, real-time locations, and other facts to increase the average purchase size. For example, Amazon employs collaborative filtering to generate "you might also want" prompts for each product bought or visited. The company was a pioneer in this area and began to offer collaborative filtering more than a decade ago. It uses a precalculated **item similarity matrix** to make real-time recommendations for customers it recognizes through cookies. Amazon has reported that one-third of sales are due to its recommendation engine. Another example of this lever is using AI to optimize in-store promotions that link complementary items and bundled products.

COMPANY EXAMPLE: eBay

Cross-selling initiatives, like those employed by eBay, demonstrate how these strategies can drive significant revenue. eBay uses sophisticated recommendation algorithms that analyze browsing history, transaction records, and user interactions with products to make tailored suggestions. More than 35 percent of eBay's sales are attributed to cross-selling, which boosts the average order value and improves customer satisfaction by helping users discover relevant products they might not have found otherwise.

The company continuously refines the algorithms using machine learning and data analytics to ensure that product recommendations remain relevant and engaging. eBay employs recommendation algorithms that analyze user behaviors—including browsing history, transaction records, and interactions—to provide personalized product suggestions. This strategy is to enhance the shopping experience by helping users discover relevant items they might not have found otherwise. In 2024, eBay introduced a new recommendation model, resulting in a 3.76 percent increase in purchases, a 2.74 percent rise in clicks, and a 4.06 percent boost in ad revenue compared to the previous model.

QUESTIONS

1. How does eBay's use of machine learning for cross-selling enhance the customer shopping experience? What factors contribute to its success?
2. What types of user data are most critical for building effective recommendation algorithms, and how does eBay ensure that these recommendations remain relevant over time?
3. How do cross-selling strategies like eBay's affect key business metrics such as average order value, click-through rate, and ad revenue?
4. What ethical considerations arise when companies use detailed customer behavior data to drive personalized product recommendations?
5. How might smaller e-commerce businesses replicate the benefits of AI-driven cross-selling without the same scale of data or technical resources as eBay?

Online distribution has made it possible for retailers to offer a much broader range of products. However, it can be difficult for consumers to navigate through a complex product array. Many online retailers have discovered that they can ease this process for consumers by offering recommendation systems. Most retailers have not developed this capability themselves but have licensed it from software providers

The recommendation systems have become very sophisticated. They can recommend purchases for specific occasions or events based on criteria the customer specifies. Overstock.com uses such an application called GiftFinder to help customers find the best gifts for certain occasions. The application asks consumers to specify the occasion, the age of the gift recipient, the relationship to the recipient, and the interests of the recipient. It then recommends specific products. Similarly, Tesco uses a Bayesian system called Tesco Direct in its online business. The recommendation system recommends specific grocery items that customers may want and replaces a previous set of manual recommendations. In addition to Tesco Direct, the company has Tesco Phone Shop and Tesco Mobile, using different channels to reach the customer.

Cross-selling is increasingly becoming commoditized in online retail. Basically, everyone is doing it. Netflix has built its own recommendation engine, called Cinematch, and has sponsored the Netflix Prize for anyone who can improve its recommendation algorithm. In 2009, the company awarded a $1 million grand prize to the team that created an algorithm that improved Cinematch by 10.6 percent. In recent years, Netflix has continued to advance its recommendation technologies. As of March 2025, the company has been developing a foundation model for personalized recommendations. This new system aims to centralize member preference learning, making it more accessible and useful across various models. By leveraging comprehensive interaction histories and extensive content data, Netflix seeks to improve the efficiency and effectiveness of its recommendation processes.

Companies should try to determine how their specific recommendations will be differentiated from other providers. Differentiation may be difficult, for example, when technology is licensed from providers who sell to multiple customers in an industry. Recommendation technology is important in online environments but will be even more important in mobile technology settings, where small screens and lower bandwidth make it more difficult to search through a wide range of products.

4.3.3 In-Store Behavior Analysis

Another way of gathering information on customer behavior is looking at their in-store behavior. Analyzing data on in-store behavior can help improve store layout, product mix, and shelf positioning, which ultimately serve to increase sales (Figure 4.4). Recent innovations have enabled retailers to track customers' shopping patterns, such as footpath and time spent in different parts of a store. They can draw real-time location data from smartphone applications, shopping cart transponders, or passively monitoring the location of mobile phones within a retail environment.

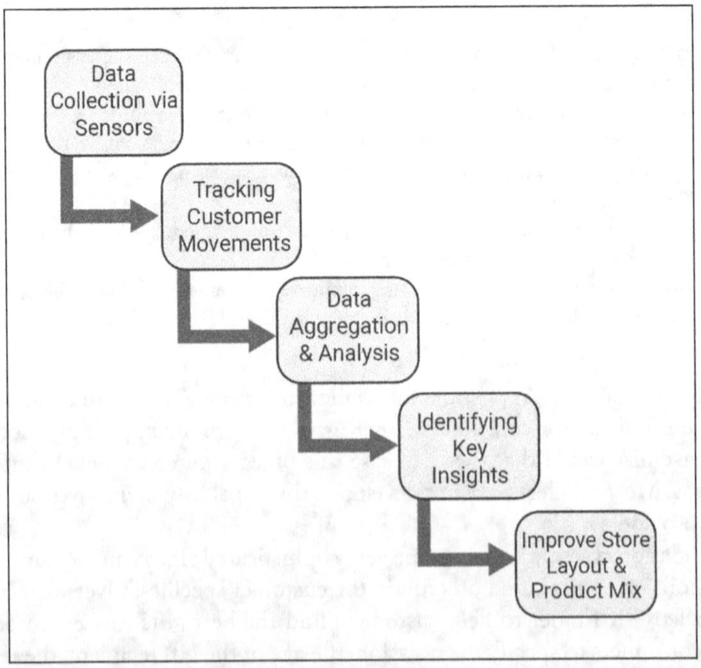

Figure 4.4 In-Store Behavior Analysis Data Flow

Many retailers are using sophisticated—and discreet—technology to monitor in-store customer behavior. Many include facial recognition cameras hidden in the eyes of mannequins to track shoppers, learn their behaviors, and increase sales.

This type of surveillance system, called Shopperception, is being utilized by Walmart. The system links motion sensors to cameras that then track a shopper's product choice on a particular shelf, as well as the time that it takes for the shopper to make a decision. Some retailers use sophisticated image-analysis software connected to their video-surveillance cameras to track in-store traffic patterns and consumer behavior. In-store behavior analysis provides data that can help improve store layout, product mix, and shelf positioning.

Although this can gather valuable information, you should note that it has been controversial and may infringe on some privacy issues. In fact, retailers may leverage *not* using such technologies as part of their marketing campaigns, while they gather customer information through other equally effective means. As we have seen in this section, there are numerous opportunities to learn about customers. Investing in every technological capability is not possible, and avoiding those that are highly controversial may be wise strategically.

4.3.4 Sentiment Analysis

Sentiment analysis is a powerful tool for understanding customer opinions expressed online—from emails, forms, surveys, internal files, and reports. It leverages these voluminous streams of data generated by consumers in the various forms of social media to help inform a variety of business decisions. The tool combines statistical learning with advanced linguistics methods to locate and analyze digital content in real time to home in on the expressed sentiment.

For example, retailers can use sentiment analysis to gauge the real-time response to marketing campaigns and adjust course accordingly. Nike, for instance, employs sentiment analysis to monitor customer opinions across online platforms, analyzing more than 1 million social media posts and customer reviews daily. This system allows Nike to respond promptly to feedback, contributing to a 15 percent increase in customer satisfaction.

Another example is the Gap, which famously relied on sentiment analysis to monitor reactions to a new logo design. When negative responses quickly surfaced on Facebook, Twitter, and blogs, the company promptly reverted to its original logo within 48 hours, demonstrating the power of real-time sentiment monitoring to guide brand decisions effectively.

4.4 Strategic Placement & Pricing

4.4.1 Price Optimization

Price optimization uses AI to determine the optimal pricing of products and services along all channels and markets. Historically, pricing had been more of an art than a science. Managers used to set prices based on intuition, judgment, and experience. The same was true for setting promotions and markdowns and deciding on which items to offer discounts and how much. The success of these old approaches pales in comparison with the results that are obtained through AI.

For example, Accenture found that roughly one-third of retailers using the old approach have at least 10 to 25 percent of their merchandise left over at the end of a season. By contrast, companies that implement AI price optimization typically see huge improvements.

A case in point is New York area grocer D'Agostino Supermarkets, which found implementing price optimization to yield a rise in gross profit of 16.1 percent and net profit by 2 percent. Another example is drugstore retailer Duane Reade, which used price optimization to increase sales of baby formula by 14 percent and of disposable diapers by 10 percent in a 20-store trial. The trial was so successful that the company expanded the use of price optimization in all its stores for nonpharmaceutical products.

Price optimization models have been available for decades. Consider revenue optimization for airline seats and hotel rooms, which have been used for years by online companies such as Priceline.com. It is AI and analytics that have given these models new capabilities.

Today, a variety of data sources can be used to optimize pricing decisions in near real time. The increased granularity of customer data on sales and pricing, coupled with the higher computing power, can provide much richer insights. Price optimization software can gather point-of-sale information and seasonal sales data at the store level. This data can then be used to determine probability distributions and create a set of demand curves for each SKU (stock keeping unit) in a particular store or cluster location. The demand curves identify the products that are the most and least price sensitive. Additional optimization routines leverage these demand curves to determine optimal recommended pricing. Retailers can use the resulting data to analyze promotional events and evaluate sources of sales lift and any underlying costs that these might entail.

Price optimization also results in an increase in sales and margins. A study by the Yankee Group found that returns on price optimization investments typically approach 20 percent. The challenge, however, is that price optimization normally requires a substantial up-front investment. This has been the primary obstacle for small and medium-sized companies. Further, the benefits of price optimization are greatest when undertaken in an integrated fashion with other AI analytics efforts, such as merchandising optimization, shelf space, and assortment management. This is the direction in which companies are moving as they look to optimize the entire marketing mix holistically.

4.4.2 Merchandising

Merchandising is another element of marketing that has been profoundly changed by AI. Merchandising is about making the items available and appealing. It refers to the variety of products available for sale and the display of those products in such a way that it stimulates interest and entices customers to make a purchase.

Two aspects of merchandising are especially affected by AI.

The first is **assortment optimization**, which involves deciding which products to carry at which stores based on local demographics, buyer perception, and other factors. These decisions not only affect sales but also have a huge influence on supply-chain management. Supply-chain management ensures the deliveries of the right quantities at the right location to meet product and quantity requirements; otherwise, there would be no products to sell.

One example is IKEA, the global furniture retailer, which leverages AI to optimize its product assortment and in-store layout across its locations worldwide. By gathering data on customer preferences, shopping behaviors, and store traffic patterns, IKEA identifies the most popular products and tailors its store layouts to enhance the customer experience. Through real-time sales, data analysis, and customer feedback, IKEA makes continuous adjustments to its product offerings and shelf layouts, resulting in a 15 percent increase in sales revenue and a 20 percent reduction in excess inventory levels.

The second aspect of merchandising is **shelf space allocation**. By gathering customer data on the retail store level and observing customer behavior patterns, such as what visually draws or repels the consumer, companies are increasingly learning the types of merchandising strategies that maximize sales. IKEA uses predictive analytics to adapt continuously. By anticipating shifts in consumer trends and adjusting inventory levels and product displays in real time, IKEA has achieved a 10 percent improvement in inventory turnover.

Product inventory, shelf space, and displays have always been a valuable resource for retailers. They have competed for the best shelf space such as end displays. Now AI can be used to determine exactly which merchandising strategies provide the highest level of sales and profits. Think about all the customer behavioral data that companies are currently gathering, including in-store behavior analysis. Certainly all this information provides clues as to how best to visually display the items on store shelves.

Historically, assortment and shelf space optimization have been a seasonal activity. With increased competition and awareness of shelf space utilization, it has evolved into a continuous process. Space optimization was often done by manufacturers as category captains in grocery retailing. Today it is evolving toward a joint process between retailers and key suppliers. This is especially true through initiatives such as **vendor-managed inventory (VMI)**' an approach in which suppliers or vendors manage the inventory levels of their customers. **Planogramming**, creating and implementing visual merchandising tools to optimize store layouts, has increasingly become more computerized with the use of computer-aided design (CAD) tools. However, it has only been through the use of optimization techniques that the real effect on the profit of different planograms and store layouts can be realized. Automated planogram generation uses optimization techniques to make store-level trade-offs between shelf space and assortment depth while considering desired product display blocking.

AI also analyzes the financial effect of different shelf space assortments, specific preferences of shoppers in the store area, and the likely responses to pricing and promotions. These tools can provide information on profit contributions of each brand and SKU and even identify cannibalization effects. AI now optimizes assortments at the micro level, looking at brands, packaging, and sizing. It then optimizes macro-store layouts, such as space allocation, and then integrates them with the micro-store analysis.

4.4.3 Location-Based Marketing

Location-based marketing is an AI application that is based on personal location data, such as mobile devices, and can enable real-time responsiveness to customers. It can then be used to target consumers who are close to stores or already in them. As a consumer approaches a store, for example, that store may send a special offer for an item to the customer's smartphone. In fact, it is estimated that nearly 50 percent of smartphone owners use or plan to use their phones for mobile shopping. The company Placecast, specializing in location-based marketing, claims that more than 50 percent of its users have made a purchase as a result of such location-based ads called ShopAlerts.

An example is McDonald's, which has leveraged AI to optimize its menu offerings and engage customers through location-based marketing. Using its mobile app and loyalty program, McDonald's gathers data on customer preferences, order history, and location to tailor promotions and introduce new, region-specific menu items. By analyzing this data, McDonald's has successfully introduced new menu items aligned with local tastes, contributing to a 12 percent increase in sales within the first year of launch. Data-driven promotions and personalized offers have also led to a 20 percent boost in customer loyalty and repeat visits.

There is a huge amount of information on where people are in the world. An early source of personal location data came from individuals' credit and debit card payments. This information was linked to personal identification data from cards swiped at POS terminals, which are typically in fixed locations. The advancement of technologies such as GPS allowed us to quickly locate a device as small as a mobile phone within a few dozen meters. This type of personal location data can pinpoint accurately, within a few city blocks, where a person—or really the device—is in real time.

Geo-Targeted Advertising. Geo-targeted mobile advertising is one of the most common ways organizations can use personal location data. Consumers can sign up to receive geo-targeted ads. Personalized ads can then appear on the consumer's smartphone when they are close to that store. This could also include coupon offers from restaurants or coffee shops. These geo-targeted advertisements are very effective, as the ads go to the customer precisely when the purchase decision is about to be made. This is a fast-growing area of AI, and companies are investing at increasing rates in this service for advertising.

COMPANY EXAMPLE: Nike

An example of geo-targeted advertising in action is Nike's innovative location-based marketing campaign. Nike uses geofencing technology to reach customers with personalized promotions and messages when they're near a Nike store or attending sports events. Through its Nike App, Nike tracks users' running routes, providing rewards and incentives to encourage store visits. This allows the company to send personalized offers and promotional messages to users when they are near physical store locations, encouraging visits and purchases. Inside the retail stores, Nike employs geofencing to detect when a customer enters, prompting the Nike App to update with relevant offers and content. This integration allows customers to book one-on-one appointments, scan mannequins for product information, and have specific sizes delivered to dressing rooms, thereby enhancing the in-store shopping experience.

This strategy has resulted in a 20 percent increase of in-store visits from app users receiving geofenced notifications. By integrating location data with user preferences and purchase history, Nike can offer highly relevant content and product recommendations, leading to a 15 percent uplift in conversion rates. By harnessing the power of location-based marketing, Nike not only boosts store traffic but also strengthens brand engagement, driving increased sales across its retail network.

By integrating geofencing technology with user preferences and purchase history, Nike delivers highly relevant content and product recommendations, strengthening brand engagement and driving increased sales across its retail network.

QUESTIONS

1. How does Nike's use of geofencing through the Nike App enhance both the digital and in-store customer experience, and what makes this approach effective?
2. What role does customer data—such as running routes, purchase history, and location—play in Nike's ability to deliver personalized promotions, and how does this affect conversion rates?

3. What are the key benefits and potential privacy concerns associated with using geolocation and app-based tracking in retail marketing?
4. In what ways does integrating digital tools like the Nike App with physical store experiences (e.g., scanning mannequins, booking appointments) improve customer satisfaction and loyalty?
5. How can other retailers adopt similar geo-targeted marketing strategies, and what challenges might they face in terms of technology, scale, or customer adoption?

Enabling Technologies. Personal location data has tremendous value in understanding customer shopping patterns. Technologies such as GPS and RFID enable tracking shoppers to identify buying patterns. None are perfect, and each has certain benefits. GPS signals, for example, often do not penetrate indoors so companies need to use other technologies to track shoppers. One option is RFID with tags placed on shopping carts.

Today, RFID tags are cheap and accurate. They enable companies to monitor customer movements in real time as they move about the store. The problem is that they monitor the cart and not the shopper, so often they provide wrong information. Shoppers, for example, often leave carts in aisles while they go about their shopping.

The retailer Target exemplifies how advanced technology can be used effectively in tracking shopper behavior. By equipping shopping carts with RFID tags, Target captures real-time data on shopper traffic and in-store movements. This approach provides granular insights into high-foot-traffic areas and customer engagement with specific products or displays. Complementing RFID, Target also employs video cameras strategically placed to enhance its understanding of general traffic flow. They can be an excellent source of information for understanding general traffic flow.

Simply looking at customer location information can tell companies about foot-traffic density—where most customers are going. It also provides detailed insights about where shoppers slow down, such as looking at a particular type of display. This data can be studied to see responses to promotions and advertising, and then these patterns can be linked with data on product purchases, customer demographics, and historical buying patterns. Such detailed information provides intelligence to offer better product choices, better product assortments and merchandising, and improved store layouts.

Aggregated Location Data. As the availability of personal location data becomes more common and awareness of its value more widespread, new business opportunities are emerging that take advantage of both aggregated and raw data. One striking example is Google Maps' real-time traffic insights, which rely on anonymized, aggregated location data from millions of mobile devices. When many users' phones report slower movement along a route, Google's algorithms identify congestion and instantly update traffic conditions, helping drivers choose faster paths and cities manage traffic flow more intelligently. This collective use of location data demonstrates how individual digital traces, when aggregated responsibly, can generate powerful public and commercial value.

4.5 Customer-Centric Marketing

In this concluding section, we put together the big picture from this chapter. We have seen that AI has changed the marketing function by enhancing traditional tools and creating a host of new tools with new capabilities. Here are the key aspects.

4.5.1 Optimizing Marketing Spend

An old marketing complaint goes like this: "Half of the money I spend on advertising is wasted; the trouble is, I don't know which half."

With AI, this is no longer true. To achieve the greatest impact on the marketing mix, companies have many possible media and promotional approaches from which to choose. The number of possibilities is growing all the time as technology rapidly advances. Further, companies can optimize the marketing mix—product, price, place, and promotion—together in one bundle rather than as separate entities, determining which marketing investments work and which are less effective.

Resources are finite. Marketing mix models help companies determine where they should spend their marketing resources for maximum effect. These models allow companies to determine which marketing vehicles provide the greatest sales lift. Using econometric models, they can determine how much lift a particular promotional approach generates for which target market. Variations on and combinations of variables can be tested for maximum effect. For example, they can examine how often a particular campaign should be run, with what frequency, and through which medium.

The rise of marketing mix models is one factor behind the decline of traditional print and broadcast advertising and the rise of online media. In general, marketing mix analyses have determined that the traditional, nonaddressable media are less effective overall at lifting sales, and retailers and other advertisers have responded by reallocating their resources.

In response to the evolving advertising landscape, there has been a pronounced shift toward digital channels. Projections for 2025 indicated that digital advertising would constitute nearly 61 percent of global advertising expenditure, while television's share was expected to decline to less than 22 percent. This transition is driven by the expansive reach of digital platforms and the increasing number of social media users worldwide. As of January 2025, approximately 5.24 billion individuals, or 63.9 percent of the global population, are active on social media. This figure is projected to grow to more than six billion by 2028. The surge in social media usage underscores the importance of online advertising, offering companies enhanced capabilities for precise targeting and real-time consumer engagement.

This paradigm shift reflects the ongoing evolution in marketing strategies, with digital platforms increasingly favored for their ability to deliver personalized content and measurable outcomes. The integration of data analytics and algorithmic strategies further amplifies the effectiveness of digital advertising, aligning with broader trends identified through marketing mix models.

4.5.2 Linking the Marketing Mix

Numerous AI applications exist for companies to optimize elements of the marketing mix, from understanding customer behavior to price optimization and merchandising. However, these elements are all related. Merchandising affects pricing, which affects store behavior. What is really new and important is to **optimize holistically** as a bundle and not separately.

An excellent example is that assortment and space optimization are also closely linked with price optimization. It is difficult to optimize prices on the wrong product inventory. Further, assortment and shelf space optimization affect one of the broadest sets of business processes among the categories available to retailers. Limited Brands, for example, uses a randomized testing approach and control groups to improve store assortments, as well as segmentation of stores by brand and customer.

A prominent example of the holistic marketing approach is demonstrated by Kroger, the grocery retail chain. Kroger uses AI to implement highly personalized marketing strategies, including analytically targeted coupons that achieve redemption rates well above the industry average, with some campaigns seeing rates more than 10 percent. By leveraging customer data to tailor promotions and offers, Kroger has increased overall sales by more than 5 percent annually. This data-driven approach has also yielded a 10 percent boost in customer satisfaction scores and a 15 percent increase in customer retention rates, fostering stronger customer relationships and brand loyalty. Additionally, Kroger's segmentation efforts have driven a 20 percent increase in transaction values among targeted segments, proving the power of integrated marketing strategies.

An added complication for companies is the growth of the **omnichannel**—the seamless interface of different channels to create one unifying experience for the customer. For the customer, this means availability, ease, and the ability to purchase across channels. For the retailer, this involves huge coordination of inventories, variations in demand, and pricing. Companies are aware that to drive sales, customer satisfaction, and loyalty, they need to enhance the omnichannel experience. Retailers can use AI to seamlessly integrate promotions and pricing for shoppers, whether those consumers are online or in store. Williams-Sonoma, for example, has integrated customer databases with information on some 60 million households. The company tracks information such as household income, housing values, and number of children. It then uses this information to target emails. The response based on this information is up to 18 times the response rate of emails that are not targeted.

It is important to remember the linkages between the elements of the marketing mix. These linkages then need to be tied to the other supply-chain levers. The advances in AI provide an unprecedented view of consumer behavior. Marketing, however, needs to work with operations to ensure that product with the exact characteristics needed are produced; sourcing needs to ensure that supplies are available in a timely and cost-effective manner; logistics needs to ensure deliveries of exactly what is needed throughout the supply chain; also, finance needs to be involved in securing appropriate funding. The AI advances in marketing cannot bring success if they are not linked to fulfillment. This involves coordination across the entire supply chain.

Chapter Review

Key Points

1. The Sell lever of the supply chain is Marketing interfacing with the customer and concerned with the downstream part of the supply chain.
2. AI has enabled optimization of all the Four Ps of Marketing: Product, Price, Place, and Promotion.
3. Personalization at scale can be achieved with tools such as micro-segmentation, cross-selling, in-store behavior analysis, and sentiment analysis, enhanced with AI.
4. Traditional tools such as price optimization, merchandizing and location-based marketing have new capabilities when enhanced with AI.
5. Location-based marketing AI tools use geo-targeted advertising, providing new insights into customer shopping patterns.
6. Successful outcomes with AI require optimizing elements of the marketing mix holistically as a bundle, and not separately.

Discussion Questions

1. What do you think truly drives customers to make a purchase today? How have these motivations changed over the past 10 years?
2. How is AI transforming the way companies understand and reach customers? Are there any risks or downsides to this transformation?
3. Which element of the marketing mix do you think has the most influence on customer decisions, and why? Can this change, depending on the product or market?
4. Why is it important for marketers to think beyond the average customer? What problems could arise from marketing to the average?
5. How does micro-segmentation create more effective marketing campaigns? Can companies go too far with personalization?
6. What are some good examples of cross-selling you have experienced personally? Did it work on you? What made it feel helpful versus pushy?
7. In what ways can in-store behavior analysis enhance the customer experience? Is there a point where data tracking in stores becomes invasive?
8. How can companies use sentiment analysis to make better business decisions? What are the limitations of relying on sentiment data?

Case Studies

TARGET

Fifty billion dollars—that is a rough estimate of how much money marketers are putting into understanding their customers. So how can advanced technologies like Modern AI & big data analytics enhance these efforts? Consider the case of Target. Target's sophisticated computer system is able to guess that a female shopper is pregnant—and in the second trimester, no less.

So how does Target know this?

The company tracks customer purchases and applies sophisticated analytics to the data. It is based on a purchase of an assortment of 25 items that pregnant women buy. Some of those items include vitamins, zinc, and magnesium supplements. A few extra items are added to the mix—for example, cocoa butter, large clothing, or a rocking chair—and the computer model says the customer is pregnant. Target assigns every customer an ID number that is tied to their name, email address, and credit card number. The company can track purchases and send coupons based on the customer's preferences. The company even knows that if it can lure mothers-to-be in their second trimester, they will maintain the highest loyalty rates. That is a lot of customer knowledge.

In the case of Target, however, an irate father scolded a Target manager because his teenage daughter received baby coupons in the mail. As it turned out, his daughter was indeed pregnant. *AI was right.*

In the modern retail landscape, it is not just about knowing your customers; it is about anticipating their preferences before they even recognize them. Modern AI is paving the way for this proactive approach. For instance, Tesco has harnessed AI-driven analytics to dynamically adjust their inventory and marketing strategies. By analyzing vast amounts of data, their AI systems predict customer purchasing behaviors, enabling Tesco to stock products tailored to anticipated needs. This forward-thinking strategy not only enhances customer satisfaction but also streamlines operations, leading to sustained growth and a competitive edge in the fast-paced retail sector.

Target is pushing the boundaries of generative AI even further. As of June 2024, Target launched a transformative initiative to embed generative-AI technology in the workflows of store team members across its entire chain. This empowers employees with real-time predictive tools to optimize customer service and streamline inventory management. AI tools now assist in making proactive suggestions based on customers' past purchasing patterns and anticipated product needs.

For example, if AI detects an increase in demand for specific products, it alerts team members to reorder or reallocate stock promptly. This system continually learns from interactions, sharpening its ability to provide personalized suggestions and timely inventory insights—elevating customer experience while supporting inventory control.

Questions

1. How can predictive analytics and AI be used ethically to anticipate customer needs without violating personal privacy? (Consider the balance between personalization and intrusion.)
2. What are the potential risks and benefits of companies like Target being able to predict major life events such as pregnancy based on shopping patterns?
3. How does the integration of generative AI into employee workflows (as seen at Target in 2024) change the role of human workers in retail operations? Does it augment their work or begin to replace decision making?
4. In what ways can AI-driven inventory and marketing optimization lead to improved efficiency and customer satisfaction? Are there limits to this automation?
5. Should customers be explicitly informed when their purchasing data is being used to make predictions about them? Why or why not? Where is the line between smart marketing and manipulation?
6. What kind of regulations or policies should be in place to govern how retailers collect, analyze, and act on customer data?
7. How might public trust in a brand be affected if customers discover just how much the company knows about them, before they share it themselves?
8. Could there be value in customers having more control over their own data, such as being able to opt-in to receive predictive offers or insights? How would that change the dynamic?

FAST FASHION

Fast fashion has revolutionized the apparel industry by dramatically shortening the product life cycle—from concept to store shelf—while maintaining affordability and trend relevance. This model thrives on rapid inventory turnover, responsive supply chains, and, increasingly, AI-driven insights. Industry giants like Zara and Shein exemplify how **Lean principles** and **digital tools** can be combined to deliver fashion at unprecedented speed and scale.

Lean Principles in Fast Fashion

At its core, fast fashion aligns closely with Lean manufacturing principles—eliminating waste, streamlining processes, and maximizing customer value. Companies like Zara apply Lean thinking to nearly every aspect of their supply chain, from design to distribution. Zara, part of the Inditex Group, is known for vertically integrating its supply chain. Unlike traditional retailers that plan collections months in advance, Zara designs, manufactures, and delivers new styles to stores in as little as two to three weeks. Lean practices such as just-in-time (JIT) production, minimal inventory holding, and tight feedback loops from stores to designers allow Zara to minimize unsold inventory and react quickly to customer preferences.

In contrast to bulk ordering, Zara produces small initial batches and relies on real-time sales data to determine which styles to restock. This reduces overproduction—a key waste in Lean methodology—and encourages frequent visits from consumers seeking the latest styles.

Fast Inventory Turnover: Churn as a Strategic Asset

High inventory turnover, or "churn," is a hallmark of fast fashion. Instead of seasonal collections, companies launch new products weekly or even daily. This rapid product cycle fosters a sense of urgency among consumers and helps businesses stay agile in responding to changing trends.

Shein, a Chinese-based e-commerce giant, has taken churn to a new level. Shein lists thousands of new items each day, enabled by a decentralized and digitized supply network. Its agile production model breaks down traditional mass production by manufacturing in micro-batches (as few as 100 pieces) and scaling up only when demand is proven.

The churn not only creates scarcity and demand but also reduces the need for deep discounting, a margin-eroding necessity in slower-moving retail operations. Both Zara and Shein rely on limited stock and frequent refreshes to keep customers returning.

The Role of AI and Data-Driven Design

AI is now a key driver in fast fashion's evolution. From predicting trends to optimizing logistics, AI helps brands act on data faster and with greater accuracy.

Zara utilizes AI for demand forecasting, stock management, and store layout optimization. Machine-learning models analyze sales patterns, weather data, and customer feedback to guide restocking decisions. AI-driven tools also assist in product recommendations and personalization on their e-commerce platforms. Shein takes a data-first approach to design itself. Using AI algorithms, Shein monitors search trends, social media, and online behavior to determine what styles are gaining popularity. Its in-house design teams then quickly translate these insights into new products. This closed-loop system in which customer data informs design reduces risk and speeds up innovation. AI also powers Shein's operational efficiency, including warehouse automation, demand-based pricing models, and real-time supply-chain adjustments.

Sensors and Real-Time Feedback Loops

Fast fashion retailers increasingly leverage IoT sensors and RFID technology to close the gap between physical and digital operations. Zara has deployed RFID tags across its inventory, allowing it to track garments from warehouse to store in real time. These sensors feed data into centralized systems, giving store managers and corporate decision-makers live updates on stock levels and customer interaction.

This data helps automate replenishment, minimize stockouts, and enable "buy online, pick up in-store" services more efficiently. With real-time insights, Zara can even relocate inventory across stores based on regional demand patterns—another Lean tactic that reduces waste and maximizes sales.

Fast fashion is a prime example of Lean thinking supercharged by technology. Brands like Zara and Shein have reshaped consumer expectations by delivering trend-responsive fashion with speed, agility, and data intelligence. Their success lies in combining Lean principles—waste reduction, flexibility, customer focus—with cutting-edge tools like AI, IoT sensors, and real-time analytics. However, this model is not without criticism, particularly concerning sustainability and labor practices. As the industry evolves, integrating environmental accountability with operational efficiency will be the next frontier for fast fashion.

Questions

1. How do lean principles apply differently in fast fashion compared to traditional manufacturing industries? What unique adaptations have Zara and Shein made to fit Lean methods into fashion retail?
2. What are the advantages and risks of rapid inventory churn in the fast fashion model? How does this affect both profitability and sustainability?
3. How does vertical integration support Zara's ability to respond quickly to changing consumer preferences? Could this model be applied in other industries?
4. In what ways does Shein's use of AI go beyond simple trend forecasting? How does it influence product development, marketing, and customer experience?
5. What role do sensors and RFID technology play in optimizing Zara's supply chain? Can these technologies help traditional retailers compete with fast fashion brands?
6. How do data-driven feedback loops transform the role of designers and buyers in fast fashion companies? Is creativity compromised or enhanced by reliance on data?
7. Compare the customer experience of shopping at Zara versus browsing on Shein's digital platform.
8. How do their business models shape consumer behavior? What ethical and environmental concerns are associated with fast fashion, and how do they relate to the Lean principle of reducing waste? Is fast fashion inherently unsustainable, or can lean and green coexist?
9. Should other industries (e.g., tech, food, home goods) adopt similar models of fast turnover and AI-driven design? What lessons can be transferred, and what pitfalls should be avoided?
10. If fast fashion companies were required to internalize the environmental cost of their operations, how might that change their use of technology, supply-chain design, or pricing strategies? Could AI help them balance speed with sustainability?

Chapter 5

Impact on Operations, or "Make"

In This Chapter You Will Learn

1. that operations is the function responsible for creating goods and services we sell;
2. why it is critical to coordinate Make and Sell;
3. about decisions involved in Make and how AI affects them;
4. how AI improves the production process, including inventory and quality management;
5. about the power of AI in running the digital factory;
6. how operations connects the entire value chain.

People think operations is just about making things, like baking a cake. In reality, it is more like running a pop-up bakery during a street festival, with unpredictable crowds, a temperamental oven, and a dozen custom orders flying in every hour.

Operations is not just about making products; it is aligning production schedules with fluctuating demand, coordinating equipment maintenance, managing labor shifts, ensuring quality standards, and tracking expiration dates on organic ingredients. A single misstep in operations—like a broken filling machine or a delayed ingredient—can cascade into missed shipments, unhappy customers, and expensive bottlenecks.

Consider WildGold Botanicals, a sustainable skincare startup known for its minimalist, eco-friendly formulations. All went well in the beginning. But as orders grew, the founder struggled to scale production while maintaining quality and avoiding costly waste. They used spreadsheets to manage batch schedules and inventory. Staff often overproduced, underproduced, or spent time idle, waiting for raw materials. Then WildGold adopted an AI-powered operations platform that connected sales forecasts with production planning, inventory, and equipment status in real time. Within six months, the company cut production delays by 30 percent, reduced ingredient waste by 20 percent, and improved on-time order delivery to 98 percent. The operations function shifted from firefighting mode to a finely tuned engine of efficiency.

This is the power of AI in operations.

5.1 Making the Things We Sell

Operations management (OM), or operations - is the business function that produces the company's goods and services. In business jargon it is often referred to as **Make** and is responsible for orchestrating the efforts needed to transform inputs into finished products. Without operations, there would be no products to sell, whether tangible goods or services (Figure 5.1).

Operations is responsible for planning, organizing, and managing all the resources needed to produce a company's products. This includes raw materials, people, equipment, technology, and

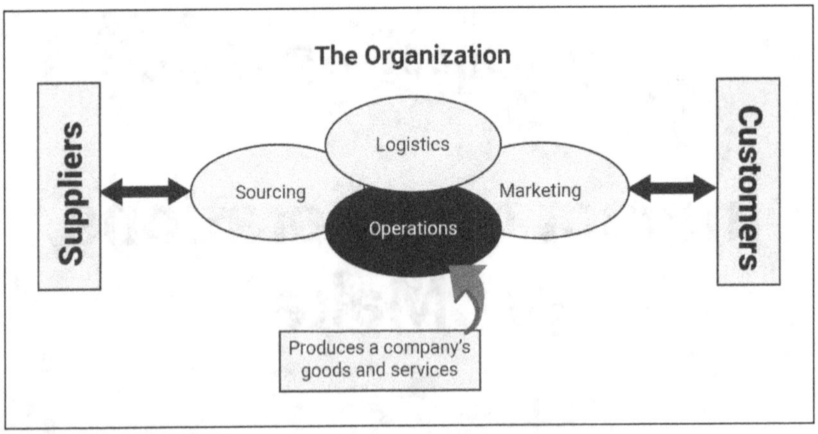

Figure 5.1 The Operations Function

information. It is a core function of every company, whether large or small, service or manufacturing, profit or nonprofit.

Consider Gap Inc. Gap's marketing function identifies characteristics of products desired by customers for each of its three brands—Banana Republic, Gap, and Old Navy. However, it is the operations function that must design and produce the merchandise, from t-shirts to jeans. It must also ensure that the right goods—variety, colors, and styles—are available in the right quantities at various retail locations across the globe. And it must be done cost effectively. Operations management can be described as having **a transformational role** in the supply chain by transforming resources into finished goods and services (Figure 5.2).

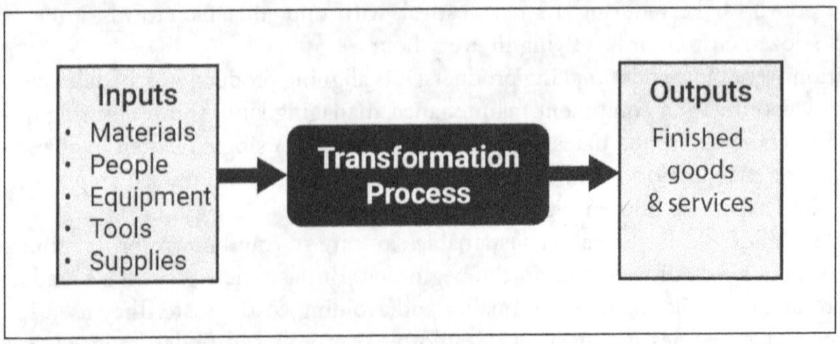

Figure 5.2 The Transformational Role of Operations

COMPANY EXAMPLE: Apple

Apple's operations strategy is one of the key pillars of its sustained success, consistently setting benchmarks across the tech and retail industries. At the heart of this strategy lies a finely tuned balance of innovation, efficiency, and agility—qualities that enable Apple to deliver premium products at scale while maintaining strong financial performance.

Some of Apple's inputs include product design ideas (e.g., engineering talent), raw materials (e.g., aluminum, cobalt, copper, and rare earth minerals), various components (e.g., chips, displays, batteries, magnets, and other electronic parts), and recycled materials. Outputs include finished products such as iPhones, iPads, Macs, Apple Watches, AirPods, and accessories. Getting from the inputs to the outputs requires tremendous coordination. This is shown in Figure 5.3. The transformation process takes all the inputs and converts them into products consumers buy. This is operations management.

A prime example of operational excellence is the iPhone, Apple's flagship product. The iPhone continues to deliver substantial profitability, with gross margins hovering around 38 percent, despite increasing component costs and global economic fluctuations. Behind this success is a sophisticated global supply chain, which spans over 300 suppliers across more than 40 countries. This extensive network allows Apple to source the best components while maintaining flexibility and resilience in production.

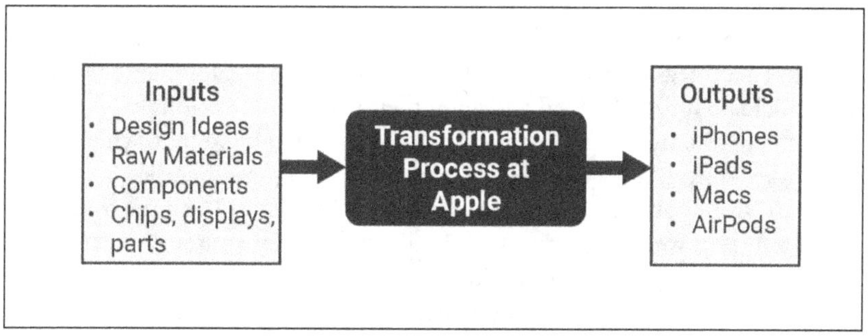

Figure 5.3 The Transformation Process at Apple

Inventory management is a critical area. Even during periods of high demand and supply-chain disruptions, Apple maintains remarkably low inventory levels. Its just-in-time model ensures that devices are available to customers without the burden of overstocking inventory.

New product launches are also a study in operational precision. The iPhone 15 Pro models significantly contributed to Apple's robust sales in fiscal year 2024. In the first three quarters alone, Apple sold 151.3 million iPhones worldwide. The iPhone 15 Pro Max emerged as the best-selling smartphone globally in the first half of 2024, with 21.8 million units shipped.

Apple's capability to scale up production and distribution rapidly is the key to enabling this, getting the products as quickly as they are demanded. Apple's supply chain is so efficient that it takes only nine days to transform raw materials into a new iPhone, ready for sale. This quick turnaround highlights Apple's impressive operational efficiency, moving products swiftly from manufacturing to retail shelves.

QUESTIONS

1. How does Apple achieve a balance between innovation and operational efficiency in its transformation process? Why is this balance critical to its competitive advantage?
2. In what ways does Apple's just-in-time inventory model contribute to both cost savings and environmental sustainability? What risks might this model introduce?
3. Apple's global supply chain spans over 300 suppliers in more than 40 countries. How does this global footprint support Apple's operations strategy, and what challenges might it pose in times of geopolitical or economic instability?
4. Apple's ability to turn raw materials into finished iPhones in just nine days is a testament to its operational efficiency. What systems or capabilities must be in place to support such rapid transformation? How might AI or automation be involved in enabling this speed?

5.1.1 Optimizing Operations

Operations management is complex. There are many decisions that are involved in transforming inputs into finished products. For operations to be able to design and produce the products customers want, specific customer preferences, such as likes and dislikes, need to be translated into tangible characteristics.

When describing products, customers typically speak in general terms, such as "this is pretty" or "it feels good." But these are subjective statements. It is up to operations to translate these sentiments into specific, and measurable, product features. Operations then must decide on the type of process needed to produce the specific product. This includes types of raw materials, level of technology and automation, skill levels and job requirements of workers, capacity of facilities, scheduling work, and complying with Environmental, Social, and Governance (ESG) requirements.

The role of AI is pivotal here, bringing valuable insights to each step in the process. With AI, operations management has new tools for predictive modeling, decision optimization, and quality control. By analyzing historical and real-time data, AI can predict demand trends, highlight areas for efficiency gains, and reduce production costs. AI can also assess quality metrics in real time to minimize defects.

Table 5.1 lists some key decisions for which operations is responsible, and how AI enhances the tasks. The first is **forecasting** (covered in Chapter 3), which is a foundational task in enabling businesses to anticipate future demand and align their resources accordingly.

In the past, forecasting relied on historical data and human judgment, which often struggled with sudden market shifts. AI transforms this process by using machine-learning algorithms that continuously learn from both structured data (such as sales history) and unstructured data (such as social media trends or weather patterns). This allows for more accurate, dynamic forecasts that adjust in real time, reducing stockouts, minimizing overproduction, and improving service levels.

Another decision is **capacity planning**, which is establishing exactly how much production capability is needed. Then there is **inventory management**, determining how many goods to order, produce, and store. Managing the right inventory at the right time is a delicate balancing act. AI systems use real-time data and predictive analytics to determine optimal inventory levels. By recognizing patterns in customer purchasing behavior and supply variability, AI can suggest when to reorder stock, identify excess inventory, and even detect fraud or shrinkage. Companies, including Amazon and Walmart, now use AI to automate inventory replenishment decisions across all product categories.

Table 5.1 Operations Management Tasks & AI Improvements

Operations Management Task	Definition	How AI Can Improve This Task
Forecasting	Predicting future demand for products or services to ensure proper planning of resources, inventory, and capacity	AI improves accuracy using machine-learning algorithms on historical and real-time data, reducing forecast errors.
Capacity Planning	Determining the production capacity needed by an organization to meet changing demands	AI dynamically adjusts capacity plans based on predictive analytics and demand variability.
Inventory Management	Supervising the ordering, storage, and use of materials and products to avoid stockouts or excess inventory	AI optimizes inventory levels through real-time tracking and predictive restocking algorithms.
Scheduling	Allocating resources over time to perform tasks or jobs efficiently, including labor and machine usage	AI automates complex scheduling, optimizes timelines, and adapts to disruptions in real time.
Quality Management	Ensuring that products and services meet defined quality standards through inspection, testing, and continuous improvement	AI enables real-time defect detection, predictive quality control, and anomaly detection using computer vision.
Process Design	Structuring workflows and production methods to maximize efficiency and reduce waste	AI supports simulation and optimization of workflows using digital twins and reinforcement learning.
Product and Service Design	Developing new products or services that meet customer needs effectively	AI uses customer data and simulations to co-design with human teams, accelerating innovation cycles.
Lean and Continuous Improvement	Identifying and eliminating waste to improve overall efficiency (e.g., Lean, Six Sigma)	AI detects inefficiencies, predicts bottlenecks, and continuously monitors KPIs for process improvement.
Workforce and Labor Management	Planning and managing human resources in operations, including shift planning, training, and performance	AI optimizes staffing, identifies training needs via analytics, and improves safety through monitoring systems.
Sustainability and ESG Compliance	Ensuring that operations are environmentally and socially responsible while adhering to regulatory and ethical standards	AI tracks carbon footprints, monitors regulatory compliance, and models sustainable alternatives in real time.

Scheduling, especially in manufacturing and service environments, is often complex due to variables like machine availability, labor shifts, and customer deadlines. AI uses optimization algorithms to create efficient schedules that minimize downtime and maximize throughput. When disruptions occur, such as machine failure or worker absence, AI can rapidly reschedule tasks, reducing the human effort needed, and ensure smooth operations.

Quality management has evolved from reactive inspections to proactive, AI-driven control. Computer vision and anomaly-detection systems can identify defects in real time on production lines, often with higher accuracy than human inspectors. Predictive quality analytics can help anticipate failures even before they occur.

Another area is **sustainability and ESG compliance**, where AI is making a measurable impact. By monitoring resource usage, emissions, and supplier compliance through IoT sensors and analytics platforms, AI helps companies reduce their environmental footprint. It also ensures alignment with regulatory requirements and social governance standards, offering transparency to both regulators and consumers.

Understanding the scope of these operational decisions allows organizations to ask the right questions and focus on AI applications that align with their strategic goals. After all, while almost any process can be optimized, the real question is:

Does optimizing a process support the strategic goals of the business? If not, it does not matter.

COMPANY EXAMPLE: Birch Coffee

Birch Coffee is a small specialty coffee company founded in New York City in 2009. It operates a handful of cozy neighborhood cafés across the city and has developed a loyal following. Birch is known for ethically sourced beans, in-house roasting, and a welcoming ambiance for reading and remote work. Alongside in-store service, it also offers a subscription service for home delivery of coffee beans.

Birch primarily targets urban professionals, coffee enthusiasts, and remote workers aged 25 to 45 who value high-quality, ethically sourced coffee and a comfortable workspace. Many are health-conscious, tech-savvy, and socially aware consumers, seeking premium experiences.

Let's look at some key operations decisions and how Birch Coffee has handled them:

Forecasting. Birch must predict daily and weekly customer demand at each location. This includes forecasting peak hours, seasonal changes (e.g., more hot coffee in winter, iced drinks in summer), and the demand for beans via its online store. They use historical sales data and local event calendars to improve accuracy.

Capacity Planning. With limited café space and equipment, Birch plans seating, barista headcount, and espresso machine capacity to avoid long wait times during rush hours. During COVID, they had to reevaluate capacity for social distancing and implement mobile ordering.

Inventory Management. Birch manages inventory of perishable goods like milk, pastries, and fresh beans. They must avoid stockouts that affect service while minimizing waste. Inventory is tracked with POS systems that alert managers when to reorder. For roasting beans, they must balance green bean purchases with roasting schedules and freshness guarantees.

Scheduling. Employee scheduling is based on traffic patterns, balancing labor cost, and service levels. More-experienced baristas are scheduled during peak hours. Roasting schedules must align with delivery timelines for subscriptions and wholesale orders.

Quality Management. Quality control is essential. Birch trains baristas to ensure consistency in drink preparation, maintains roasting standards for flavor profiles, and evaluates customer feedback. They perform regular calibration on espresso machines and conduct taste tests for every roast batch.

Even though Birch Coffee is small, it also uses AI. The company integrates AI into its operations through a partnership with Odeko, a technology platform designed to support independent coffee shops. Odeko provides tools that automate inventory ordering and offer predictive analytics, enabling Birch Coffee to streamline its supply-chain management and operations. This collaboration allows Birch to focus more on delivering exceptional customer experiences while maintaining efficient back-end processes

Birch Coffee is a great example of how small businesses must strategically manage operations across multiple functions to deliver a premium, consistent customer experience. Despite limited resources, decisions around forecasting, inventory, scheduling, and quality directly influence customer satisfaction and profitability.

QUESTIONS

1. How does Birch Coffee's use of AI through Odeko support its operations strategy? What advantages does this provide compared to traditional small-business management methods?
2. What operational challenges does Birch face due to its size and limited resources, and how has it adapted its forecasting and capacity planning to address these challenges, especially during unpredictable events like COVID-19?
3. In what ways does Birch Coffee balance inventory efficiency with quality and freshness, especially given its use of perishable goods and a commitment to high-quality standards?
4. How do employee-scheduling and quality-management practices at Birch contribute to delivering a consistent customer experience across its cafés and online subscription service?
5. What lessons can larger organizations learn from Birch Coffee's approach to integrating AI, managing operational complexity, and staying aligned with its brand promise of quality and customer satisfaction?

Now let's contrast this with a large company example.

COMPANY EXAMPLE: Tesla's AI-Driven Operations

Tesla, Inc., exemplifies the future of advanced operations management through its tightly integrated, customer-focused, and AI-powered strategy. The company redefines automotive innovation by translating real-time consumer preferences into dynamic product features, such as over-the-air software updates, personalized in-car experiences, and autonomous driving capabilities through its Full Self-Driving (FSD) system. This responsiveness to consumer demand is enabled by a highly agile product development cycle along with a software-first approach rarely seen in traditional automotive firms.

At the heart of Tesla's operations are its gigafactories—massive, vertically integrated facilities designed to consolidate production, reduce supply chain dependencies, and drive economies of scale. These factories leverage state-of-the-art robotics, machine learning, and AI-driven automation to optimize processes from battery cell production to final vehicle assembly. The result is faster throughput, lower costs, and consistent product quality across global markets.

AI plays a pivotal role across Tesla's entire value chain. Machine-learning models analyze vehicle telemetry from millions of miles driven to improve performance, safety, and predictive maintenance. AI also supports real-time supply-chain visibility and production planning, enabling Tesla to react quickly to material shortages and demand fluctuations. The company has even deployed AI to design custom silicon chips for its FSD computer system.

Let's look at specific operations decisions and how Tesla uses AI:

Forecasting. Tesla uses AI and machine learning to predict vehicle demand by region and model and integrates real-time market data, order patterns, and macroeconomic indicators to refine forecasts.

Capacity Planning. Tesla designs flexible gigafactories with vertically integrated production for rapid scaling. Modular production is used to add or reconfigure capacity as new models (e.g., Cybertruck, Model Y) ramp up. Capacity is also built ahead of demand (e.g., Gigafactory Texas) to minimize future bottlenecks. Battery and vehicle assembly lines are co-located to optimize throughput.

Inventory Management. Tesla operates with low finished goods inventory; vehicles are often made to order or held briefly before delivery. They use real-time analytics and RFID tracking across the supply chain to monitor parts, batteries, and vehicle locations. Inventory of parts is adjusted dynamically based on production plans and supplier risk assessments.

Scheduling. Tesla uses AI-based production scheduling to optimize line performance and reduce downtime. Worker shifts and resources are adjusted in real time based on supply availability and delivery targets. They coordinate delivery scheduling with customers and logistics, to ensure high customer satisfaction and low idle inventory.

Quality Management. AI-powered visual inspection systems are integrated on production lines to detect defects in real time. They implement over-the-air (OTA) updates to fix software issues after delivery—unique among automakers. They also track vehicle performance and service data to improve quality feedback loops.

Tesla has integrated AI and digital technologies to streamline its operations and set a new benchmark for the future of manufacturing and mobility. In 2024, Tesla reported revenues surpassing $96.8 billion, with more than 1.8 million vehicles delivered globally. Despite rising competition in the electric vehicle (EV) space, Tesla remains an industry leader in operational efficiency.

QUESTIONS

1. How does Tesla's software-first approach and ability to deliver OTA updates transform traditional operational models in the automotive industry?
2. In what ways do Tesla's gigafactories reflect advanced capacity planning and vertical integration, and how do these design choices enhance operational agility and scalability?
3. How does Tesla's use of AI in forecasting, scheduling, and inventory management contribute to its ability to maintain low finished goods inventory and respond quickly to demand shifts?
4. What are the advantages and risks of Tesla's real-time AI-powered quality-management systems, especially when compared to traditional post-production inspection models?
5. Given Tesla's use of AI across the value chain, from chip design to customer delivery, how might its operations strategy evolve as global competition in the EV market intensifies?

Intense global competition has placed even greater importance on the operations function of the supply chain. Organizations within the supply-chain network must operate at the highest level of efficiency and utilize the most current operations methods. Many inroads that are made in competitive markets come from innovations in operations.

Let's look at Build Your Dreams (BYD), a competitor to Tesla, and see how they use AI along the Make lever.

COMPANY EXAMPLE: BYD

A good example of an emerging EV leader is BYD, which produces some of the most affordable EVs on the global market. Originally a battery manufacturer, BYD has successfully leveraged its core expertise to become a dominant player in the EV space. In 2023, the company gained attention for selling fully electric cars for as little as $11,000 USD, significantly undercutting competitors like Tesla. While Tesla emphasizes high-end performance and premium branding, BYD focuses on affordability, practicality, and accessibility, making EV adoption more feasible for the mass market, particularly in developing economies.

BYD's cost advantage stems largely from its vertically integrated supply chain, a strategic choice that minimizes dependency on third-party suppliers. The company produces its own batteries, electric motors, semiconductors, and even some raw materials, allowing it to tightly control quality, reduce lead times, and keep production costs low. This level of integration also improves resilience against supply-chain disruptions and allows faster innovation cycles.

BYD has also invested heavily in AI-powered manufacturing and operational analytics. AI is used to optimize everything from battery chemistry and energy-management systems to production scheduling and quality control. Smart sensors and machine-learning algorithms monitor manufacturing lines in real time, detecting anomalies, predicting equipment maintenance needs, and minimizing downtime. AI also enhances vehicle design through simulation and testing, reducing development cycles while improving reliability and performance.

Here are some of their Make decisions:

Forecasting. Utilizes AI-driven predictive models to anticipate demand for EVs, batteries, and energy products.

Capacity Planning. Employs modular factory designs to scale production efficiently across various vehicle types and battery product. Invests in vertical integration, producing batteries, semiconductors, and other components in-house to reduce reliance on external suppliers.

Inventory Management. BYD implements material requirements planning (MRP) systems to manage inventory levels and reduce waste. They strategically locate parts warehouses to ensure timely delivery and service to customers.

Scheduling. They use MRP systems for production scheduling, aligning manufacturing operations with demand forecast.

Quality Management. The company adopts total quality management (TQM) practices to eliminate manufacturing errors, particularly in battery production and leverages AI in battery manufacturing to reduce defects, ensuring longer battery life and superior performance.

AI applications extend to all aspects of operations, allowing BYD to balance just-in-time manufacturing with the need for regional customization. As a result, the company not only maintains competitive pricing but also scales efficiently across different global markets. Through this combination of vertical integration, cost leadership, and AI-driven operational excellence, BYD has positioned itself as a formidable global competitor capable of challenging incumbent automakers and reshaping the economics of electric mobility.

How does BYD compare with Tesla along these Make decisions?

Forecasting. Both companies use AI for demand forecasting; however, Tesla emphasizes real-time data analytics, while BYD integrates broader market trends and policy changes.

Capacity Planning. Tesla focuses on rapid scaling with gigafactories and flexible production lines, whereas BYD emphasizes modular factory designs and vertical integration to enhance scalability.

Inventory Management. Tesla operates with minimal finished goods inventory, relying on JIT principles, while BYD uses MRP systems to manage inventory and strategically places warehouses for efficient distribution.

Scheduling. Tesla employs AI-based production scheduling for real-time adjustments, while BYD utilizes MRP techniques for structured scheduling.

Quality Management. Tesla integrates AI-powered visual inspections and OTA updates for continuous quality improvement, whereas BYD focuses on TQM practices and AI in battery manufacturing to enhance product quality.

QUESTIONS

1. How does BYD's strategy of affordability and vertical integration contrast with Tesla's premium branding and flexible production approach? What does this suggest about the different ways companies can compete in the EV market?
2. In what ways does BYD's use of AI in battery production and manufacturing processes contribute to its ability to offer low-cost, high-quality electric vehicles?
3. What are the operational advantages and potential limitations of BYD's reliance on MRP systems compared to Tesla's real-time, AI-driven production scheduling?
4. How does vertical integration strengthen BYD's resilience against global supply-chain disruptions, and what role does this play in its success in emerging markets?
5. As BYD expands globally, what challenges might it face in maintaining quality and cost advantages across diverse regulatory environments and customer expectations? How might its operational model need to evolve?

5.1.2 Coordinating Make and Sell

Without Make, there would be no products to Sell. However, operations cannot succeed without coordinating with other business functions. Most importantly, Make must work with the Sell lever to understand and design the exact products customers want. Operations (Make) must then create the production process to efficiently produce the exact products. This is where AI capabilities—such as micro-segmentation, sentiment analysis, customer behavior analysis, and location-based marketing—come in. These capabilities shift understanding of customers from mere guesses to a fact-based understanding.

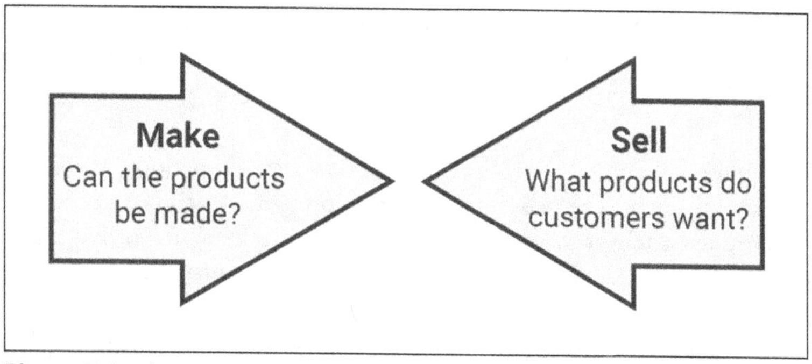

Figure 5.4 Coordinating Make and Sell

The Sell lever must also understand the capabilities of Make, such as the types of products operations can, and cannot, produce. This may sound trivial, but it is very important. Otherwise, companies will sell products the company cannot make.

Consider a furniture manufacturer in North Carolina where marketing pushed for the selling of laminated versus customized furniture only to find that the operations requirements were completely different for one versus the other. The company was simply not able to deliver. Based on a thorough understanding of the Make and Sell levers, an organization's supply-chain manager must coordinate Make and Sell activities (Figure 5.4).

COMPANY EXAMPLE: Amazon

Amazon shows effective coordination between the Make and Sell levers, achieved through advanced analytics and a robust supply chain. By leveraging AI data, Amazon personalizes its offerings through tools like micro-segmentation, driving more than 35 percent of sales through its recommendation engine alone. This data-driven approach ensures that Amazon's offerings align with customer preferences in a precise and targeted manner.

Amazon uses advanced operations management through an expansive but highly integrated global supply chain. As of 2025, the company operates more than 200 fulfillment centers worldwide, supported by a logistics network made up of more than 120,000 trucks, vans, and planes, along with 1.275 million associates and partners. This extensive infrastructure enables Amazon to deliver products rapidly and efficiently to customers across the globe. Embracing a systems-thinking approach, Amazon aligns its production capabilities with sales targets to ensure seamless execution. This strategic alignment has been pivotal in Amazon's continued success, with the company reporting net sales of $638 billion in 2024, marking an 11 percent increase from the previous year. By effectively coordinating its Make and Sell, Amazon optimizes its supply chain, demonstrating the power of synchronizing operations with sales goals in a dynamic marketplace.

Amazon is also investing in AI to enhance operational efficiency. Amazon is developing more than 1,000 generative-AI applications and has introduced its Trainium chips, custom-built machine-learning chips designed by Amazon Web Services (AWS). These chips are part of a family of AI chips that provide 30 to 40 percent better price-performance and are scalable, meaning they can support large-scale AI systems. These advancements are expected to further streamline Amazon's operations and improve customer experiences.

QUESTIONS

1. How does Amazon's integration of AI-driven personalization with supply-chain operations illustrate effective coordination between the Make and Sell functions?
2. What roles do Amazon's global fulfillment and logistics network play in supporting its dynamic sales strategy, and how does this contribute to customer satisfaction and loyalty?
3. How does Amazon's systems-thinking approach help align production and fulfillment with customer demand, and what challenges might arise in maintaining this alignment at scale?
4. In what ways do Amazon's investments in custom AI chips and generative-AI applications give it a competitive edge in operations management and innovation?

5. Considering that Amazon drives more than 35 percent of sales through its recommendation engine, how might AI-driven selling strategies shape operational planning and inventory management in the future?

It is not enough for a company to manage its own operations functions. These must be coordinated across the supply chain. The reason is that each company depends on other members of its supply chain to be able to deliver the right products to its customers in a timely and cost-effective manner. Processes operate as a system, and every system has a **constraint**, or a **bottleneck**, that must be managed.

On the upstream part of its supply chain, a company depends on its suppliers for the delivery of raw materials and components in time to meet production needs. If deliveries of these materials are late, or are of poor quality, production is delayed, regardless of the efficiency a company's operations process.

On the downstream side, a company depends on its distributors and retailers for the delivery of the product to the final customer. If these are not delivered on time, are damaged in the transportation process, or are poorly displayed at the retail location, sales will suffer. Also, if the operations function of other supply-chain members is not managed properly, excess costs will result, which will be passed down to other members of the supply chain in the form of higher prices.

Figure 5.5 illustrates the organization between the upstream and downstream parts of the supply chain and some potential risks.

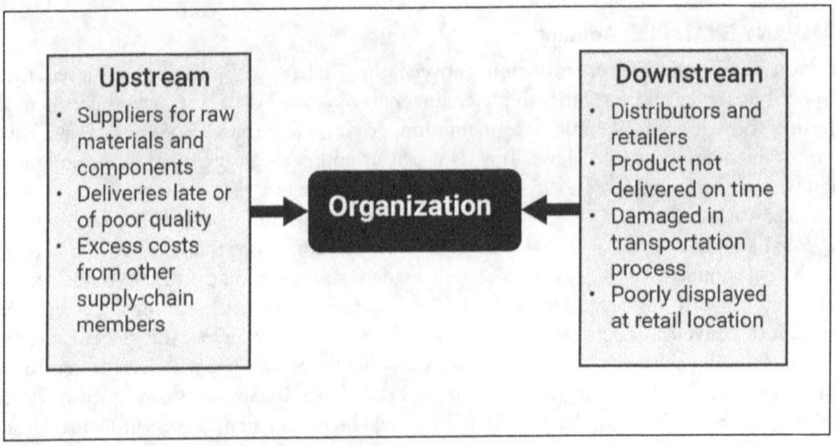

Figure 5.5 Potential Upstream & Downstream Risks

COMPANY EXAMPLE: Miso Robotics & CaliBurger

A great example of using AI to match supply and demand is CaliBurger. CaliBurger is a regional burger chain based in California that blends fast-casual dining with cutting-edge technology. While it's a relatively small business compared to national chains, it gained attention for being one of the first to integrate AI and robotics into its daily operations through a partnership with Miso Robotics, a startup specializing in AI-powered kitchen assistants.

At the heart of CaliBurger's innovation is Flippy, an AI-enabled robotic system that handles frying and grilling tasks. Flippy uses computer vision and machine learning to monitor cooking times, food temperature, and kitchen demand in real time. By linking Flippy to point-of-sale and inventory systems, CaliBurger dynamically adjusts prep timing based on customer traffic, so when there's a lunch rush, for instance, the kitchen stays efficient without over-prepping or wasting ingredients.

The AI system also forecasts demand based on weather data, foot traffic predictions, and historical sales trends. For example, on a hot weekend, the system may anticipate a spike in milkshake sales and prompt staff to stock up in advance. These insights not only minimize food waste but also streamline labor scheduling, ensuring optimal staffing levels without overspending.

By using AI in this focused, practical way, CaliBurger has reduced operational costs, improved order consistency, and boosted customer satisfaction, all while keeping its footprint small and agile.

QUESTIONS

1. How does CaliBurger's integration of AI and robotics through systems like Flippy enhance operational efficiency and help balance supply and demand in real time?
2. In what ways does connecting AI tools to point-of-sale and inventory systems create a competitive advantage for a small regional chain like CaliBurger?
3. How does CaliBurger's AI forecasting, which incorporates weather and foot traffic data, differ from traditional demand-forecasting methods in the restaurant industry?
4. What are the potential risks or limitations of relying on AI and robotics in a fast-casual kitchen environment, and how might a company like CaliBurger address them?
5. How could CaliBurger's AI-driven operational model scale to larger restaurant chains, and what adaptations might be needed to support broader deployment?

5.1.3 Manufacturing and Service

Manufacturing organizations were early users of AI to drive quality and efficiency. Since the dawn of the computer era, manufacturing firms have been using information technology and automation to design, build, and distribute products. In the 1990s, manufacturing companies achieved large productivity gains due to new operational improvements. These were focused on increasing operational efficiency through Lean systems and improvements in quality such as TQM. These advancements in manufacturing were quickly followed by applications in service organizations that began to achieve similar benefits. The same is true today with applications of big-data analytics. However, these applications need to be modified to the unique needs of services.

Manufacturing and service organizations each pose unique challenges for the operations function. There are two primary distinctions between them. First, manufacturing organizations produce physical, tangible goods that can be stored in inventory before they are needed. By contrast, service organizations produce intangible products that cannot be produced ahead of time. Second, in manufacturing organizations, most customers have no direct contact with the operation. For example, a customer buying a bottle of Pepsi never sees the bottling factory. However, in service organizations, such as hospitals, colleges, theaters, and barber shops, the customer is typically present during the creation of the service.

COMPANY EXAMPLE: UnitedHealth Group

UnitedHealth Group, one of the largest healthcare and insurance service providers in the United States, showcases how AI can transform operations in the service sector. With more than 150 million individuals served globally, the company uses AI and advanced analytics to improve patient care, reduce administrative burdens, and enhance operational efficiency across its healthcare and insurance divisions.

One of UnitedHealth's key uses of AI lies in **predictive analytics** for health outcomes. By analyzing massive volumes of patient data—including electronic health records, lab results, and claims history—AI models can identify individuals at high risk for chronic conditions like diabetes or heart disease. These insights are then used to proactively engage members with preventive care programs, virtual consultations, or care coordination services, which can lead to better outcomes and reduced costs.

On the administrative side, UnitedHealth employs AI-driven claims processing and fraud detection systems. Natural language processing (NLP) helps automate the review of medical documentation, reducing manual workload and speeding up approvals. At the same time, AI models flag unusual billing patterns or inconsistencies, helping to curb fraudulent activities and minimize revenue leakage.

AI is also used in personalizing member services. Virtual assistants powered by conversational AI guide patients through insurance plans, scheduling appointments, and offer tailored wellness advice, delivering both cost savings and a more engaging customer experience. UnitedHealth's approach demonstrates how AI, when thoughtfully integrated into service operations, can drive not only internal efficiencies but also meaningful improvements in human health and satisfaction.

QUESTIONS

1. How does UnitedHealth Group's use of predictive analytics to identify high-risk patients illustrate the potential of AI to shift healthcare from reactive to preventive care?
2. In what ways does AI-driven automation in claims processing and fraud detection improve operational efficiency and financial performance in the healthcare service sector?
3. What are the ethical considerations and potential privacy risks associated with using AI to analyze patient data for health predictions and personalized care?
4. How does the integration of conversational AI in customer service enhance the member experience, and what limitations might still exist in fully automating patient interactions?
5. What lessons can other service-based industries learn from UnitedHealth Group's approach to embedding AI across both customer-facing and back-end operations?

The differences between manufacturing and service organizations are not as clear-cut as they might appear, and there is much overlap between them. Most manufacturers provide services as part of their business, and many service firms manufacture physical goods they use during service delivery.

For example, a manufacturer of furniture may also provide shipment of goods and assembly of furniture. A barber shop may sell its line of hair care products. Even in pure service companies, some segments of the operation may have high customer contact while others have low customer contact. The latter can be thought of as backroom or behind-the-scenes segments. Examples would include the kitchen segment at a fast-food restaurant, such as Wendy's, or the laboratory for specimen analysis at a hospital.

It is important to remember that AI applications exist for both manufacturing and service organizations, although some are more applicable to one industry versus the other.

As tangible goods become more commoditized, manufacturers are increasingly developing integrated product-service offerings that differentiate products and deliver additional value. This is called **servitization**. An example is Rolls-Royce selling not just aircraft engines but also complete service solutions that include all operations and maintenance functions. Similarly, tool manufacturer Hilti expanded its offering from only the products to the broad service capabilities that the tools deliver. These companies need a broad set of applications that address both the tangible and intangible aspects of Make.

5.1.4 How AI Affects Operations

Companies are using AI to improve every aspect of their operations, from quality to efficiency to labor utilization.

COMPANY EXAMPLE: Western Digital

Western Digital is an American data storage company and a leading global manufacturer of hard disk drives. The company has relied on leveraging AI and advanced analytics to enhance their product quality and operational efficiency. By integrating SAS Asset Performance Analytics into its manufacturing processes, the company can monitor equipment sensors and analyze machine-to-machine data to identify hidden patterns that predict potential failures early in the production process. This proactive approach allows engineers to make timely decisions to avoid yield excursions, ensuring that hard drives meet the highest quality standards.

The implementation of AI-driven quality-control systems enables Western Digital to perform real-time analytics on the production floor while the product is being made. This capability facilitates the scanning, recording, testing, and tracking of every disk drive during manufacturing. If a defect is detected at any stage, the system can promptly remove the faulty unit from the production line or inventory, preventing defective products from reaching customers. Such measures have contributed to a significant reduction in warranty return rates, bolstering customer trust and loyalty.

Western Digital also employs AI for visual inspections and defect pattern recognition, aiming for a 100 percent pass rate on production lines. By predicting quality failures and test escapes, the company minimizes machine idle time and enhances throughput. This integration of AI into manufacturing not only improves product quality, but also streamlines operations, positioning Western Digital as a pioneer in utilizing AI for industrial excellence.

QUESTIONS

1. How does Western Digital's use of AI and predictive analytics improve product quality and reduce the risk of production failures?
2. What advantages does real-time monitoring and quality control during the manufacturing process offer compared to traditional post-production testing methods?
3. In what ways does integrating AI for visual inspection and defect detection contribute to operational efficiency and customer satisfaction?
4. How does Western Digital's AI-driven approach to yield management affect cost savings and competitive advantage in the data storage industry?
5. What challenges might arise when implementing AI in complex manufacturing environments, and how can companies like Western Digital address them effectively?

Like Western Digital, numerous companies now use AI for inventory management, optimization of stock levels, maintenance optimization, and quality control. Some are also using AI in workforce productivity evaluation, as well as the study of capacity constraints. Productivity and quality have huge applications. Companies can now run real-time analyses of operational performance, aggregating and reporting by location, SKU, department, and employee. They can also track quality and alert companies of defects and other problems. Although companies have been using technology to evaluate performance for many years, what is different now is *timing and scale*. AI enables real-time monitoring at the highest level of detail, and many decisions can be automated into a "digital factory."

AI significantly enhances the Make phase of manufacturing by improving efficiency in product design and production, improving product quality, and enabling the development of products that better meet customer needs. AI can reduce product development time by more than 50 percent. This enables bringing innovative products to market more rapidly. Additionally, AI-driven simulations and testing can proactively identify potential defects before production begins, contributing to higher-quality finished products.

Predictive maintenance is another area where AI can reduce downtime by up to 50 percent. For example, at BMW's manufacturing plant in South Carolina, AI-powered predictive maintenance is used to monitor the robotic arms on the entire assembly line. By analyzing sensor data in real time, the system predicts potential failures before they happen, allowing it to intervene.

Labor optimization is yet another area with large AI applications. Companies can now maintain service levels through optimizing labor, automating and tracking employee attendance, and improving labor scheduling. Amazon uses AI in their fulfillment centers to analyze worker performance and operational efficiency. This enables Amazon to optimize task allocation, reduce unnecessary movements, and align workforce schedules with real-time demand, enhancing productivity and reducing delivery times.

Figure 5.6 shows a summary of AI applications in operations, or Make.

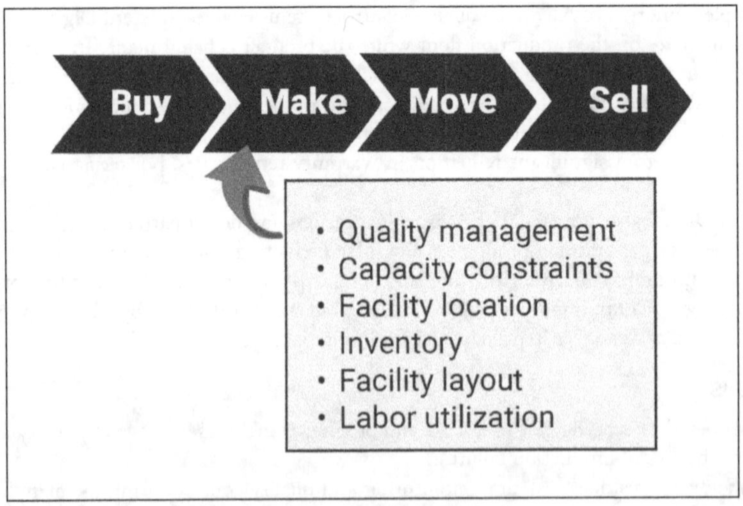

Figure 5.6 AI applications in Operations

5.2 Product Design & Innovation

AI can affect the Make lever not only along tactical decisions, such as inventory stock levels, but it can also answer strategic questions such as what exactly the company should make in the first place. AI can help foster innovation by building cross-functional research and development (R&D) and product design databases along the supply chain to enable concurrent engineering, rapid experimentation and simulation, and co-creation of products.

5.2.1 Product Design

Product design is the process of specifying the exact features and characteristics of a company's product. Ideas for which features to include are driven by the Sell lever—marketing—whose job is to understand the products that customers want. Marketing understands that customers respond to a product's appearance, color, texture, and performance as well as packaging, delivery, and service. Sell and Make then work together to create the product bundle that meets these consumer preferences.

Product design decisions are one of the most important in operations management as they drive other operations decisions. One of these is process design. **Process design** is the design of the production process needed to produce the desired product. This includes decisions regarding technology and automation, implementation of Lean systems, and performance measurement. Product design defines a product's characteristics and then translates them into measurable dimensions the process can use to produce the product.

Product and process design affect every aspect of the product, from quality to cost to customer satisfaction.

Consider Apple's iPhone 4, manufactured in 2010. The product design specified one of the color options to be white. However, after white iPhones were produced, consumers found the white casing to leak too much light from the phone unit, washing out pictures and creating other problems. The white idea was great. The production process, however, could not deliver. If the product is not well designed, or if the production process is not capable of producing the exact product design features, the quality of the product will suffer. Therefore, for a product to succeed in the marketplace, it must have a good design *and* a good production process. These decisions are all intertwined.

Where does AI come in?

AI identifies exactly what the customer wants and helping design the process to make that product.

5.2.2 What the Customer Wants

Historically, product design decisions were made based on guesswork and instinct, using traditional marketing tools to identify what customers want. The use of AI and analytics changes this. AI enhances the understanding of customers and offers opportunities to improve and accelerate product design in several ways.

First, on the Sell side it helps identify the most important and valuable features based on concrete customer inputs rather than hunches.

AI can

- identify designs that meet customer criteria while minimizing production costs,
- harness consumer insights to reduce development costs through processes such as open innovation,
- improve the efficiency and effectiveness of the design process, and
- increase the value-added content of products and services at a lower cost.

These efforts provide insights that drive product design more precisely. However, gaining access to comprehensive data about customers to achieve true insights can be a significant barrier. The reason is that many companies—from distributors to retailers, to other players—are often unwilling to share such data. *Why?* Data is a competitive asset.

COMPANY EXAMPLE: Dell

Not long ago, Dell was famous for its "configure-to-order" model, where customers could personalize their computers directly on the company's website. Options ranged from memory size and screen type to software packages and design features. While this flexibility was a major win for customers, it created a logistical nightmare for operations, producing more than seven septillion potential product configurations. That kind of complexity posed serious challenges for inventory planning, fulfillment speed, and cost control.

To address this, Dell transitioned to a data-driven operations model. The goal: maintain personalization while optimizing inventory and production costs.

Dell's analytics team began by analyzing historical sales data and customer orders. Using cluster analysis, they identified patterns and trends in the most frequently selected configurations. What they discovered was surprising. Despite the theoretical variety, most customers were choosing a small subset of popular configurations.

This insight allowed Dell to dramatically simplify its production strategy. The company reduced the effective number of configurations from septillions to just a few million, and among those, a small number stood out as top sellers. Dell began pre-building these high-demand models, stocking them in advance for rapid shipping.

The result? Customers still enjoy tailored options, but Dell can now ship common configurations as quickly as the next day thanks to streamlined inventory. The new system not only improved customer satisfaction with faster delivery but also generated an additional $40 million in revenue by reducing manufacturing inefficiencies and inventory costs.

In short, Dell found a way to marry customization with scalability, turning a complex challenge into a competitive advantage.

QUESTIONS

1. How did Dell's use of data analytics help the company overcome the operational complexity of its original configure-to-order model?
2. What are the trade-offs between full customization and operational efficiency, and how did Dell strike a balance between the two?
3. How can identifying high-frequency customer preferences through cluster analysis improve both production planning and customer satisfaction?

4. In what ways did Dell's transition to pre-building popular configurations enhance inventory management and speed to delivery?
5. What lessons can other companies with complex product offerings learn from Dell's approach to simplifying operations without sacrificing customer choice?

5.2.3 Service Design

AI can greatly enhance the product design of services. Product design in the service industry has an added complication as the product is intangible and there is a high degree of customer interaction. Service design requires the design of the entire service concept. As with a tangible product, the service concept is based on satisfying customer needs. However, there are aesthetic and psychological benefits the service provides that must also be considered.

Services are experienced by the customer. Details that enhance this experience are all part of the service concept. An example of this is seen in the retail sector with the trend of shopper marketing. Here, retailers focus on creating the entire shopping experience rather than just focusing on features of one product.

Companies can leverage data to design products that better match customer needs. Consider an example of smart phones where the phone "learns" its owner's habits and preferences. It holds applications and data tailored to the user's needs. As a result, it is more valuable than a new device that is not customized to a user's needs. Also, we can now embed sensors in products that can "phone home." These devices can then serve to generate data about how products are actually used and monitor for emerging defects. Companies can then respond immediately by making changes in real time to their current production process. R&D and operations can share this data as they engage in redesign and new product development.

AI also enables service companies to understand and anticipate customer needs and preferences with incredible precision. By aggregating and analyzing customer data from various touchpoints—social media interactions, transaction histories, and service usage patterns—companies can tailor their services to meet individual preferences and enhance overall customer satisfaction.

Netflix provides the real-life application of using AI to revolutionize its service design by customizing viewing experiences. The streaming giant collects data on viewing habits, search histories, and ratings provided by its users to feed algorithms that recommend shows and movies. This personalization makes the service more engaging and improves user retention, as each recommendation is tailored to the viewer's established preferences.

COMPANY EXAMPLE: Hilton Hotels

Hilton Hotels uses AI to enhance its service design by personalizing the guest experience across its global properties. Through its Connected Room initiative, Hilton leverages AI and IoT technologies to allow guests to customize and control various in-room features, such as lighting, temperature, TV content, and even digital artwork, via a mobile app.

Behind the scenes, Hilton's AI systems analyze guest preferences from previous stays, loyalty program data, and app usage to anticipate needs and tailor the environment even before a guest arrives. For instance, if a guest prefers a specific room temperature or frequently orders room service at a certain time, the system can proactively adjust these services.

This AI-enhanced service design not only improves comfort and satisfaction but also streamlines operations, reducing manual interventions by staff and increasing efficiency in housekeeping and maintenance scheduling. By embedding AI into the core of its service design, Hilton redefines luxury and personalization at scale.

QUESTIONS

1. How does Hilton's Connected Room initiative illustrate the integration of AI and IoT to enhance the customer experience in the hospitality industry?

2. In what ways does AI-driven personalization improve guest satisfaction, and how might it influence brand loyalty for a global hotel chain like Hilton?
3. What operational efficiencies can be gained by using AI to automate housekeeping, maintenance scheduling, and other behind-the-scenes services?
4. What are the potential privacy concerns related to collecting and analyzing guest data for personalization, and how should Hilton address them?
5. How might Hilton's AI-enhanced service model be adapted for different cultural contexts or guest expectations across its global properties?

5.2.4 Co-Creation

Co-creation is when a company involves external stakeholders, such as suppliers, in the design and development of a product. The process can enhance design, testing, and experimentation. Designers and manufacturing engineers can share data. They can quickly create simulations to test different designs, the choice of parts and suppliers, and compute associated manufacturing costs. This is especially useful because decisions made in the design stage typically drive 80 percent of manufacturing costs.

In co-creation, AI enables more dynamic interactions among all parties involved in product design and manufacturing. By leveraging data from various collection points, customer feedback, and real-time market analytics, companies can gain valuable insights that drive the co-creative process. This integrated approach allows for rapid prototyping, testing, and iteration, significantly shortening development cycles and enhancing the adaptability of the products to meet customer expectations.

Again, consider BMW, which has pioneered the use of AI in co-creation. Through its Customer Innovation Lab, BMW collects and analyzes vast amounts of data from vehicle sensors and driver feedback. This data informs the development of new features and improvements in car designs. BMW also collaborates closely with technology suppliers to integrate the latest innovations into its vehicles, such as advanced driver-assistance systems (ADAS) and electric powertrains, ensuring that the vehicles not only meet, but also anticipate, customer needs and preferences. Toyota, Fiat, and Nissan have all used collaborative design and data experimentation methods. The result has cut new-model development time by 30 to 50 percent, and Toyota claims to have eliminated 80 percent of defects prior to building the first physical prototype.

5.2.5 Innovation

AI enables innovation and creation of new business models. For example, in the auto industry sensor data has enabled innovative after-sales services. Let's look at BMW again. Their Connected-Drive offers drivers directions based on real-time traffic information, automatically calls for help when sensors indicate trouble, alerts drivers of maintenance needs based on the actual condition of the car, and feeds operation data directly to service centers. The ability to track the use of products at a micro level has also made possible monetization models that are based not on the purchase of a product but instead on services priced by their usage, as described previously.

Companies are relying increasingly on outside inputs to drive innovation and develop new products for emerging customer needs. They are also using innovative channels to access this talent. This is called **open innovation**, where organizations utilize both internal and external sources of ideas, knowledge, and technology to drive progress. Many companies are inviting external stakeholders—consumers, suppliers, and other third parties—to submit ideas for innovations or even collaborate on product development. For example, consumer goods companies such as Kraft and Procter & Gamble (P&G) invite ideas from their consumers. They also collaborate with external experts in new product development, including academics and industry researchers. Half of the new products at P&G now have elements that originated outside the company.

Open innovation has extended to other industries as well. Again, BMW shows how this is done by creating an idea management system to help evaluate ideas submitted through its virtual innovation agency. This has cut the time taken to identify high-potential ideas by 50 percent. It has also reduced the time in assessing the feasibility of an idea. The process has resulted in the incorporation of multiple ideas from open innovation into models every year.

5.3 Improving the Production Process

Inventory management, quality management, and labor utilization are three major areas where AI can bring about improvements in the production process.

5.3.1 Inventory Management

Inventory management is one of the most important decision areas for Make. Not having the right goods when and where they are needed means lost sales and poor customer service. Too much inventory means high inventory carrying costs. In fact, carrying costs can be as high as 50 percent of item costs. Managing this trade-off is a constant problem for Make.

Inventory management has one of the largest applications of AI and digital technologies. RFID tags, for example, are now affixed to items and allow tracking and monitoring of stock levels, location status, and in-transit shipments and can trigger automatic reorder quantities. These sensors enable real-time visibility for in-transit inventory. They also enable the monitoring of environmental conditions, such as temperature control for perishable items, and provide a time stamp on how long inventories spend in different stages of transit. Coupled with POS data on the Sell side, they can create intelligent inventories that know exactly how much stock needs to be placed in a specific location and automatically generate orders.

With AI applications, there is the term called **adaptive inventory**. This is where the system dynamically adjusts inventory levels and replenishment in response to changing customer demand. The inventory levels essentially *adapt*. By combining multiple data sets, such as sales histories, POS data, weather predictions, and seasonal sales cycles, forecasting accuracies are improved. Algorithms then automatically connect forecasts to current stock levels at a granular level. This results in optimized inventories.

AI applications permit supplies to be more tightly coupled with demand signals, called **demand sensing**. Sensing the real-time demand signal can make inventory more productive. Here, lower levels of stock can offer higher customer-service levels. Real-time location sensing can create a better supply-demand match. This highly accurate information can feed into all levels of inventory calculations, from production planning levels to stocking levels to ordering levels.

COMPANY EXAMPLE: Walmart & Amazon

Walmart exemplifies the use of AI in its inventory management on a large scale. The company employs RFID tags on inventory items, enabling real-time tracking and monitoring of stock levels, in-transit shipments, and item locations, thus reducing stockouts by as much as 30 percent. With **adaptive inventory** practices, Walmart tracks stocking levels in real time, making it possible to optimize inventory while also lowering carrying costs by approximately 20 percent.

Additionally, Walmart's use of **demand sensing** allows it to respond to real-time demand signals, making inventory more productive. By tightly aligning inventory with actual demand, Walmart improves customer-service levels, maintains lower stock levels, and reduces overall inventory costs. With these capabilities, Walmart has transformed its inventory management, minimizing stockouts and reducing carrying costs, while ensuring high availability of products.

Similarly, Amazon employs its anticipatory shipping model, where it predicts future purchases using data from consumer buying patterns and browsing behaviors. Products are shipped to distribution centers closer to the customer's location even before the purchase is finalized, streamlining the delivery process and enhancing customer experience by reducing wait times. These examples highlight how AI can be used strategically to refine inventory management and improve the overall efficiency of production processes in retail.

QUESTIONS

1. How do Walmart's use of RFID tags and real-time tracking capabilities enhance inventory visibility and reduce stockouts across its large-scale operations?
2. What are the advantages of adaptive inventory practices and demand sensing, and how do they contribute to both cost savings and improved customer satisfaction at Walmart?
3. How does Amazon's anticipatory shipping model illustrate a proactive approach to inventory management, and what risks or challenges might come with predicting customer purchases?
4. In what ways do Walmart and Amazon demonstrate the role of AI in transforming traditional inventory management into a strategic asset for competitive advantage?
5. How can smaller retailers apply lessons from Walmart and Amazon to improve their own inventory practices, even with fewer technological resources?

5.3.2 Quality Management

AI and machine learning are reshaping quality management, enabling companies to monitor and enhance quality with unprecedented accuracy. **TQM**, a holistic approach to improving quality at every level of the organization, traditionally relied on manual tools to detect defects as early as possible. However, AI now allows companies to anticipate and even prevent quality issues, reducing reliance on reactive measures.

Two key advancements have emerged.

First, AI enables companies to define customer-specific quality with precision. In the past, quality expectations were inferred through surveys or focus groups, giving only approximate insights. Today, advanced AI extracts detailed and actionable insights from real-time customer data. It reveals specific quality aspects that matter most to consumers. AI-powered sentiment analysis, customer segmentation, and behavior analytics transform the Voice of the Customer from guesswork into measurable, data-driven understanding.

All the Sell tools discussed in Chapter 4, from customer micro-segmentation to in-store-buying analysis, allow companies to know what customers are buying with great accuracy. There is no need to try to guess what customers want. This information is then used to inform Make, Move, and Buy with great precision.

Second, traditional quality management focuses on identifying quality defects as early as possible in the production process. TQM tools have traditionally included the **seven tools of quality**. These tools include **checklists, fishbone diagrams, histograms, Pareto analysis, scatter diagrams, flowcharts,** and **control charts**. Control charts, for example, rely on random sampling to test whether a process is operating within expectations for specific dimensions (e.g., product weight, dimensions, or the number of customer-service complaints). These tools are meant to be used in tandem to help companies detect and root out causes of quality. They are, however, quickly becoming ancient history.

Why?

We now have sensors and tags attached to products that can monitor performance in real time. There is no need to do randomized product sampling. Yes, sensors are costly, but they are also much more accurate. They are also rapidly becoming affordable for everyday use. Consider that readings from a heat sensor on a factory machine can detect problems before they occur. Sensors can detect that an engine is prone to a breakdown based on the heat or vibrations that it produces. Sensors can be affixed to bridges and buildings to watch for signs of wear and tear. Sensors are also used in large chemical plants and refineries, where a piece of broken equipment could bring production to a standstill; the cost of collecting and analyzing the data that indicates when to take early action is lower than the cost of an outage. All this changes the traditional concepts and standards of quality.

Let's look at some examples in the auto industry that leads the way in applying AI.

COMPANY EXAMPLE: Honda

Honda continues to lead in leveraging AI and analytics to enhance vehicle quality and customer satisfaction. By 2025, the company has significantly advanced its early-warning systems to detect and address quality issues proactively. Honda's approach involves mining data from various sources, including warranty service records, mechanics' notes, and customer call center transcripts. This comprehensive data collection allows for the identification of emerging trends and potential problems before they become widespread. These insights are then reviewed by human analysts who make informed risk management decisions, exemplifying how AI can augment human judgment.

A notable advancement is Honda's partnership with SAS Analytics, which has streamlined the analysis of warranty claims. Previously, identifying noncompliant claims was a time-consuming process, but with AI integration, the detection time has reduced to under a minute, with a 76 percent accuracy rate in flagging issues. This efficiency not only reduces costs but also accelerates the resolution of potential defects. Honda also utilizes AI to forecast parts demand and service needs, ensuring that dealerships are well equipped to meet customer requirements. This predictive capability enhances customer satisfaction by minimizing service delays and maintaining vehicle reliability.

Overall, Honda's integration of AI and analytics represents a significant shift in quality management, moving from reactive measures to proactive strategies that prioritize customer-defined quality and real-time defect identification.

QUESTIONS

1. How does Honda's use of AI to mine diverse data sources—like warranty records and customer-service transcripts—enhance its ability to detect quality issues early?
2. In what ways does Honda's partnership with SAS Analytics improve the efficiency and accuracy of warranty claim analysis, and what operational benefits result from this?
3. How does Honda balance AI-driven insights with human judgment in its risk management processes, and why is this hybrid approach valuable?
4. What effect does AI-driven forecasting of parts demand and service needs have on customer satisfaction and dealership operations?
5. How does Honda's shift from reactive to proactive quality management redefine traditional approaches to product reliability and customer experience in the automotive industry?

COMPANY EXAMPLE: Volvo

Volvo continues to be a leader in leveraging AI to achieve its bold vision: zero fatalities, zero serious injuries, and eventually zero collisions in its vehicles. As of 2025, the company has made significant strides toward this goal by embedding advanced safety technologies, real-time data collection, and AI-powered diagnostics across its global fleet.

Today, every Volvo vehicle is equipped with hundreds of sensors that continuously monitor driving behavior, vehicle health, and environmental conditions. When a car visits a dealership or connects to Volvo's cloud-based system via OTA updates, all accumulated sensor data, ranging from brake performance to steering anomalies, is uploaded to the company's centralized AI system. This data warehouse now exceeds five petabytes, integrating inputs from millions of vehicles around the world.

Volvo's AI systems analyze this data in real time to detect patterns, predict maintenance issues, and identify early signs of component failure or design flaws. This has dramatically reduced the defect detection and resolution cycle, from an average of six to eight months in the past to just two to four weeks today. The company's ADAS and Driver Understanding System (DUS), which use AI to monitor driver alertness and behavior, have contributed to a measurable decline in serious accidents involving Volvo vehicles. By integrating predictive maintenance, machine learning, and behavioral AI, Volvo has positioned itself as one of the safest and most data-driven automotive brands on the road.

QUESTIONS

1. How does Volvo's use of real-time sensor data and centralized AI systems support its vision of achieving zero fatalities and zero collisions?
2. What operational advantages arise from reducing the defect detection and resolution cycle from several months to just a few weeks using AI?
3. In what ways do Volvo's ADAS and DUS enhance driver safety, and how might they influence the broader automotive industry's approach to accident prevention?
4. How does Volvo's ability to collect and analyze more than five petabytes of vehicle data reflect the growing role of big data and machine learning in modern automotive operations?
5. What challenges might Volvo face in managing the ethical and privacy concerns related to continuous data collection from vehicles, and how can these be addressed while maintaining safety goals?

Another important quality-management application is using sensor data from products that are in use to improve quality and service offerings. For example, analyzing the data reported by sensors embedded in complex products enables manufacturers of aircraft, elevators, and data-center servers to create proactive, smart, preventive-maintenance service packages. A repair technician can be dispatched before the customer even realizes that a component is likely to fail. Other manufacturers have been able to use this capability to transform the commercial offering from a tangible product to one that is focused on a service. Good examples are traditional manufacturers of jet engines such as Rolls-Royce, General Electric, and Pratt & Whitney who now sell "power-by-the hour" and similar services.

5.3.3 Labor Utilization

Labor is often the most significant expense after cost of goods. For many labor-intensive organizations—retailers and distribution centers—workforce scheduling is one of the most pressing problems. Tools for labor scheduling can not only reduce management time spent on this task but can also do a better job of scheduling workers with the right skills when customers are most likely to visit their stores. The most advanced systems use sophisticated forecasting to consider changes in demand due to upcoming promotions, arriving inventory, and local macroeconomic data. Current software can also identify the most productive employees. These workers can be scheduled during peak sales hours in retail or have the highest-value orders routed to them in a warehouse. Including these types of factors in workforce scheduling can result in significant cost savings from workforce efficiency.

COMPANY EXAMPLE: Starbucks, Walmart, REI

An example of data-driven labor scheduling can be found at Starbucks. The coffee giant leverages AI to maintain optimal staffing levels across its stores, ensuring that employees are available when customer demand is highest. The system incorporates factors such as customer traffic, seasonal trends, and employee skill sets to create optimized schedules. By analyzing sales patterns and customer behavior, Starbucks dynamically adjusts staffing to meet real-time demand, lowering labor costs while maintaining high service standards.

Another is Walmart, which has implemented a workforce management system that schedules workers based on predictions of when customers will be most likely to shop. This approach is currently being adopted by other retailers as well. One criticism of the software is that it can assign workers to split shifts and hours they do not want.

Other companies are using applications that consider employee preferences. Employee preferences are important as they also affect productivity. Outdoor retailer Recreational Equipment, Inc. (REI) is using a workforce management application that supports labor forecasting and workforce optimization. Resulting improvements in customer service have led to significantly higher sales. Also, the employees feel that the system leads to greater equity in scheduling.

QUESTIONS

1. How does Starbucks use AI to balance labor-cost efficiency with customer-service quality, and what key data inputs drive its scheduling decisions?
2. What are the potential ethical and employee satisfaction concerns associated with predictive labor-scheduling systems like those used by Walmart?
3. How can incorporating employee preferences into scheduling algorithms, as seen at REI, affect workforce morale, retention, and customer experience?
4. What trade-offs do companies face when trying to optimize labor schedules using AI, and how can they ensure both operational efficiency and employee well-being?
5. As more retailers adopt AI-driven workforce management systems, what best practices should be considered to ensure fair and effective implementation across different organizational cultures?

5.4 The Digital Factory

This is an exciting time for operations. A wave of new technologies is emerging, maturing, and converging in a way that is reshaping product design and manufacturing. It is shifting from a world defined by hardware and production constraints to one that is defined by software and data. The increasing deployment of the Internet of Things is allowing manufacturers to use real-time data from sensors to track parts and monitor machinery, software, and AI to completely reshape operations.

5.4.1 Transformative Technologies

Research conducted by IBM has identified three important technologies that are transforming manufacturing. They are 3D printing, a new generation of intelligent assembly robots, and the rise of open-source hardware. Individually, each of these trends is transformative. However, together, and driven by AI, their power is multiplied.

First is 3D printing, a technology that is changing manufacturing as we know it. Three-dimensional printers use technology similar to inkjet and laser-jet printers. They deposit materials like plastics and metals in thick layers, one atop the other. The process gradually builds up one layer at a time until the object is produced. This means that a solid object is created from a software design with just a click of a button. This technology makes desktop manufacturing possible and requires no economies of scale. Just a few years ago, three-dimensional printers were considered useful tools in prototype studios. They are now becoming essential machine tools on the production line. Just consider the opportunities this offers in manufacturing flexibility. Now imagine coupling that with real-time demand data. It is changing the Make lever as we know it.

Second is the next generation of intelligent assembly robots. Past robotic systems required huge, complex installations. These typically started at around $250,000 per assembly station, well above the cost of what most manufacturers could afford. This new generation costs a fraction of that price—at around $25,000 per robot. They can also be installed in less than a day. Suddenly, efficient, effective manufacturing automation is within the reach of even small companies. Coupled with AI to drive performance, super production capability is possible.

Third is the rise of open-source hardware. Open-source hardware are physical devices, such as computer hardware, whose designs are publicly available for anyone to study, modify, distribute, and make. This information can include documentation, schematic diagrams, construction details, parts lists, and logic designs—all the information needed to build the product. No reverse engineering is required.

Imagine the collective potential of these technologies coupled with AI. Collectively, these technologies are removing all the physical constraints of the Make lever—building molds, ordering parts, reconfiguring assembly machinery—and turning them into processes that can be run through software. This is the AI-driven supply chain.

5.4.2 Sensor-Driven Operations

Taking inputs from product development and historical production data (e.g., order data, machine performance), manufacturers can apply advanced computational methods to create a digital model of the entire manufacturing process. Such a **digital factory**—including all machinery, labor, and equipment—can be used to design and simulate the most efficient production systems. For any specific product, the simulation can create the facility layout and sequencing of steps. Manufacturers are now using this to optimize production of new plants.

COMPANY EXAMPLE: Siemens and BP

A prime example of sensor-driven operations can be seen in Siemens' approach to manufacturing. Siemens leverages AI, collecting vast amounts of data from sensors installed across its machinery. These sensors provide real-time insights into equipment performance, enabling Siemens to identify potential issues before they escalate into costly problems. This proactive monitoring not only reduces downtime but also ensures a consistent level of product quality.

In addition to sensor data, Siemens employs **digital twin technology**, a digital replica of its factories and production processes. By creating these virtual models, Siemens can simulate various production scenarios, optimize schedules, and assess the potential affect of adjustments without disrupting actual operations. This technology has become a core part of Siemens' strategy to boost efficiency and product quality.

Another example of how high-sensor monitoring can improve quality is seen at BP's Cherry Point Oil Refinery in Blaine, Washington. Managers use AI to detect trends in oil corrosiveness, which can be highly damaging to pipes. The company uses multiple sensors across the plant to generate vast amounts of information. This serves to compensate for the data fuzz created by the intense operating activity in the environment (such as noise, heat, and vibrations).

This is how it works. Wireless sensors are installed throughout the plant to collect vast amounts of data in real time. The environment of intense heat and electrical machinery can distort the readings. However, this is compensated for by the huge quantity of information generated from both wired and wireless sensors. Increasing the frequency and number of locations of sensor readings serves to offset the distortion. By continuously measuring the stress on pipes, BP learned that some types of crude oil are more corrosive than others, a problem it could not spot prior to relying on big data.

QUESTIONS

1. How does Siemens' use of real-time sensor data and digital twin technology enhance operational efficiency and reduce production risks?
2. What are the advantages of simulating production changes using digital twins before implementing them in real-world operations?
3. How does BP's approach to high-frequency sensor monitoring help mitigate the effects of data fuzz in harsh industrial environments?
4. What are the challenges and limitations of relying heavily on sensor-generated data in manufacturing or refinery operations, and how can companies address them?
5. How can the insights gained from sensor data, such as identifying more corrosive crude oils, lead to long-term strategic improvements in product quality and cost savings?

5.5 Connecting the Value Chain

AI applications have an unbounded potential to generate value for the Make lever of the supply chain, from improvements in product design to labor utilization and capacity planning. Data and analytics applications integrated across the extended enterprise raise productivity by increasing efficiency and improving the quality of products. For example, in emerging markets, operational improvements can build a competitive advantage that goes beyond relatively low labor costs. In developed markets, companies can use AI to reduce costs and deliver greater innovation in products and services.

5.5.1 Shared Data

Achieving meaningful operational improvements depends on the seamless exchange of data across the entire supply chain. No single company can unlock the full potential of its operations in isolation. The challenge lies in the fragmented nature of data ownership; each supply-chain participant holds a different piece of the puzzle.

While advanced AI systems now monitor activities across extended enterprises, the data they collect often remains siloed within the organization where it originated. To realize the full benefits of real-time insights and collaborative decision making, however, this data must be shared and integrated across supply-chain partners. This is a barrier to overcome.

Then there is the product itself, which is constantly generating data. Sensors embedded in machinery track real-time production metrics, while post-sale product monitoring systems gather performance feedback from the field. For instance, during a single transcontinental flight, a modern Boeing 737 can produce more than 240 terabytes of data, including engine performance, maintenance needs, and safety checks. As manufacturers embed more sensors and integrate the digital tools—such as CAD, CAE (computer-aided engineering), CAM (computer-aided manufacturing), PLM (product lifetime management) systems, and digital twin platforms—the volume and complexity of data will continue to rise. The opportunity for supply chains is to share this across organizational boundaries for end-to-end visibility.

An example is Unilever's collaboration with Microsoft and SAP to create a Virtual Ocean Control Tower that consolidates real-time data from suppliers, manufacturing plants, logistics providers, and even weather services. This shared data ecosystem allows Unilever to make agile decisions around production planning, shipping routes, and inventory levels, resulting in increased sustainability, reduced waste, and improved customer service.

5.5.2 Matching Supply and Demand

Sales and operations planning (S&OP) is an organizational process that aligns demand and supply plans with a company's financial and strategic objectives. It is a plan that bridges the gap between Make and Sell, enabling businesses to align production with demand.

S&OP is a highly data-driven process. It relies on cross-functional collaboration to balance supply and demand, ensuring that resources are efficiently allocated while customer needs are met. This integrated approach allows leaders to anticipate market changes but avoid reactive responses. When S&OP is executed effectively, it drives strategic focus and functional alignment, enabling an organization to operate in a synchronized, forward-looking approach.

S&OP hinges on reliable, clean, and accurate data, which often requires gathering and consolidating information from various departments. In this data-rich environment, companies can better synchronize their planning cycles with real-time production data, enabling them to adjust supply dynamically. This means using real-time data to adjust schedules and collaborate with marketing to shape demand through tactics like time-sensitive promotions. With the help of AI and big data, S&OP essentially enables all functions, including Make and Sell, to respond rapidly to change.

COMPANY EXAMPLE: P&G

P&G, the consumer goods leader, uses a highly integrated S&OP process to align demand forecasts with production and supply capabilities across its diverse product lines. For example, when launching a new shampoo line, the marketing team develops demand projections based on historical data, promotional plans, and market trends. These forecasts are reviewed and adjusted during the monthly S&OP cycle to reflect changing market dynamics or competitor actions.

P&G's operations, finance, and supply-chain teams collaborate during the S&OP meetings to ensure that manufacturing capacity, raw material procurement, and inventory levels align with the projected demand. If marketing expects a spike in sales due to a new national ad campaign, supply-chain managers might increase production and adjust distribution plans in advance.

Finance evaluates the cash flow and working capital implications of these decisions. By synchronizing plans across departments, P&G minimizes stockouts and overstocks, improves customer service, and increases profitability. This end-to-end visibility and coordination enable the company to respond quickly to changes in demand while maintaining operational efficiency, making S&OP a core part of its competitive strategy.

QUESTIONS:

1. How does P&G's integrated S&OP process help align marketing forecasts with supply-chain and manufacturing capabilities, especially during new product launches?
2. What role does cross-functional collaboration play in ensuring the success of P&G's S&OP process, and how might this model benefit other organizations?
3. How does end-to-end visibility in the S&OP cycle help P&G respond more effectively to changes in market demand or competitor activity?
4. In what ways does P&G's S&OP process support both customer-service improvements and cost control, and how does finance contribute to balancing these goals?
5. What challenges might arise in maintaining accurate and agile S&OP processes across P&G's global operations, and how can these be addressed through technology or organizational practices?

Another example of effective S&OP implementation is Coca-Cola. By leveraging integrated data and aligning functions across the organization, Coca-Cola has achieved substantial improvements in supply-chain efficiency. Through its S&OP process, Coca-Cola has reduced stockouts and excess inventory by 20 percent, bolstered forecast accuracy by 15 percent, and enhanced responsiveness to market fluctuations. These efficiencies have led to a 10 percent increase in overall operational efficiency, a 5 percent improvement in customer-service levels, and optimized cash flow.

AI enables S&OP by improving forecast accuracy, providing real-time decision making, and fostering collaboration across all functions. This process has allowed companies like Cisco Systems, Unilever, and Nestlé to synchronize production with real market demands and align their supply chains. By integrating consumer insights and production data, these companies can then swiftly adapt to changing market conditions.

Chapter Review

Key Points

1. Make is the supply-chain lever that is operations management and is responsible for orchestrating the transformation process of inputs into finished products.
2. Coordinating Make with Sell is critical for a company to produce precisely the goods and services customers want, in the quantities needed.
3. AI has had a tremendous effect on improving decisions along the Make lever, including product design, innovation, quality management, efficiency, and labor utilization.
4. AI is equally important in the creation of both manufactured goods and services.
5. Today we can create the digital factory—including all machinery, labor, and equipment data—where AI can be used to design and simulate the most efficient production systems.
6. AI enhances the S&OP process, which relies on having reliable, clean, and accurate data.
7. AI allows operations management to connect the entire value chain through shared data and S&OP.

Discussion Questions

1. How does the Make lever differ from the other supply-chain levers (Buy, Move, Sell), and why is its coordination with Sell essential in today's customer-driven market?
2. In what ways can AI optimize the transformation process from raw materials to finished products? Can you provide examples from both manufacturing and services?

3. How does the use of AI in product design and innovation accelerate time-to-market and affect customer satisfaction?
4. Discuss the concept of a digital factory. How do simulations, machine data, and AI-driven models contribute to operations efficiency?
5. What are the key data challenges in implementing AI in the Make lever, and how can companies ensure the accuracy and reliability of this data?
6. Can AI fully replace human decision making in operations management, or is human oversight still necessary? Provide arguments for both sides.
7. How does AI improve labor utilization without necessarily leading to job losses? What examples demonstrate AI as a tool for augmentation rather than automation?
8. In what ways has AI changed the traditional S&OP process? How does it enhance forecasting and planning accuracy?
9. Compare and contrast the application of AI in the creation of physical products versus digital services. What are the unique challenges and opportunities in each?
10. How can AI help integrate and optimize the entire value chain, from sourcing raw materials to delivering products to customers? What role does visibility and data flow play in this process?

Case Studies

The Great Toilet Paper Buy

In March 2020, the world changed overnight. The COVID-19 pandemic triggered a global crisis with cascading effects across health systems, labor markets, and, unexpectedly, supply chains. Among the most surprising and iconic consequences was the massive shortage of toilet paper in stores across the United States. Consumers, driven by fear and uncertainty, began panic-buying toilet paper in bulk, leaving shelves empty for weeks. What had long been a routine, predictable product became the poster child for supply-chain disruption.

Surge in Demand

Prior to March 2020, toilet paper demand in retail outlets was steady and well forecasted. On March 9, sales were brisk but not alarming. But by March 12, daily sales skyrocketed by more than 2,300 percent. In just one week, nearly a year's worth of toilet paper was sold across 1,200 grocery stores in the US Southern and Midwestern regions. On March 13 alone, sales were 10,128 percent higher than the same day the previous year. Consumer panic buying cascaded into logistic chaos.

By the end of March, more than 70 percent of US stores were out of stock, including giants like Amazon and Costco. As images of empty shelves went viral, they only fueled more panic, creating a self-reinforcing cycle. The item that had once ranked 20th to 30th in grocery spending became the number-one-selling product in America.

Supply-Chain Structure: Lean but Fragile

Toilet paper supply chains were built for efficiency, not flexibility. They are characterized by

- low margins and high volumes,
- Lean inventory (1–2 weeks of demand at retail and DCs),
- factories operating at 92 percent capacity pre-pandemic, and
- tight coordination and demand-driven replenishment.

This system, while cost-efficient, had little buffer to absorb demand shocks. When the surge hit, manufacturers ramped up production to 98 percent capacity. Yet, due to production constraints and labor shortages from social distancing, this translated to only an 8 percent increase in output, far short of demand.

Manufacturing Challenges

Most plants produced either consumer-grade or away-from-home (AFH) toilet paper. With offices, schools, and restaurants closed, demand for AFH products collapsed, while demand for consumer products soared. Yet the two types were made with different materials (virgin pulp vs. recycled paper), and shifting production required significant capital investment and time.

Adding to the burden were

- shortages of packaging material and pulp,
- rising pulp prices (up 10 percent per ton), and
- reduced labor availability due to COVID safety protocols.

Operations Strategy Pivot: From Lean to Hybrid

Facing this unprecedented challenge, both manufacturers and retailers were forced to rethink their operational strategy. Several key shifts occurred.

SKU Rationalization. Companies like Kimberly-Clark reduced SKUs (stock-keeping-units) by up to 50 percent, focusing on fast-moving, standard products (e.g., mega 6-packs instead of 12-packs). This minimized downtime due to machine changeovers, increasing output.

Retailer Rationing and Prioritization. Retailers like Kroger and Publix imposed purchase limits. Algorithms were deployed to flag demand surges and automatically initiate rationing.

Enhanced Information Sharing. Real-time sales and inventory data were shared across suppliers, distribution centers, and stores to optimize shipments. Retailers prioritized shelf space for essential items and used advanced forecasting to prioritize restocking.

Repurposing Capacity. Manufacturers invested in converting AFH capacity to produce consumer-grade tissue, though slowly and at cost.

Increased Automation. P&G and others expanded use of automation and contactless delivery to keep operations running with fewer workers.

Supplier Diversification and Expansion. Manufacturers added new suppliers for pulp and packaging. Temporary distribution centers were opened to speed up fulfillment.

Strategic Dilemma

Toilet paper is a functional product with predictable demand and low profit margins, ideal for a Lean supply chain. But the pandemic exposed the fragility of this model in crisis. Should manufacturers and retailers now invest in greater supply-chain flexibility, traditionally reserved for high-margin, innovative products like fashion or electronics?

Investing in flexibility means higher operating costs, added inventory, more capacity slack, and redundancy. But it also means resilience in the face of uncertainty. Is the new normal one in which even basic products must be protected against black swan events?

Questions

1. What operational practices made the toilet paper supply chain so vulnerable to a sudden spike in demand? Consider inventory levels, capacity utilization, and SKU variety.
2. How did the bullwhip effect manifest in this case, and what could have been done to mitigate its effect? Discuss information flow, demand visibility, and communication across the supply chain.
3. Toilet paper is considered a functional product. Do you agree that its supply chain should remain lean, or has the pandemic shifted this thinking? Why?
4. Compare and contrast the challenges faced by consumer-grade toilet paper producers versus away-from-home producers during the pandemic. What does this tell you about demand segmentation?
5. What trade-offs were made when companies shifted from efficiency to flexibility in this case? Discuss costs, capacity reallocation, product variety, and customer service.
6. How did automation and real-time data sharing help stabilize the supply chain during the crisis? Could these changes become permanent fixtures in operations strategy?
7. If you were an operations manager at a major toilet paper manufacturer in early March 2020, what immediate actions would you have taken? What long-term changes would you recommend for future resilience?
8. What lessons can other industries learn about supply-chain design and preparedness for sudden demand disruptions from this case?

BYD versus TESLA

Two global EV giants, Tesla and BYD, represent divergent strategies in the race to dominate the EV industry. While both companies leverage advanced technologies and AI to drive innovation, they differ fundamentally in their business models, target markets, and supply-chain approaches.

Company Overviews

Tesla: Founded in 2003, USA
Focus on premium EVs with performance and branding
Price range from $40,000–$ 100,000
Target market: Brand-driven, innovation-led, early adopters
Global reach: North America, Europe, China, expanding globally
AI & tech integration is through FSD, OTA updates, and battery optimization

BYD: Founded in 1995 (as battery co.), EV production since 2003, China
Focus on affordable, mass-market electric cars
As low as $11,000 USD
Target market: Value-driven, broad accessibility
Global reach: Dominant in China, rapidly expanding to Latin America and Europe
Smart manufacturing, battery optimization, predictive maintenance.

Supply-Chain Strategy

1. **Battery Supply**
   Tesla: Initially outsourced (Panasonic), now vertically integrating (Gigafactories)
   BYD: Fully integrated; produces its own batteries and even battery raw materials
2. **Manufacturing**
   Tesla: Assembly-focused with some vertical integration (Gigafactories)
   BYD: Deep vertical integration; produces motors, semiconductors, batteries, chassis
3. **Flexibility & Speed**
   Tesla: High speed to market with focus on innovation cycles
   BYD: High production scalability and cost control through integration
4. **AI Usage in Operations**
   Tesla: AI used for FSD, predictive analytics, product development
   BYD: AI used for manufacturing optimization, quality control, logistics

Strategic Positioning

Tesla's value proposition lies in its cutting-edge innovation, sleek design, and autonomous driving capabilities, appealing to tech-savvy, affluent consumers. It uses software and AI to deliver a continuous stream of updates and enhancements to vehicle performance and features.

BYD, on the other hand, pursues cost leadership and operational efficiency, targeting middle-income and price-sensitive consumers. Its strategy is grounded in vertical integration and AI-enhanced Lean manufacturing. With affordable EVs and expanding global reach, BYD challenges Tesla not by imitation but through a fundamentally different model.

Future Dilemma

Tesla continues to innovate with FSD ambitions and energy products (solar, Powerwall), while BYD aggressively expands internationally, investing in EV buses, trucks, and emerging markets. Both companies exemplify the disruptive potential of AI and automation in reshaping global mobility, albeit through contrasting lenses.

Questions

1. Does vertical integration give BYD a competitive edge in cost control and supply-chain stability? If so, how?
2. Why has Tesla chosen to pursue a premium strategy, and what are the risks and benefits of this positioning?
3. How do AI applications differ in Tesla's and BYD's operations, and what does this suggest about their strategic priorities?
4. Could BYD's model work in North America, or is it better suited to emerging markets? Why?
5. What lessons can traditional automakers learn from Tesla and BYD's contrasting approaches to supply-chain design?
6. To what extent does brand matter in the EV industry, and how do Tesla and BYD differ in building brand equity?
7. How might geopolitical tensions (e.g., US–China relations) affect the global expansion of both Tesla and BYD?
8. What are the long-term sustainability implications of Tesla's innovation-led model versus BYD's volume-led model?
9. Which company is better positioned to lead in autonomous driving, and why?
10. How might AI reshape vehicle design, production, and customer service across the automotive industry in the next decade?

Chapter 6

Impact on Logistics, or "Move"

In This Chapter You Will Learn

1. that logistics is responsible for moving goods through the supply chain;
2. the decisions involved in delivering the perfect order;
3. about the costs involved in the infrastructure of transportation and logistics;
4. how Artificial Intelligence (AI) optimizes Move, from inventories, warehousing, to reverse logistics; and
5. how logistics activities are integrated across the supply chain.

What do everyday products like Tide detergent, Nike sneakers, and Dove soap have in common?

The companies that sell these products—from Walmart, Procter & Gamble, Unilever, and Nike—rely on Maersk to deliver these goods. Maersk is one of the world's largest shipping and logistics companies that transport goods for a wide range of clients, including major global retailers and manufacturers. Without Maersk's timely and reliable shipping services, store shelves would sit empty, delivery times would increase, and consumer access to these familiar goods would be disrupted.

The work of Maersk is invisible to the average consumer, yet the company serves thousands of businesses across industries like automotive, electronics, pharmaceuticals, and agriculture. Its work enables global trade through ocean freight, inland transport, and end-to-end logistics services. Maersk's optimized logistics operations ensure that everyday consumer goods—like electronics, clothing, and food—arrive on retail shelves more quickly and reliably.

How does Maersk do this?

Maersk relies on AI to enhance the efficiency and reliability of its global supply-chain operations. Through its multiple platforms, Maersk applies AI to process real-time data from more than 600 ports, terminals, and carriers worldwide. AI algorithms optimize shipping routes by analyzing weather, port congestion, and vessel conditions, thereby reducing delays by up to 30 percent and fuel consumption by 10 percent. AI processes Internet of Things (IoT) sensor data to provide real-time visibility into the location, condition, and estimated time of arrival (ETA), of containers, thus improving inventory planning and customer service. AI automates bills of lading and customs processes, reducing human error and processing time by 40 percent.

Through these applications, Maersk has streamlined its logistics function, cut operational costs, and improved on-time delivery performance globally.

6.1 Moving the Things Will Sell

Logistics is the business function responsible for transporting and delivering products to the right place at the right time throughout the supply chain. In business jargon it is often referred to as **Move** and plays a critical role in the functioning of both the organization and the supply chain. Without logistics, materials would not arrive when and where they are needed.

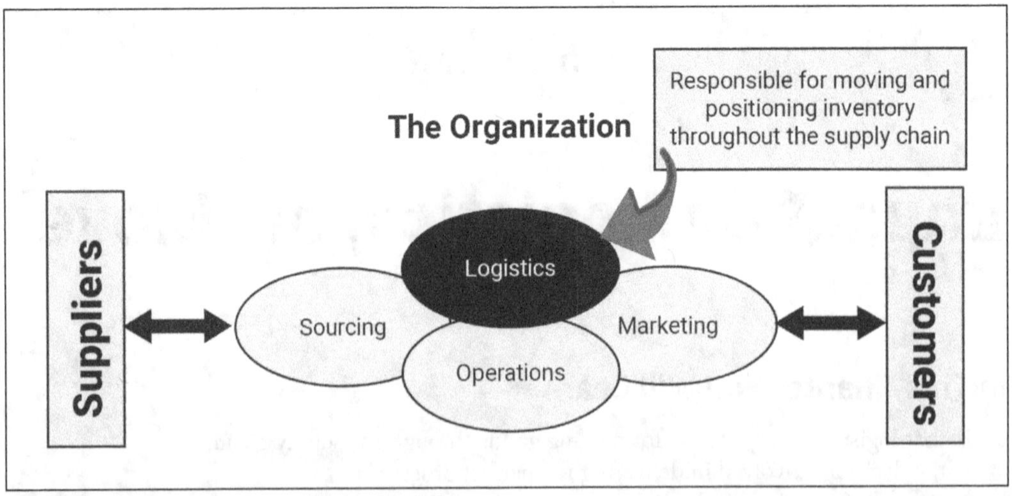

Figure 6.1 The logistics function

To achieve this, logistics must plan and coordinate all material flows from source of supply to all users in the chain (Figure 6.1). There are many factors to evaluate and optimize in a supply-chain network. As such, AI technology is an essential enabler.

6.1.1 Delivering the Perfect Order

Logistics involves much more than movement and storage of goods throughout the supply chain.

On the upstream side of the supply chain, logistics works with sourcing to link the organization to its suppliers. It works with operations to ensure that sourced materials are transformed into finished products. On the downstream side of the supply chain, it works with marketing to ensure distribution and delivery of outbound products to external customers.

Logistics is responsible for delivering what is known as **the perfect order** to every customer. The perfect order is based on the five rights of customers. These rights are receiving the **right product** at the **right place** at the **right time** in the **right condition** for the **right cost.**

As customers we all assume that when we place an order it will be delivered as promised. The definition of the perfect order is one that meets the following conditions of delivery:

- **Complete:** All items are included in the order, in the quantity requested.
- **On time:** The order arrives on the customer's request date, using the customer's definition of on-time delivery.
- **Accurate:** All documentation is included, such as packing slips and invoices.
- **Perfect condition:** The order is in the correct configuration, is damage free, and is faultlessly installed.

Achieving the perfect order requires optimization and coordination of all activities of the supply chain. As the pace of commerce has increased, so have customer expectations. "Better, faster, and cheaper" is no longer good enough. Today's customers are demanding perfect orders, shipped and delivered on time, at a cost that leaves little margin. It is up to logistics to make this happen. And it is true for all companies, regardless of size of industry.

Let's look at an example of how this works in a small company.

COMPANY EXAMPLE: Bella's Custom Tea Co.

Bella's Custom Tea Co. is a small specialty tea company based in Philadelphia that sells handcrafted loose-leaf tea blends through its online store. The company prides itself on personalized customer service, sustainable packaging, and quick delivery. To ensure customer satisfaction and repeat business, Bella's logistics operations are designed to consistently deliver the perfect order, fulfilling the five rights of logistics:

Right Product. Bella's uses an integrated order management system that syncs online customer orders with inventory. When a customer orders a Lavender Earl Grey blend, the system cross-verifies that the SKU matches the selected product. Each order is then hand-packed and labeled to ensure accuracy.

Right Place. The company uses address-verification software at checkout to reduce errors. Orders are shipped via USPS or local courier depending on the delivery address. For local customers, same-day delivery is available.

Right Time. Orders placed before 3:00 PM EST are guaranteed to ship the same day. Bella's uses regional warehouses and predictive carrier data to select the shipping option that meets the promised delivery window, typically one to three business days.

Right Condition. The tea is packed in eco-friendly, moisture-resistant pouches, placed in insulated boxes, and cushioned with recyclable filler. Packages are sealed, tracked, and quality-checked to ensure that the tea arrives fresh, untorn, and undamaged.

Right Cost. Bella's offers flat-rate shipping within the United States and free delivery for orders over $50. Because the company maintains tight inventory control and uses zone-based carrier optimization, it keeps costs low without sacrificing service quality.

Result Is the Perfect Order. Thanks to this logistics setup, Bella's achieves a 97 percent perfect order rate, meaning that nearly every customer received exactly what they ordered, where and when they expected it, in pristine condition, and at a fair cost. This high level of fulfillment accuracy boosts customer loyalty and positive reviews, which are critical for a growing small business.

The Technology. Bella's uses a machine-learning-based recommendation system integrated with its e-commerce platform. This AI system analyzes customer purchase history, seasonal preferences, and feedback data to suggest personalized tea blends and package sizes for returning shoppers. It also forecasts demand for specific ingredients and blends to optimize sourcing and minimize stockouts. The AI tracks inventory levels in real time and triggers reordering of ingredients before they run low, ensuring that each order meets the "five rights" of logistics: right product, place, time, condition, and cost.

QUESTIONS

1. You are in charge of logistics at Bella's. How would you use AI to balance cost and speed in shipping decisions? What prompts would you use? What data would you need to make smart choices? List at least two trade-offs you would have to make. How would you integrate issues like supplier delays or unpredictable trends?
2. How could AI help Bella's team anticipate customer preferences beyond order history? How would you integrate external data like weather, holidays, or customer location?
3. What risks or challenges might arise from relying on AI for inventory management in a small business?
4. How would you improve Bella's recommendation engine to enhance personalization? Would you incorporate surveys, social media data, or customer reviews?
5. How can Bella's use customer-feedback data to continuously improve the AI system? Propose a feedback-loop strategy between AI and customer service.

6.1.2 Optimum Cost-Service Mix

As we see, the goal of logistics is customer service to deliver the perfect order. This is achieved by all the logistics activities working in tandem. Customer service can be defined as a customer-oriented philosophy that integrates and manages all elements of the customer interface within a predetermined optimum cost-service mix.

In measurable terms, the perfect order means the following:

- ship all items ordered (measured as **percent order complete**)
- without error (measured as **percent order accuracy)**
- on time to the customer (measured as **percent on-time delivery)**.

This raises an important question: *What does the customer really mean by "on time"*?

The recent trend has been moving toward a standard of a minimum of next-day shipping. However, an increasing number of retailers are moving toward offering same-day shipping, and a few are offering same-hour shipping for selected SKUs in selected geographic areas.

Here is the problem:

Such high levels of service are expensive. Retailers that have a multichannel distribution network can possibly provide same-hour shipping and delivery from a brick-and-mortar store. However, this is only feasible within a certain geographic area. And it is still costly. It is for this reason companies need to ask some important questions:

- Do all our customers really require next-day shipping?
- Is this necessary to achieve high customer service?

For some market segments, this will certainly be true. Many customers, however, may prefer a discount on the shipping price in exchange for later delivery. Offering these options provides distribution operations with flexibility that can help with smoothing order demand and balancing staffing. Also, relaxing next-day service levels requirements, even to 48-hour delivery, significantly reduces the capital investment needed as flexible labor can be used for added capacity.

AI is the solution to identifying which segments respond to which elements of customer service. Micro-segmentation—using software applications and analysis—reveals exactly the customer service requirements that are order winners in each segment. It can create segments of customers for which such high delivery standards are critical. It then permits allocating distribution costs to those segments rather than averaging it across segments and including those that are indifferent to such a costly proposition. This then provides a better sense of true costs.

For example, Greenleaf Electronics, an online retailer specializing in smart home devices, faced rising logistics costs as customer expectations for rapid delivery grew. To deliver the "perfect order" —complete, accurate, and on time—they initially offered free next-day shipping across all orders. However, analytics revealed that only 30 percent of their customers, mainly business clients, truly valued immediate delivery. The majority preferred lower shipping costs over speed, a trade-off that is very important.

Using micro-segmentation tools, Greenleaf redesigned their logistics strategy: premium customers continued receiving next-day service, while others were offered discounted shipping for two or three-day delivery options. By aligning service levels with customer preferences, Greenleaf reduced their distribution costs by 189 percent, improved order accuracy, and maintained high customer satisfaction across all segments.

6.1.3 Coordination

Moving goods requires coordination. Without this, there is no perfect order. This includes managing the entire distribution network, location of warehouses, distribution centers and plants, and coordinating the modes of transportation between them. It also includes design and management of operations throughout the network for efficient storage and quick movement of goods.

There are many logistics tasks involved in coordinating movement of goods along the supply chain. Figures 6.2 illustrates some of these tasks within the flow of a supply chain. Table 6.1 then list these tasks and their definitions.

Let's look at implementation of these specific tasks at Walmart and see how an AI platform coordinates them.

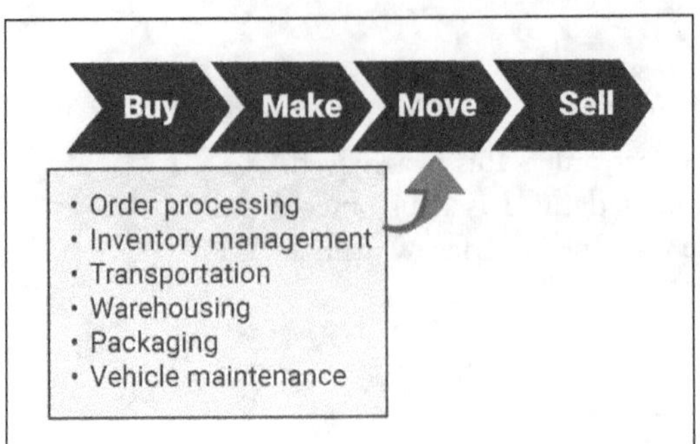

Figure 6.2 Logistics Tasks

Table 6.1 Logistics Tasks and Definitions

Task	Definition
Order Processing	Managing customer orders from receipt to fulfillment
Inventory Management	Tracking and managing stock levels to meet demand
Transportation Management	Planning and executing the movement of goods
Warehousing	Storing goods safely until they are needed
Packaging	Preparing goods for safe and efficient transport
Material Handling	Moving goods within warehouses and production facilities
Customs Clearance	Managing documentation and compliance for cross-border shipments
Demand Forecasting	Predicting future customer demand to optimize logistics planning
Returns Management	Handling the return of goods from customers efficiently
Route Optimization	Planning the most efficient delivery paths to save time and costs

COMPANY EXAMPLE: Walmart

Walmart operates one of the most digitally integrated logistics and supply-chain systems. The platform is called Retail Link and is driven by real-time data exchange between the company and its suppliers. It allows near real-time visibility into sales, inventory, forecasts, and replenishment needs. At the core is a seamless flow of information powered by technologies such as barcoding, radio frequency identification (RFID), IoT, machine learning, and blockchain.

This system enables automatic replenishment. In its retail stores, Walmart uses bar code scanners and smart shelf sensors to capture real-time point-of-sale (POS) data, which is immediately transmitted through its centralized system to both internal logistics teams and external suppliers. As products are sold, inventory counts are updated in real time, and restocking orders are triggered algorithmically without the need for manual intervention. Suppliers receive this demand data instantaneously and use it to optimize their production and shipping schedules. As of 2025, Walmart's supply chain can process more than 265 million item-level transactions daily, a capability made possible by cloud-based analytics.

In parallel, Walmart receives continuous updates from its suppliers on shipment status, arrival times, order quantities, temperature-controlled transport data, and electronic invoicing. Suppliers benefit from this transparency. They can track how their products are performing in the market in near real time and receive faster payment cycles, which significantly improves their cash flow. Walmart's newer initiatives, like its Supplier Quality Excellence Program and blockchain pilot with IBM, further enhance traceability, reduce invoice disputes, and ensure food and pharmaceutical safety.

Walmart has deployed RFID tags on more than 90 percent of its apparel and general merchandise SKUs and is expanding RFID use to home goods and electronics by 2026. This allows Walmart to track the exact quantity and location of goods in its stores and distribution centers with high precision. Their distribution centers are automated, powered by robotics and AI, ensuring that restocking happens faster and with lower error rates. While Retail Link itself is primarily a data and analytics platform, it plays a key enabling role in managing Walmart's broader transportation and logistics strategy.

Here's how Retail Link, powered by AI, affects specific logistics tasks at Walmart:

TRANSPORTATION

- Retail Link enables suppliers to access transportation data, including shipment tracking, delivery performance, and compliance scores.
- It supports Walmart's Vendor Managed Inventory by allowing suppliers to plan shipments based on real-time demand.
- Walmart uses this visibility to consolidate loads and optimize freight costs, especially through its private fleet and third-party carriers.

ROUTING

- Retail Link integrates with Walmart's routing systems, such as Global Logistics System and Carrier Logistics Management tools.
- It helps coordinate route optimization by providing SKU-level data, demand forecasts, and store-level sales trends so that products are routed efficiently to DCs or stores.
- Routing decisions are improved by predictive analytics, reducing unnecessary miles and emissions.

SCHEDULING

- Suppliers use Retail Link to book delivery appointments at Walmart's distribution centers via the Scheduler 2.0.
- Walmart uses historical data from Retail Link to optimize dock scheduling and reduce bottlenecks at high-volume facilities.
- This improves truck turnaround times, enhancing overall transportation efficiency.

DIRECT AND INDIRECT DELIVERY

- Most goods flow through Walmart's regional DCs. Retail Link provides visibility into DC-level inventory needs and enables precise inbound planning.
- Some products (like perishable goods) go straight to stores. This direct-delivery Retail Link helps suppliers track store-level sales and inventory so they can schedule timely direct deliveries. This improves delivery timing and inventory accuracy, minimizing out-of-stocks and overstocks.

VEHICLE MAINTENANCE

- Walmart uses advanced fleet management technology that incorporates AI and advanced analytics to manage its massive private fleet, one of the largest in the US, with more than 10,000 drivers and thousands of tractors and trailers.
- Walmart vehicles are equipped with GPS tracking and telematics that provide real-time data on fleet performance and optimize driving routes.
- AI machine learning is used to optimize delivery routes in real time, reduce empty miles and fuel consumption, adapt to traffic patterns and weather conditions, and store schedules.
- AI is used to forecast demand, helping logistics managers match shipments to capacity.
- AI-driven predictive maintenance tools analyze vehicle sensor data to detect early signs of mechanical issues and schedule repairs before breakdowns occur.
- AI systems assess driver behavior using telematics (e.g., hard braking or rapid acceleration).

Walmart uses AI-driven logistics technology to optimize its supply chain through real-time demand forecasting, intelligent route planning, and predictive maintenance of its private fleet. These systems enhance delivery efficiency, reduce transportation costs, and support the company's sustainability and service-level goals. With this, Walmart continues to be a benchmark in logistics excellence. Currently it is maintaining its position as the world's largest retailer with more than 10,500 stores across 19 countries and an e-commerce business that surpassed $100 billion in annual revenue in 2024.

QUESTIONS:

1. You are Walmart's logistics manager. How would you use Retail Link data to reduce transportation costs while maintaining service levels? Consider how real-time sales and inventory data might influence routing, load planning, or shipment frequency.
2. In what ways does Retail Link enable better coordination between Walmart and its suppliers for inbound scheduling and delivery? What role does the appointment system play in reducing bottlenecks at distribution centers?
3. How can Retail Link data improve the efficiency of direct-to-store versus distribution-center delivery strategies? How would you use Retail Link to decide what gets shipped directly to stores versus routed through DCs?

4. What kind of performance metrics should Walmart managers monitor in Retail Link to evaluate carrier and supplier logistic performance? Consider factors such as on-time delivery, fill rates, and compliance scores.
5. If you were a logistics manager at Walmart, how would you use Retail Link insights to proactively manage disruptions like delays, inventory shortages, or overstocking? Describe the decision-making process using data visibility across the supply chain.
6. How can AI-powered fleet management systems help Walmart balance cost efficiency with sustainability goals? How can route optimization, predictive maintenance, and electric vehicle integration impact both operational and environmental performance?
7. List potential risks or limitations of relying heavily on AI and telematics in fleet operations.

6.1.4 Logistics Costs

Logistics is costly. It is a function that requires large investments in infrastructure, such as transportation vehicles, material-handling equipment, and information technology. For this reason, most companies outsource this function to outside firms. Companies such as UPS, FedEx, and DHL have made it their business to provide logistics services to their clients. These companies have invested in the needed infrastructure, such as their own fleet of airplanes and trucks and the latest technology to track packages from point of origin to destination using bar code technology and scanners. In addition to moving products, these companies can be contracted to perform a range of logistics functions, including designing an entire distribution network for their clients. They can also provide services such as management of inventory, warehouse management, and even customer interface.

Controlling the logistics costs is one of management's top concerns. As a result, efficient and effective control of logistics is an imperative. The aspect that adds substantial cost is the need for technology, both information and processing technology.

Let's look at a small company example to see how they use AI technology to reduce their logistics costs.

COMPANY EXAMPLE: GreenCrate Organics

GreenCrate Organics, a regional distributor of farm-fresh produce in the Pacific Northwest (name modified) faced rising transportation and logistics costs due to fuel price volatility, underutilized delivery trucks, and last-mile inefficiencies. As a small business with a network of local farms and urban customers, transportation made up nearly 40 percent of its total operating costs.

To address these challenges, GreenCrate took several strategic steps:

Route Optimization with AI. They implemented a cloud-based route optimization system that uses real-time traffic, weather, and order volume data to dynamically plan daily delivery routes. This resulted in a reduction of total miles driven by 18 percent, saving significantly on fuel and driver labor costs.

Collaborative Shipping with Other Local Vendors. GreenCrate partnered with nearby small businesses, like bakeries and dairies, to share truck space for overlapping delivery zones. This increased vehicle load capacity from 55 percent to 85 percent, dramatically improving cost efficiency per shipment.

Hub-and-Spoke Distribution Model. Instead of dispatching from a single central warehouse, GreenCrate set up small micro-fulfillment hubs in suburban areas closer to customer clusters. This cut last-mile delivery costs by 25 percent and reduced delays.

Sustainable Packaging and Delivery Incentives. The company introduced reusable crates with a deposit system and offered customers discounts for flexible delivery windows, allowing better route batching and fewer missed deliveries.

By rethinking logistics through technology, collaboration, and smarter routing, GreenCrate Organics cut transportation and logistics costs by 22 percent within a year, while improving on-time delivery and customer satisfaction. The implementation followed a phased approach.

First, they began integrating a cloud-based AI route optimization system, training dispatchers and drivers to use the platform, and analyzing initial delivery patterns. Then, GreenCrate initiated partnerships with local vendors, established shared delivery protocols, and used AI to identify overlapping routes to improve truck utilization. The company used delivery data to choose optimal locations for reducing last-mile costs and congestion. These changes not only made the business more profitable but also aligned with its eco-friendly mission.

The Technology. GreenCrate Organics leveraged AI-driven route optimization, collaborative shipping partnerships, and a decentralized hub-and-spoke model to cut logistics costs and enhance delivery efficiency. By integrating real-time data and aligning technology with sustainability goals, the company improved operational agility while staying true to its eco-conscious mission.

QUESTIONS

1. How did AI-powered route optimization help GreenCrate reduce costs? What types of data inputs were crucial to its success? How would you incorporate factors like real-time traffic, weather, and order volume?
2. List the operational and strategic benefits of collaborating with other local vendors for shared shipping? How might AI tools help manage and coordinate such partnerships effectively? Specify how you would use AI to do this.
3. How does the shift to a hub-and-spoke model complement the use of AI in route planning and last-mile delivery? Discuss the logistic and cost implications of this change.
4. What challenges might GreenCrate face in maintaining or scaling its collaborative shipping model, and how could AI help address them? Explain how you would coordinate complexity, data sharing, and scheduling.
5. In what ways did GreenCrate align its logistics changes with its sustainability mission, and how does AI support those goals? Do you think there is a link between reduced emissions, packaging, and eco-conscious delivery? If so, what is it?
6. If you were advising GreenCrate, what other AI tools or data sources could further enhance their logistics operations? Would you suggest AI tools to focus on demand forecasting, customer behavior analytics, or predictive maintenance?
7. How would you use AI to evaluate the effectiveness of customer incentives for flexible delivery windows? What prompts would you use? What data would be needed and how would insights be applied?
8. What lessons can larger companies learn from GreenCrate's AI-enabled logistics strategies, especially regarding agility and cost efficiency? Could these strategies be adapted at scale, or are they unique to small business models?

6.2 AI Technologies in Logistics

Logistics is a highly data- and technology-enabled function that demands real-time visibility, security, speed, and precision. There is little room for error as mistakes mean not delivering the perfect order and incurring higher costs. To achieve this precision, companies must track the movement of goods across global networks, optimize delivery timing down to the hour, and ensure the security of products against theft, tampering, or spoilage. These operational demands cannot be met with traditional methods alone; state-of-the-art technologies, driven by AI, are now essential to managing modern logistics systems efficiently and at scale.

Organizations are investing heavily in AI technologies to reduce logistics costs, improve customer service levels, and enhance operational resilience. These initiatives extend beyond transportation and warehousing and include data protection, cybersecurity, disaster recovery planning, and data warehousing.

6.2.1 Core Technologies

The logistics technology landscape has expanded significantly, blending traditional supply-chain tools with sophisticated AI-powered innovations. A summary of these technologies is listed in Table 6.2.

Table 6.2 AI Logistics Technologies

Technology	Definition	How It's Used in Logistics
AI and ML	AI simulates human intelligence; ML allows systems to learn from data autonomously	Optimizes demand forecasting, route planning, warehouse operations, and fleet management
IoT Sensors	Networked sensors embedded in goods, vehicles, and facilities to monitor and transmit data	Tracks real-time location, condition, and movement of goods; enables predictive maintenance
Blockchain for Supply-Chain Security	Decentralized digital ledger for secure, transparent tracking of goods and transactions	Authenticates products, combats counterfeiting, and provides secure end-to-end cargo tracking
Advanced Barcode Scanning and Next-Gen RFID	Upgraded scanning technologies offering faster, more reliable item tracking and inventory visibility	Automate inventory check-ins/check-outs, reduces manual errors, and triggers real-time stock updates
Vision Systems for Automated Inspection	Computer vision AI that interprets visual data to automate inspections and verifications	Automates packaging inspections, damage detection, and load verifications in warehouses
Real-Time Satellite and Drone Data Transmission	Integration of real-time satellite imagery and drone feeds into logistics operations	Monitor remote logistics sites, assesses infrastructure conditions, and enables dynamic rerouting
AMRs and AGVs	Robots that navigate warehouse floors (AMRs) or follow programmed paths (AGVs) to move goods	Automate picking, packing, and internal warehouse transport, enhancing efficiency and accuracy
Edge Computing Devices	Local (on-device) processing of data to enable faster decision making without relying on cloud servers	Enables real-time analytics at the source (e.g., warehouses, vehicles) for faster corrective actions

A more detailed overview of these key technologies, how they work, and how they are being used today is given here.

1. Artificial Intelligence and Machine Learning

Definition: As defined in Chapter 1, AI refers to the ability of machines to mimic human intelligence, while machine learning (ML) allows systems to automatically learn from data without being explicitly programmed. AI and ML are the backbone of the majority of technologies used in logistics as described below.

Use in Logistics

- Demand forecasting and inventory optimization through predictive analytics
- Route optimization using real-time traffic, weather, and delivery constraints
- Autonomous decision making for dispatching, fleet management, and warehouse automation

Example: UPS uses AI in its ORION platform to dynamically reoptimize driver routes midday based on changing delivery volumes.

2. Internet of Things (IoT) Sensors

Definition: IoT devices are networked sensors embedded into products, vehicles, and infrastructure to collect and transmit data.

Use in Logistics

- Real-time tracking of goods in transit (temperature, humidity, location)
- Predictive maintenance of trucks and warehouse equipment based on usage patterns

Example: Maersk deploys IoT-connected containers to monitor cargo conditions across its global shipping network.

3. Blockchain for Supply-Chain Security

Definition: A decentralized digital ledger that securely records transactions across multiple parties.

Use in Logistics

- Authentication of goods to combat counterfeiting
- Transparent tracking of products from origin to customer

Example: Maersk's blockchain platform, described in the chapter introduction, improves transparency across global shipping logistics by recording immutable cargo movement data.

4. Advanced Barcode Scanning and Next-Gen RFID

Definition: Modern barcode scanners and radio frequency identification (RFID) systems are wireless technologies that use radio waves to identify and track objects. They offer faster, more detailed item scanning with greater range and reliability.

Use in Logistics

- Automated check in/check out of goods at warehouses
- Inventory visibility across large facilities without manual counting

Example: Zara uses RFID tags across its stores and warehouses to monitor inventory movements and trigger replenishment in real time.

5. Vision Systems for Automated Inspection

Definition: Computer vision uses AI and cameras to interpret and act on visual data.

Use in Logistics

- Automated quality checks on packaging integrity
- Counting and verification of pallet loads and container contents

Example: Amazon uses vision systems in its fulfillment centers to inspect outgoing packages for damage or errors before shipping.

6. Real-Time Satellite and Drone Data Transmission

Definition: Integration of satellite imagery and drone feeds into logistics operations.

Use in Logistics

- Monitoring large outdoor storage yards or shipping ports
- Dynamic rerouting based on real-time infrastructure damage (e.g., after natural disasters)

Example: DHL uses drones for last-mile deliveries in remote areas and to monitor warehouse yards for efficiency.

7. Autonomous Mobile Robots (AMRs) and Automated Guided Vehicles (AGVs)

Definition: AMRs navigate dynamic warehouse environments independently, while AGVs follow preprogrammed paths to transport goods.

Use in Logistics

- Picking and packing automation inside warehouses
- Material transport between production lines and loading docks

Example: Walmart uses thousands of AMRs in its fulfillment centers to speed up item retrieval and reduce labor costs.

8. Edge Computing Devices

Definition: Edge computing processes data closer to where it is generated rather than sending it to centralized servers.

Use in Logistics

- Faster decision making in warehouse robotics and autonomous vehicle navigation
- Real-time processing of sensor data for immediate corrective action

Example: FedEx uses edge computing to enable faster routing decisions for its autonomous delivery robots.

On the software side, the landscape has also evolved. AI-driven warehouse management systems (WMSs), AI-enhanced transportation management systems (TMSs), predictive inventory optimization platforms, advanced enterprise resource planning (ERP) systems with embedded AI, dynamic supply-chain planning suites, AI-powered global trade management platforms, and intelligent labor management systems are now central to logistics operations.

Additionally, digital twin technology, discussed next, is gaining adoption for simulating supply-chain scenarios in real time. Each of these software solutions not only provides independent data streams but also creates a connected ecosystem. Let's look at this next.

6.2.2 Emerging Technologies & Connected Ecosystems

Technological advancements are happening rapidly, resulting in numerous emerging technologies in addition to core capabilities. One example is the use of **digital twins**, essentially virtual simulations of logistics networks. DHL, for example, uses digital twins of its warehouses to simulate operations, test different layouts, and optimize workflows in real time, improving efficiency and reducing bottlenecks without disrupting actual operations.

Another technology is **AI-powered supply-chain control towers**—centralized, real-time platforms that use artificial intelligence to provide end-to-end visibility, predictive insights, and automated decision making across the entire supply chain. For instance, Dell uses an AI-powered supply-chain control tower to monitor global shipments in real time, predict potential disruptions like port delays, and automatically reroute inventory to ensure on-time customer deliveries.

Companies are not just implementing these technologies in isolation; they are building fully connected ecosystems in which hardware, software, and AI analytics work together to maximize visibility, automate decisions, and optimize outcomes across the entire logistics value chain. By linking hardware devices with advanced software applications through cloud platforms, AI orchestration engines, and digital twins, organizations can achieve levels of agility, efficiency, and resilience that were previously unattainable. The future is not just smarter individual technologies; it is the intelligent connection between them.

One example is **cloud platforms**. They are internet-based systems that provide scalable computing resources, applications, and data storage to manage and optimize logistics operations, such as inventory tracking, transportation management, warehouse operations, and real-time supply-chain visibility. These platforms allow companies to access, share, and analyze data anytime, anywhere, improving speed, flexibility, and decision-making.

Example: DHL uses Microsoft Azure as a cloud platform to power its Resilience360 system, which provides real-time supply-chain risk management. By leveraging Azure's cloud infrastructure, DHL collects and analyzes vast amounts of global logistics data, offering predictive analytics on potential disruptions like natural disasters, port closures, or political instability. This enables DHL and its customers to reroute shipments quickly and maintain continuous operations.

Another example is **AI orchestration engines**. These are AI-powered systems that automatically coordinate and manage multiple AI models, data sources, and decision-making processes to optimize logistics operations—such as routing, warehouse automation, inventory management, and demand forecasting—in real time. They act like a conductor, ensuring that all AI tools and technologies work together efficiently across the supply chain.

Example: Maersk uses an AI orchestration engine through its supply-chain management platform called NeoNav. NeoNav integrates AI models for vessel tracking, container optimization, customs clearance predictions, and real-time disruption alerts. The orchestration engine continuously prioritizes and reassigns resources, such as rerouting shipments due to weather, by balancing delivery speed, cost, and carbon footprint, without human intervention.

Yet another example is **integrated data lakes.** These are centralized, cloud-based repositories that store vast amounts of structured and unstructured data from across the entire supply chain—including orders, shipments, inventory levels, weather, sensor (IoT) data, customer interactions, and more—and connect them so they can be easily accessed, analyzed, and used by different systems, including AI and analytics tools. The integrated part means the data lake pulls data from multiple sources (e.g., warehouses, transportation systems, ERP software) into one unified environment for better end-to-end visibility and smarter decision making, which allows organizations to combine all of these technologies into a seamless, agile, and data-driven logistics system.

Example: FedEx built an integrated data lake on Amazon Web Services to centralize package tracking data, vehicle telematics, weather patterns, and customer service interactions. By integrating these different data sources, FedEx enables real-time package tracking, predictive maintenance of delivery trucks, and AI-driven delivery route optimization, improving delivery accuracy and customer satisfaction.

Let's look at an example of connecting these hardware and software technologies.

COMPANY EXAMPLE: DHL

DHL has developed a fully integrated logistics network with advanced hardware and software technologies that work in seamless coordination. In their smart warehouses, IoT sensors track the location, temperature, and condition of goods in real time. AMRs equipped with vision systems and next-generation RFID scanners autonomously retrieve and transport goods to packing stations. Wearable devices, such as smart glasses, assist workers with picking instructions and navigation, improving both speed and accuracy. All of these hardware systems are connected through 5G networks and feed real-time data into an AI-enhanced WMS.

Meanwhile, TMS powered by AI analyze this warehouse data alongside external inputs such as satellite weather data, traffic patterns from connected vehicle fleets, and customer order dynamics. Using digital twin technology, DHL creates real-time simulations of warehouse and transportation operations, allowing it to model and optimize various shipping scenarios and identify the most efficient routes and delivery times. Blockchain secures and verifies all data exchanges across the network, ensuring integrity, particularly for sensitive or high-value shipments.

At the heart of this system lies an integrated data lake, which consolidates data from warehouses, vehicles, customer orders, IoT sensors, weather feeds, and external partners into a unified, cloud-based repository. This integration enables advanced analytics, machine-learning models, and predictive algorithms to continuously learn from, and optimize across, the entire supply chain.

Overseeing the flow of data and operations is DHL's AI-driven control tower, a centralized, real-time command center that monitors logistics activities end to end. The control tower uses insights drawn from the integrated data lake and live operational feeds to detect disruptions early, orchestrate rapid rerouting, dynamically allocate resources, and continuously optimize performance.

Here are the steps for how this works at DHL:

- **Smart warehouse** collects real-time IoT data and feeds into the **AI-enhanced WMS.**
- **AI-enhanced WMS** provides warehouse operational data to the **AI-powered TMS.**
- **TMS** integrates external data (weather, traffic, customer orders) and sends to the **data lake.**
- **Blockchain** secures the data exchanges between systems.
- **Integrated data lake** aggregates all inputs.
- **AI-Driven control tower** oversees everything in real time, making dynamic optimizations.

Rather than each system operating in isolation, DHL's synchronized ecosystem, powered by integrated data lakes and control towers, allows logistics decisions to be made automatically, intelligently, and in real time. This transforms traditional logistics operations into intelligent, self-optimizing supply networks that deliver near-perfect order fulfillment, predictive delay management, lower costs, and unparalleled supply-chain transparency.

QUESTIONS

1. How does the integration of IoT sensors, AMRs, and wearable devices improve operational efficiency in DHL's warehouses?
2. In what ways does real-time data from connected technologies influence transportation planning and delivery performance at DHL?
3. What role does digital twin technology play in enhancing DHL's decision making across their logistics network?
4. How does blockchain technology contribute to supply-chain security and transparency in DHL's operations?
5. What potential challenges could DHL face in maintaining the connectivity and coordination among so many hardware and software systems, and how might they address them?

As we can see, logistics is no longer a manual, disconnected function. It is a data- and AI-enabled strategic function requiring real-time visibility, speed, security, and flexibility.

This change means investing in technologies such as IoT sensors, AI-driven WMS and TMS platforms, autonomous robots, and real-time data analytics tools. Traditional methods are no longer sufficient; staying competitive now requires integrating technologies like digital twins, blockchain, and edge computing to drive faster, more accurate, and lower-cost logistics operations.

6.3 Transportation

Transportation is a major logistics activity. It involves the movement and flow of goods from point of origin to point of consumption. Transportation is often the single largest cost in the logistics process. Therefore, it is a critical component that must be managed effectively.

Transportation involves decisions such as selecting the method of shipment: air, rail, water, pipeline, truck, or intermodal. The most common is the latter, which involves selecting a mix of transportation options. Air is fastest but is expensive and has capacity constraints; water and rail are cheapest but are slow; and pipeline has geographic limitations. Transportation decisions also involve choosing the specific path (routing); complying with various local, state, and federal transportation regulations; and being aware of both domestic and international shipping requirements. AI and analytics technologies are essential to optimize the mix considering multiple criteria, including delivery times, speed, capacity, security, and cost, as well as other factors such as minimizing the carbon footprint.

Route optimization and advanced telematics systems are two key areas of transportation where AI technologies are essential. Let's look at these next.

6.3.1 Route Optimization

One of the most impactful applications of AI and analytics in transportation is route optimization, the process of selecting the most efficient, cost-effective, and sustainable routes based on a variety of real-time conditions. Smart routing has evolved dramatically from basic digital maps to sophisticated systems that continuously analyze traffic data, weather patterns, road conditions, and even driver behavior.

Modern route optimization systems integrate GPS-enabled telematics with AI and analytics to

- determine the best real-time routes based on live traffic congestion, accidents, and construction zones;
- optimize for fuel efficiency, minimizing idling and unnecessary mileage;
- schedule preventive maintenance by monitoring vehicle performance in transit;

- improve driver behavior by analyzing harsh braking, speeding, and idle times; and
- support full fleet management, balancing delivery schedules, driver hours, and load capacities across a network.

More advanced platforms use AI and machine-learning algorithms to anticipate future traffic conditions, adjust delivery schedules dynamically, and even suggest rerouting on the fly if unexpected delays occur. These systems not only optimize travel time but also reduce fuel consumption, lower carbon emissions, and improve overall asset utilization, directly contributing to sustainability goals.

Current navigation platforms (e.g., Google Maps, Waze for fleets, Verizon Connect, and Samsara) can integrate real-time feeds such as

- traffic accidents and congestion alerts,
- scheduled road closures or detours,
- hazardous weather forecasts,
- location of service stations or EV charging points, and
- delivery priority adjustments based on customer urgency or supply-chain disruptions.

In addition, a key innovation is the feedback loop. Vehicles equipped with telematics devices report movement and location data back to central servers, which update traffic conditions for the entire network in real time. This crowdsourced data improves predictive models and continually refines route suggestions, creating a self-improving logistics ecosystem.

COMPANY EXAMPLE: UPS

UPS provides a world-class example of route optimization in action. Through its proprietary ORION platform, UPS analyzes more than 250 million address data points daily to optimize routes for its delivery fleet.

KEY ACHIEVEMENTS OF USING ORION

- Route optimization has saved UPS more than 100 million miles driven annually, equating to more than 10 million gallons of fuel saved and a significant reduction in CO_2 emissions.
- The system uses real-time traffic feeds, customer delivery windows, and driver-specific performance data to adjust routes on the fly.
- UPS drivers now make fewer left turns, minimizing time spent idling at intersections, a seemingly small adjustment that saves millions in fuel costs and improves driver safety.

In 2024, UPS expanded ORION capabilities with AI-enhanced predictive analytics, allowing it to dynamically reassign deliveries across nearby drivers mid-route if sudden disruptions occur (e.g., road closures, canceled stops). By combining telematics, big data, and AI, UPS has not only slashed operating costs but also improved its on-time delivery rate to 98.5 percent, bolstering customer satisfaction in a fiercely competitive last-mile delivery market.

QUESTIONS

1. How does UPS's use of AI and real-time data through the ORION platform demonstrate the value of operational efficiency in last-mile delivery? Consider the effect on cost, fuel usage, and emissions.
2. What are the broader implications of small behavioral changes—such as reducing left turns—on overall logistics performance and sustainability? How can other companies apply similar principles to improve efficiency?
3. How does the integration of predictive analytics and real-time data into ORION change the role of human drivers in UPS's operations? Discuss the balance between automation and human decision making.
4. What risks or challenges might UPS face in relying heavily on AI-driven route optimization? Think about data accuracy, system failures, and driver adaptability.

5. How can companies in other industries adopt similar AI-enhanced logistics strategies to improve customer satisfaction and environmental outcomes? Can this be replicated in retail, healthcare, or field services?

6.3.2 Telematics Systems

Nearly all transportation and logistics carriers deploy advanced telematics systems integrating GPS technology, IoT sensors, and real-time analytics. These systems gather extensive data on vehicle location, driver behavior, traffic patterns, weather conditions, idle time, fuel consumption, and maintenance needs. Modern telematics systems extend far beyond simple GPS tracking. They provide predictive maintenance alerts, notifying drivers and fleet managers about potential mechanical issues before breakdowns occur. They can also remotely push over-the-air (OTA) software updates to vehicles and immediately locate assets during emergencies, such as airbag deployment or accident detection.

UPS, Schneider National, and FedEx have been at the forefront of using telematics to redesign their delivery networks and boost operational efficiency. UPS, for example, continues to leverage telematics as a core part of its smart logistics strategy. Its ORION platform, discussed previously, powered by telematics and AI optimizes 330,000 daily routes and saves the company an estimated 100 million miles and 10 million gallons of fuel per year.

Telematics data is not just used for vehicles; it is instrumental in labor management and safety enhancements. UPS uses its vast fleet data to not only optimize routes for fewer left turns (saving time and reducing accident risks), but also to monitor driver performance for coaching and training, contributing to a 30 percent improvement in delivery efficiency across key regions.

In addition to traditional telematics, AI-driven autonomous technologies are revolutionizing supply-chain transportation. Lesser known companies like Aurora Innovation and Waymo Via have been piloting autonomous freight deliveries. They even have some trucks now operating without safety drivers on certain US highway corridors. Autonomous vehicles, discussed in the next section, are equipped with AI to continuously analyze sensor data from cameras, LiDAR, radar, and GPS. They use this data to make real-time navigation, speed, and routing decisions. These systems enable more consistent driving patterns, leading to fuel efficiency gains of up to 15 percent and a significant reduction in human error.

Beyond operational benefits, autonomous electric trucks offer environmental and economic advantages. Einride, a Swedish company, has already deployed electric autonomous pods for commercial freight use, aiming to cut emissions by up to 90 percent compared with diesel trucks. As these technologies mature, transportation costs could eventually shrink to primarily the cost of electric charging, maintenance, and time management, freeing companies to reinvest in customer service enhancements and broader supply-chain innovations.

6.3.3 Autonomous Vehicles (AV)

AVs have rapidly moved from experimental technologies to real-world logistics solutions. AVs are reshaping how goods are transported across first-mile, middle-mile, and last-mile logistics networks. Autonomous trucks, delivery vans, and even AMRs within warehouses are being deployed to optimize speed, reduce labor costs, and enhance operational efficiency. Companies are using AVs not just for experimentation but also for live commercial operations on highways, urban delivery routes, and in distribution centers.

According to McKinsey & Company, autonomous technologies could lower logistics costs by up to 45 percent over the next decade, with the greatest near-term effect expected in middle-mile trucking and controlled environments like warehouses and ports. Recall that logistics is a huge expense for companies making this a significant projection. Key benefits driving adoption include increased vehicle utilization rates, lower accident rates, reduced fuel consumption, and smoother, more predictable delivery schedules.

Some examples of where AVs are being used today include the following:

Middle-Mile Trucking

Middle-mile trucking refers to the transportation of goods between two hubs or facilities, typically from a supplier's warehouse to a retailer's distribution center or between distribution centers. It is the middle stage between the first mile (moving goods from manufacturer to warehouse) and the last mile (delivering goods to the final customer).

Middle-mile trucking is crucial because it connects major nodes in the supply chain and often involves long-haul trucking, regional routes, or transfers between major logistics hubs.

Walmart uses middle-mile trucking to move pallets of goods from its regional warehouses to local distribution centers, where items are sorted for delivery to individual stores or for last-mile delivery to customers.

Aurora Innovation and Kodiak Robotics are companies operating autonomous freight trucks on US highways, often with limited or no human intervention.

- Aurora partners with FedEx and Uber Freight, running autonomous routes between major logistics hubs in Texas.
- Kodiak reported that its autonomous trucks completed more than 4,000 paid freight deliveries in 2024 with an on-time rate of more than 95 percent.

Last-Mile Delivery

Last-mile delivery refers to delivering goods to the final customer. Autonomous delivery robots and small vehicles are increasingly handling last-mile logistics in urban and suburban areas.

- Nuro, in partnerships with Kroger and Domino's Pizza, operates small, self-driving pods for grocery and food delivery in cities like Houston and Phoenix.
- Amazon Scout is piloting autonomous sidewalk robots for package deliveries in select neighborhoods, aiming to cut last-mile delivery costs, which can represent up to 50 percent of total shipping costs.

Warehouse and Yard Operations

Inside logistics hubs, AMRs and AGVs are streamlining warehouse workflows. Walmart uses AMRs in its distribution centers to pick, move, and sort inventory with minimal human intervention. Similarly, DHL has deployed more than 2,000 AMRs globally to support picking and sorting activities, increasing warehouse throughput by up to 30 percent.

Ports and Freight Terminals

Major ports like the Port of Rotterdam and Port of Los Angeles have deployed autonomous yard trucks and cranes to handle container movement with enhanced speed and precision.

6.3.4 The Future of Autonomous Logistics

In the future, AVs are expected to reshape logistics in the following ways:

- **Scaling Middle-Mile Operations:** It is estimated that by 2030, autonomous trucks could handle 30 percent of middle-mile freight in the United States, especially on fixed, high-volume routes between major hubs.
- **Expansion to Last-Mile Deliveries:** Improvements in AV safety, urban planning adaptations, and regulatory approval could lead to broader adoption of autonomous delivery robots in dense cities.
- **Autonomous Electric Vehicles (AEVs):** The convergence of electric powertrains and AV technology will create ultra-efficient, low-carbon logistics solutions.

- **AI-Driven Dispatching and Routing:** Autonomous fleets will be managed by centralized AI systems that dynamically assign vehicles based on real-time demand, weather, and traffic conditions.
- **Regulatory Evolution:** Governments are progressively establishing standards for AV safety, cybersecurity, and insurance, which will unlock broader commercial deployment.

Despite progress, there are huge challenges around full autonomy in urban environments, such as public acceptance, cybersecurity risks, and interstate regulatory consistency. However, with ongoing investments from companies like Waymo Via, Amazon, UPS, Daimler Truck, and Tesla, the trajectory toward broader autonomous logistics integration is clear. According to ABI Research, a global technology intelligence firm, by 2035, autonomous delivery vehicles could complete moe than 20 billion deliveries globally per year, fundamentally transforming last-mile logistics economics.

What does this mean for managers?

AVs in logistics represent both a major opportunity and a strategic imperative. Managers must proactively assess how AV technologies—whether in middle-mile freight, last-mile delivery, or warehouse operations—can be integrated into their supply chains to reduce costs, improve efficiency, and meet rising customer expectations, for faster, more reliable delivery. This means staying informed about AV developments, piloting partnerships with technology providers, and investing in digital infrastructure that supports real-time vehicle tracking, autonomous dispatching, and AI-driven route optimization.

Managers should also rethink workforce strategies, planning for the re-skilling of employees toward roles in AV maintenance, fleet management, and logistics oversight. Critically, managers must advocate for cybersecurity readiness, regulatory compliance, and scenario planning as AV adoption scales up. Those who act early to integrate autonomous logistics capabilities will position their businesses for significant competitive advantages in operational speed, cost leadership, and supply-chain resilience.

6.4 Warehousing

Traditionally, the role of warehouses has been to provide storage space for goods, as well as inbound and outbound transport. Goods arrived at the warehouse via truck or rail were unloaded and then placed in storage bins in the warehouse. When customer orders arrived, goods were picked up from their storage location—known as order picking—staged for transport, loaded on to trucks or rail, and shipped.

Contemporary warehouses are increasingly places for mixing inventory assortments to meet customer needs. They are about enhancing and expediting movement rather than lengthy storage. In fact, lengthy storage is discouraged, as it does not add value and just contributes to cost and obsolescence. Warehouses provide a centralized location that stores and organizes inventories of products before they get distributed to customers. For that reason, they are often called distribution centers.

Warehouses, or distribution centers, play a critical role in the supply chain, enabling efficient movement of goods on both the inbound and outbound side of the organization. For example, in JIT and Lean manufacturing, warehouses can be located close to the manufacturing facilities, enabling frequent deliveries of materials on a just-in-time basis. At the same time, warehouses can be utilized to create product assortments for customer shipments. A strategically located centralized warehouse, for example, can take advantage of consolidated shipments. The products can then be sorted and arranged for a particular customer and shipped. In addition to these benefits, many warehouses are increasingly performing tasks traditionally done at a manufacturing or retail site. This may include repairing items, putting garments on hangers and sequencing them to be rolled straight on to the retail floor, and adding labels and price tags. Optimizing these decisions is a role for big data.

6.4.1 Labor Utilization

Labor is the greatest expense item associated with warehousing and tasks such as picking and packing individual orders. This is especially problematic in direct-to-consumer fulfillment, generating the largest share of those labor costs. One of the most time-consuming aspects of labor is travel. Warehouses are large facilities. Associates walking or riding around the building en route to and from product storage locations is one of the most wasteful activities. Analytics can be used to create material-handling solutions that reduce travel, such as developing picking assignments that minimize distance walked. It can create pick modules and zone assignments that meet order requirements, while optimizing order picker access to SKUs. All this serves to substantially improve efficiency of labor and warehouse operations.

A prime example of optimizing labor utilization through technology is Alibaba's warehouse operations. Alibaba has significantly invested in automation, deploying more than 700 AGVs in its warehouses. This has led to a remarkable 70 percent reduction in labor costs and substantial efficiency gains. By utilizing advanced analytics, Alibaba has also improved order-processing speed by 30 percent, showcasing how strategic implementation of technology can transform labor-intensive processes. This level of automation reduces human travel within warehouses, allowing for faster, more accurate fulfilment while cutting operational costs.

6.4.2 Material Handling

Another warehouse challenge is material handling. **Material handling** is concerned with every aspect of movement of goods including raw materials, in-process inventory, and finished goods within a plant or warehouse. The objective of material handling is to eliminate handling wherever possible, minimize travel distance, and minimize goods in process. In addition, efficient material handling should provide uniform flow that is free of bottlenecks and should minimize losses from waste, breakage, spoilage, and theft.

Costs are incurred every time an item is handled. Because handling generally adds no value to a product, handling should be kept to a minimum. For items with low unit value, the proportion of material-handling costs to total product cost can be significant. Analytics applications can optimize this flow. By carefully analyzing material flows and planning efficient routes, these applications can save the organization significant amounts of money.

COMPANY EXAMPLE: Alibaba

Alibaba's warehouse operations are at the cutting edge of AI-driven logistics. At the heart of its system is the Cainiao Smart Logistics Network, which integrates AI, machine vision, IoT sensors, robotics, and digital twins to manage the storage, retrieval, and shipping of goods. As of 2025, Alibaba processes more than 1.5 billion packages daily during peak events like Singles' Day, supported by a vast network of more than 1,000 warehouses across China and globally. These include high-tech fulfillment centers equipped with thousands of AMRs that perform automated picking, packing, and sorting tasks with millimeter-level precision.

AI-powered WMS use real-time data from IoT-enabled racks, conveyor belts, and smart bins to dynamically allocate space, optimize picking paths, and reduce human travel time inside warehouses. These AI systems learn from historical data and predict demand fluctuations, allowing inventory to be repositioned proactively. This has resulted in a 45 percent reduction in stockouts, a 30 percent drop in inventory holding costs, and a 50 percent improvement in order-picking speed, compared to traditional manual warehouse operations.

Alibaba's logistics arm, Cainiao Network, extends these capabilities to international e-commerce. Using AI-enhanced route optimization and smart customs clearance algorithms, Cainiao has cut cross-border delivery times from an average of 15 days to under 9 days in many key markets, including Southeast Asia, Europe, and the Middle East. AI models forecast customs-processing durations, preschedule container loads, and balance air and sea freight usage to minimize delays and emissions.

Furthermore, Cainiao is implementing green logistics strategies powered by AI, including energy-efficient sorting systems and packaging optimization algorithms. These changes align with Alibaba's pledge to achieve carbon neutrality in its operations by 2030. Real-time tracking data is also shared with customers through AI-driven dashboards, enabling more accurate delivery estimates and improved transparency.

Through its intelligent warehouse ecosystem, Alibaba showcases how the fusion of AI and logistics can generate agility, scalability, and environmental sustainability at a massive scale. Its advancements set a new global standard for e-commerce fulfillment and smart supply-chain management.

QUESTIONS

1. How has Alibaba's use of AI in warehouse operations transformed traditional logistics models in terms of speed, cost, and scalability? Consider the role of AMRs, predictive inventory positioning, and AI-powered WMS.
2. What are the potential trade-offs or risks of relying heavily on AI, robotics, and automation in fulfillment centers? Think about resilience, workforce displacement, or technology failure.
3. How does Cainiao's integration of AI into cross-border logistics demonstrate the importance of real-time data and intelligent decision making in global e-commerce? Evaluate the impact on delivery time, customs clearance, and environmental outcomes.
4. In what ways do Alibaba's green logistics strategies align business efficiency with sustainability goals? Discuss the role of AI in reducing carbon footprint and packaging waste.
5. What lessons can other companies, especially outside of China, learn from Alibaba's logistics innovations, and what challenges might they face in replicating this model? Explore issues related to infrastructure, data access, scalability, or regulatory environments.

6.4.3 Warehouse Management Systems

A warehouse management system (WMS) is a platform designed to control or optimize warehouse operations by providing real-time information on the movement, storage, and handling of goods. Current WMS applications enable centralized management of tasks such as inventory tracking, stock location control, order fulfillment, labor management, and reporting. They can be stand-alone systems or integrated as a critical component of broader ERP solutions. Advanced WMS systems leverage data analytics, AI, and IoT to significantly improve warehouse efficiency, responsiveness, and visibility.

The early versions of WMS programs provided just basic storage functionality, such as tracking where goods were placed in the warehouse. That was very simple. Today WMS platforms are far more complex and data intensive, often requiring dedicated IT and operations teams to run them effectively. There are also high-end WMS systems that can incorporate advanced features such as RFID, machine-learning algorithms, and voice recognition technologies. Many also offer the ability to manage inventory across multiple global warehouses from a single dashboard, allowing managers to track real-time inventory, forecast demand, and support order entry. They can then monitor all this through mobile devices.

A prime example of WMS adoption is Amazon. Their fulfillment centers use a highly sophisticated WMS integrated with real-time data from IoT sensors, AI-driven robotics, and predictive analytics. Robots navigate warehouse floors autonomously, bringing shelves directly to human pickers. This significantly reduces travel time and boosts accuracy rates. Amazon's system also uses predictive AI to reorganize warehouse layouts dynamically, prioritizing fast-moving inventory closer to fulfillment zones, based on current sales data. With more than 1.5 million packages processed daily at peak times, Amazon's WMS and automation investments demonstrate how modern data-driven solutions can scale operations while maintaining high service levels.

Automated Material-Handling Systems and Data-Driven Warehousing. Another transformative innovation in warehousing is the rise of automated material-handling systems. These systems leverage real-time data and AI to automate key tasks such as picking, packing, and storage, improving both efficiency and accuracy. Modern systems dynamically slot products based on SKU velocity, placing fast-moving items near pick zones and automatically reshuffling products as their movement patterns change. Some systems even personalize workflows based on employee performance data, directing high-value or complex tasks to the most experienced workers.

Kroger was one of the first adopters, implementing the first fully automated full case picking system in its Arizona distribution center back in 2004, significantly improving operational efficiency. Now companies like Target have continued this by launching a fully automated 360,000-square-foot perishables and frozen goods distribution center in Denton, Texas, capable of operating with minimal human intervention while maintaining exceptional accuracy and throughput.

Then there are innovative companies many of us have not heard of, such as AutoStore and Ocado, that have further redefined warehouse operations. AutoStore has a cube storage system that uses small robots to retrieve bins stacked vertically, maximizing space efficiency and significantly speeding up picking processes. Ocado's highly automated grocery warehouses employ AI orchestration to manage thousands of robots moving in synchronized patterns to fulfill online grocery orders rapidly. Additionally, companies like Amazon Robotics (formerly Kiva Systems) offer material-handling solutions where workers remain in fixed stations while automated systems bring goods directly to them. This station-based picking supported by "pick-to-light" and "put-to-light" systems results in lower labor costs, shorter cycle times, and extraordinarily high picking-accuracy rates.

Current Challenges. While the introduction of AI, robotics, and real-time inventory management systems has drastically enhanced warehouse performance, one of the most pressing challenges for organizations is achieving seamless integration across the entire technology stack. In many warehouses, technologies like AI-enhanced WMS, RFID-enabled inventory tracking, and robotics picking systems operate successfully. The problem is that it often occurs in isolated silos.

To achieve full automation and optimization requires that all components, from inbound receiving, to inventory management, to outbound shipping, be interconnected. Despite these advancements, ensuring synchronization between new and legacy technologies remains a major organizational hurdle. Bridging these gaps, creating fully integrated warehouse control towers with centralized visibility and dynamic decision-making capabilities, while maintaining the agility to adapt systems as technologies evolve will be critical success factors for the next generation of smart warehouses.

6.4.4 Location Optimization

Facility location—deciding where to site warehouses, factories, retail stores, or service centers—is one of the most critical decisions. Location optimization is merely optimizing facility location. These decisions carry long-term financial implications: once capital is committed to land, buildings, and infrastructure, correcting a mistake can be costly and disruptive. A poor location can lead to higher transportation costs, labor shortages, supply-chain inefficiencies, and ultimately a loss of competitive advantage.

Historically, location decisions relied on basic geographic and demographic data and relatively simple models like load-distance analysis or factor rating systems. However, AI has fundamentally changed this process. Companies today use AI-driven algorithms, real-time market intelligence, and predictive modeling to optimize site selection.

There are two emerging shifts in location optimization.

Data Quality and Granularity. Census data was used in the past, but it is totally dated. Today's location models leverage real-time data streams, updated monthly, weekly, or even daily. Companies integrate not just demographics (age, income, population density) but also psychographics (consumer values, behaviors, brand preferences), mobile location data (foot traffic patterns), loyalty program insights, and social media sentiment.

This richer and more dynamic data allows for far more accurate modeling of market potential and risk.

Integrated Decision Making. Modern location analytics allow businesses to simultaneously optimize the location, store format, distribution integration, and even merchandising strategies. Companies now conduct what-if scenario modeling to assess whether to build a full-size store, open a micro-fulfillment center, remodel an existing facility, or deploy a hybrid format—all based on predictive ROI and customer demand projections.

McDonald's offers a great example of strategic, data-driven facility location. They operate more than 40,000 restaurants across 120 countries and serve approximately 70 million customers daily. This is quite a logistical achievement. Data-driven location strategy contributed to doubling McDonald's sales growth between 2021 and 2024. McDonald's has a precise facility location strategy, underpinned by AI and real-time analytics. McDonald's analyzes vast data sets, including economic indicators, competitor presence, pedestrian traffic, and mobile phone location patterns, to find optimal restaurant sites.

Instead of relying purely on demographics, McDonald's prioritizes markets based on growth potential, income trends, and consumer behavior analytics. New locations are chosen considering proximity to McDonald's distribution hubs, ensuring low transportation costs and fresh food delivery to maintain service quality. By integrating AI-driven site analytics, McDonald's has accelerated its global store openings by 25 percent in the past three years and maintained high profitability per site.

The future of location optimization is especially interesting. Companies are now using AI to predict market saturation points. Simulation modeling is another tool that is used. It can test multiple location and format strategies in virtual environments. These models can add other factors, such as climate risk, to select locations creating resilience to extreme weather events. They can also link to last-mile delivery optimization, to influence the design and placement of micro-fulfillment centers. In addition, many companies like Amazon and Target are including hybrid distribution-retail hubs, using AI not only to serve shoppers but also to optimize inventory placement for rapid fulfillment.

From this, you can see the further blurring of lines between logistics and retail location strategy.

Location optimization is not just about picking a spot on a map; it's about strategically positioning the business for long-term success. By leveraging AI, managers can evaluate a wide array of factors simultaneously, including real-time demographic trends, customer behavior patterns, logistics network proximity, competitive saturation, and market growth potential. A manager might use AI-powered location tools to simulate different site scenarios, assessing the projected return on investment (ROI), operating costs, and customer acquisition rates for each option. In addition, managers can optimize store formats (e.g., flagship, express, fulfillment center) based on local market characteristics. This data-driven approach minimizes the risk of costly location mistakes, enhances supply-chain efficiency, and improves responsiveness to evolving consumer demand, ultimately turning location decisions into a powerful competitive advantage.

6.4.5 Inventory Management

Inventory management is the strategic coordination of materials, parts, work-in-progress, and finished goods, to ensure that the right quantities are available at the right time and place, without tying up excessive capital. Managing inventory remains a critical balancing act: holding too much inventory incurs high costs for storage, insurance, and obsolescence, while holding too little risks stockouts and lost sales. In fact, inventory holding costs can account for 25 to 55 percent of the cost of goods sold, or more, depending on the industry.

Effective inventory management seeks to optimize this trade-off, achieving high customer service levels while minimizing holding costs. This involves careful decisions around order quantities, replenishment frequency, safety stock levels, and warehouse space utilization. The ultimate goal is to maximize availability and responsiveness without unnecessarily locking up capital.

Technologies are transforming inventory management. RFID and telematics have long enabled companies to track item locations, but the next frontier is ILC (Identification, Location, Condition) sensors. ILC sensors monitor temperature (critical for food, pharmaceuticals, and batteries). They also monitor humidity and light exposure (for perishable or sensitive goods), tilt and impact (for fragile or high-value items), and even package-opening events (for theft detection).

These sensors then transmit real-time data via cellular networks, giving companies unparalleled visibility into inventory conditions as goods move through the supply chain. This real-time insight allows immediate corrective action when conditions deviate from standards, reducing spoilage, damage, and compliance risks. Analytics platforms now harness this sensor data to drive predictive insights, optimize warehouse operations, and trigger automatic replenishment or routing adjustments when risks are detected.

A standout example is Tesla's data-driven approach. Tesla minimizes inventory holding costs through a combination of JIT manufacturing, RFID-enabled tracking, and advanced ILC monitoring across its global supply chain. Notice that these are combined in tandem leveraging the strengths of each. Tesla's warehouses and production lines use real-time dashboards fed by RFID and ILC sensors to track parts' conditions, such as battery temperature and handling shocks, ensuring that only optimal components reach the assembly line.

Also, Tesla leverages AI-powered analytics to predict parts demand, dynamically adjust order quantities, and minimize safety stock while maintaining high service levels. This not only reduces the capital tied up in inventory but also allows Tesla to scale production flexibly as market demand shifts. Tesla's integration of sensor technologies and AI exemplifies how smart inventory management can deliver both cost savings and operational excellence.

Another major advancement in inventory management is the use of AI for **vendor performance analysis and stock optimization**. By analyzing this highly detailed or granular data, companies can identify top-performing suppliers, mitigate risks from unreliable vendors, and negotiate better service-level agreements. Current AI systems monitor supplier metrics such as

- on-time delivery rates,
- defect rates,
- lead time variability, and
- responsiveness to disruptions.

AI algorithms also drive inventory optimization by factoring in seasonality trends, lead-time forecasts, macroeconomic indicators, and real-time sales velocity. This enables companies to minimize carrying costs, reduce stockouts and backorders, and dynamically rebalance inventory across multiple distribution points. Nike, for example, uses AI-based inventory optimization models to adjust stock levels across their omnichannel networks, reducing excess inventory by 15 percent while improving fulfillment speed.

6.5 Reverse Logistics

Reverse logistics refers to the process of moving products upstream from customers back to manufacturers, distributors, or suppliers. This process occurs for a variety of reasons, including damaged goods, warranty repairs, product recalls, recycling efforts, remanufacturing, or simple customer dissatisfaction. Managing reverse logistics effectively is critical, yet it remains highly expensive and complex.

According to recent data from the Reverse Logistics Association, US returns now account for more than $850 billion annually, fueled by the growth of e-commerce. Returns can cost companies between 8 and 12 percent of the total value of goods sold, and the cost of moving a returned item back through the supply chain can be up to 10 times higher than moving it downstream to the customer.

In addition to transportation and handling costs, reverse logistics brings other challenges:

- Financial complexity around refunds, restocking fees, and reverse cash flows
- Warehouse congestion, requiring dedicated space for returned goods without disrupting forward-moving operations

- Environmental compliance, especially with stricter green laws in regions like the EU, which mandates proper recycling of packaging and safe disposal of materials

Just as AI has transformed traditional logistics, it is now reshaping reverse logistics. Using data analytics, companies can better understand why returns happen, where they are most frequent, and how they can be reduced.

AI algorithms play a critical role in reverse logistics by

- predicting likely returns based on purchase patterns, product types, and customer profiles;
- helping design better products and packaging to minimize defects and dissatisfaction;
- optimizing store layouts and fulfillment processes to decrease return rates;
- managing the routing and handling of returns more efficiently, thereby reducing transportation and storage costs; and
- automating inspection and disposition decisions (repair, recycle, refurbish, or resell).

Amazon is a leader in leveraging AI for reverse logistics. By analyzing real-time customer return data, Amazon identifies patterns behind returns, such as size mismatches or misleading product descriptions, and uses these insights to adjust product listings, packaging standards, and even recommend sizing guides. Amazon's AI-powered systems also automate the sorting and disposition of millions of returned items, determining the most profitable and sustainable paths for each—whether resale, refurbishment, recycling, or donation—thus maximizing recovery value.

A lesser known company specializing in reverse logistics analytics is Inmar Intelligence, a data-driven technology company. Inmar provides real-time visibility and predictive analytics for goods flowing through the reverse supply chain. Their platforms help retailers and manufacturers monitor store activities, analyze returns at a granular level, and implement strategies to reduce return rates while improving overall supply-chain efficiency.

Inmar's engineers and data scientists gather extensive operational data from across the supply chain, analyzing everything from product damage rates to consumer feedback. These insights lead to actionable consulting studies for clients, recommending packaging improvements, process redesigns, and sales strategies that lower returns and improve customer satisfaction. For example, Inmar helped a major consumer product goods company redesign its packaging to reduce breakage during transit, resulting in a 15 percent reduction in returns within a year.

6.6 Energy & Sustainability

Most major companies recognize sustainability not just as a corporate responsibility but as a core component of long-term competitiveness. For example, Walmart, Unilever, GE, and Marks & Spencer are consistently cited as pioneers in integrating sustainability into their operations.

In recent years, Unilever has significantly expanded its Sustainable Living Plan, first launched more than a decade ago—into what it calls its Compass Strategy. Under this framework, Unilever aims to achieve net-zero emissions across its entire value chain by 2039. Key initiatives include halving the greenhouse gas footprint of its products by 2030 and ensuring that 100 percent of plastic packaging is either recyclable, reusable, or compostable by 2025. Furthermore, Unilever is transitioning all of its cleaning and laundry product formulations to use renewable or recycled carbon sources by 2030, demonstrating how sustainability can coexist with profitable growth.

What does sustainability have to do with AI and logistics?

The Move function, covering transportation, warehousing, and distribution, is one of the most energy-intensive aspects of supply chains. Optimizing these activities presents one of the greatest opportunities to reduce carbon footprints. Big data analytics empowers companies to design logistics solutions that balance multiple objectives, including cost, service level, and now, environmental impact.

From an energy perspective, companies have made significant advances. With rising fuel costs and tightening environmental regulations, sustainability initiatives now yield both ecological and financial benefits. According to the US Energy Information Administration, energy consumption in US manufacturing has declined by more than 15 percent since 2010, illustrating how continuous improvements in energy efficiency are reshaping industrial logistics.

Sustainability Metrics Using AI. Virtually every aspect of logistics can incorporate sustainability metrics through AI and analytics:

- **Facility Location Optimization:** New algorithms not only maximize market coverage but also minimize transportation distances, reducing emissions. Locating distribution centers closer to end markets reduces "food miles" and associated carbon output.
- **Ethical Supply-Chain Design:** Big data can help identify suppliers and partners that adhere to fair labor practices, local environmental laws, and social compliance standards, particularly important in developing economies.
- **Transportation Impact Analysis:** Companies now use AI to simulate the environmental effects of centralized versus decentralized warehousing models, evaluate long-term fuel consumption patterns, and plan low-emission distribution networks.
- **Green Warehousing:** Analytics-driven building designs integrate features like LEED certification, energy-efficient HVAC systems, graywater recycling, and intelligent lighting controls. Although these features may increase upfront costs, big data tools can project total cost of ownership, factoring in savings from tax incentives, lower utility costs, and reduced maintenance.

Managers can use these analytics to build strong, financially grounded business cases for investments in greener logistics operations.

A lesser known example of operational sustainability is the Rodon Group, a leader in American plastics manufacturing. Through extensive data-driven initiatives, Rodon has minimized packaging waste, transportation energy costs, water usage, and overall carbon emissions. They track energy consumption at a granular level using IoT sensors across their facilities, enabling real-time adjustments to maximize efficiency. Rodon's commitment to green manufacturing, combined with big data monitoring—has not only strengthened their environmental leadership but also significantly lowered operational costs, proving that sustainability and profitability can go hand in hand.

6.7 The Future of Logistics

Logistics, or the Move function, is a complex, interconnected system involving coordination of activities such as customer service, transportation, warehousing, inventory management, order processing, and production planning. To achieve true efficiency, these activities must be managed holistically. Companies that do not adopt a systems approach experience logistics fragmentation, where departments operate independently under separate budgets and priorities. The result is higher costs, reduced performance, and recurring issues like excess inventory stuck in wrong locations while shortages persist elsewhere.

A good example of a company that has implemented a fully integrated logistics system is 3M. They analyze more than 1.5 terabytes of data daily to drive logistics, including real-time inventory updates, customer order histories, weather forecasts, and geopolitical risk signals. This analytical capability has helped the company cut forecasting errors by 22 percent, achieve a 28 percent improvement in inventory turnover, and streamline transportation scheduling. 3M has also created a tightly connected logistics system with their S&OP process. The result was reduced excess inventory by 35 percent, improved inventory accuracy by 30 percent, and boosted customer satisfaction scores by 12 percent. This approach cut operational costs and enabled faster, more adaptive responses to market disruptions. As of 2024, the result was a 12 percent reduction in logistics costs and a rise in its on-time delivery rate to 97 percent. The company relies on a cloud-based supply-chain management platform embedded with AI-driven predictive analytics to manage more than 200 warehouses and coordinate transportation activities globally. This exemplifies

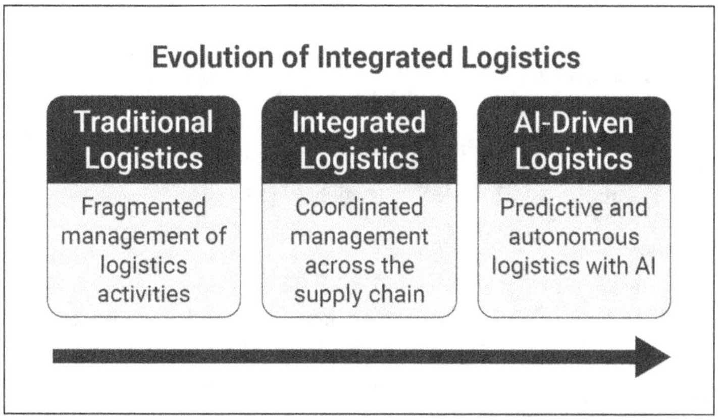

Figure 6.3 Evolution of Integrated Logistics

integrated AI-driven logistics—which is very different from traditional logistics composed of the fragmented tasks discussed at the onset of the chapter.

Looking ahead, logistics will continue to evolve rapidly with the integration of AI, autonomous transportation, digital twins, and self-healing supply chains, all the technologies we have discussed. AI will shift logistics from being primarily reactive, as we saw with traditional logistics, to an integrated and then AI-driven system (Figure 6.3). Logistics will become predictive and even autonomous, anticipating disruptions before they occur, dynamically optimizing routes, automating inventory reallocation, and coordinating global supply networks with minimal human intervention.

As AI models become increasingly sophisticated, companies will not only achieve operational efficiency but also enable new levels of supply-chain resilience, sustainability, and customer-centricity. The future belongs to organizations that can seamlessly integrate these AI-driven innovations into their logistics ecosystems.

Chapter Review

Key Points

1. Logistics is responsible for moving goods through the supply chain:
   - Logistics is the function that ensures products move efficiently from suppliers to manufacturers, from manufacturers to warehouses, and from warehouses to customers.
   - Logistics encompasses transportation, warehousing, material handling, packaging, and order fulfillment. It is central to ensuring that the right product is delivered to the right place, at the right time, in the right quantity, and at the right cost.
2. Logistics makes decisions involved in delivering the perfect order:
   - The perfect order is complete, on time, undamaged, and includes accurate documentation.
   - Key logistics decisions include transportation mode selection, route planning, carrier selection, delivery scheduling, and inventory availability.
   - Trade-offs must be made between speed, cost, service level, and reliability. Technology, such as TMS, plays a critical role in managing these decisions.
3. Large costs are involved in the infrastructure of transportation and logistics. Major cost components include fuel, labor, fleet maintenance, warehousing, packaging, and IT systems:
   - Fixed costs: warehouses, transportation fleets, automated systems.
   - Variable costs: fuel, third-party carrier fees, per-mile charges, packaging, and labor. Balancing cost with customer service is a key logistics strategy challenge.
4. AI optimizes Move, from managing inventories and warehousing, to reverse logistics:
   - AI enables predictive analytics for demand forecasting and real-time shipment tracking. In warehousing, AI-driven robotics and vision systems speed up picking, packing, and sorting.
   - AI supports dynamic routing, congestion avoidance, and real-time delivery adjustments. In reverse

logistics, AI can help identify return reasons, reduce waste, and improve product restocking or disposal.

5. Logistics activities are integrated across the supply chain:
   - Logistics must coordinate with sourcing, production, sales, and customer service to enable end-to-end visibility. Integrated logistics ensures inventory flow aligns with demand planning and production schedules.
   - Collaborative tools and shared data (e.g., through control towers or cloud platforms) help synchronize logistics partners. Integration is essential for agility, cost control, and customer satisfaction in complex, global supply chains.

Case Studies

The Harry Potter Logistics Challenge

Background

In the summer of 2007, the launch of "Harry Potter and the Deathly Hallows" was not just a major publishing event; it was a global cultural phenomenon. Anticipation for the seventh and final installment in the beloved Harry Potter series reached a fever pitch, with fans eager to get their hands on the book the moment it was released.

To meet demand and avoid spoilers leaking early, Scholastic, the US publisher, announced a synchronized release date of July 21, with 12 million copies scheduled to be available in stores and delivered to homes across the United States that very same day. This represented the largest first printing in publishing history at the time.

The Operations Challenge

Behind the excitement was a massive logistics challenge. Coordinating the simultaneous delivery of millions of books to thousands of retail outlets and online buyers across the country demanded near-flawless execution from every member of the supply chain.

Scholastic had to orchestrate a complex web of partners:

- Printers at six different locations produced the books under high security.
- Trucking companies such as Yellow Transportation and J.B. Hunt were hired to move inventory directly from printers to major retailers like Barnes & Noble and Amazon, bypassing Scholastic's own warehouses to save time.
- To ensure consistency, trailers and pallets were standardized to expedite loading and unloading processes.
- GPS tracking was installed on every trailer, allowing Scholastic to monitor movement and ensure that no shipment veered off its designated route, a key security measure to prevent leaks.

Digital Meets Physical

The challenge was especially intense for online retailers. For books to arrive on customer doorsteps precisely on July 21, they had to be shipped in advance. This required accurate forecasting and timing. For example, Barnesandnoble.com developed custom algorithms to calculate the optimal time to hand off shipments to USPS or UPS, ensuring synchronized last-mile delivery.

This operation wasn't just about logistics; it was a finely tuned orchestration of manufacturing, transportation, security, marketing, and digital strategy. It was a rare example where the book industry, not known for high-tech innovation, had to perform like a tightly integrated supply-chain organization.

Outcomes

The launch was a resounding success. Fans received their books on time, the ending remained a secret until the official release, and the event went down as one of the most impressive feats in publishing supply-chain history.

Questions

1. What were the key operational risks Scholastic faced in the synchronized launch of "Harry Potter and the Deathly Hallows?"
2. How did bypassing Scholastic's own warehouses help streamline the supply chain? What are the trade-offs of such a decision?
3. In what ways did technology, such as GPS tracking and algorithmic shipping, play a role in managing supply-chain complexity?
4. Why was standardization of trailers and pallets important for this operation? What efficiency gains did it offer?
5. Discuss the role of trust and coordination among supply-chain partners in this case. How might a breakdown in communication have affected the launch?
6. What strategies did Scholastic use to ensure both speed and security in its logistics operations?
7. From a customer satisfaction perspective, why was it critical that online orders arrived on the same day as the in-store release?
8. How did marketing and logistics need to align for this product launch to succeed? What lessons can be drawn for other industries?
9. Imagine this launch happening today. What additional digital tools (e.g., AI, real-time tracking apps, blockchain) might enhance the coordination?
10. What can other industries, such as fashion, tech, or pharmaceuticals, learn from Scholastic's approach to high-stakes product launches?

Developing Customer Service with AI Prompts

A manager can use AI-driven analytics to surface insights by crafting strategic, data-informed prompts, often as queries to a business intelligence tool, data science platform, or integrated AI dashboard. Here are examples of prompts that a manager could issue to an AI system to obtain such segmentation, service prioritization, and cost allocation insights:

I. Customer Segmentation & Service Sensitivity

Prompt: "Identify distinct customer segments based on historical order frequency, return rates, and service interactions. Which segments show the highest sensitivity to delivery speed?" *Yields micro-segments where fast delivery is an order winner.*

Prompt: "Cluster customers based on purchasing patterns and post-purchase feedback to identify which segments value premium customer service the most." *Reveals which groups respond to high-touch service.*

Prompt: "Run a correlation analysis between customer satisfaction scores and delivery performance metrics by segment." *Uncovers which customer types link satisfaction to logistics performance.*

II. Cost Allocation and Profitability

Prompt: "Estimate per-order logistics costs by customer segment based on average shipping distance, return frequency, and service-level requests." *Supports cost allocation across segments instead of averaging total.*

Prompt: "Calculate segment-specific margin impact when high-cost delivery services are included versus excluded." *Enables trade-off analysis between service levels and profitability.*

Prompt: "Which customer segments could be shifted to slower, lower-cost delivery options with minimal risk of churn?" *Allows for service differentiation without compromising loyalty.*

III. Strategic Planning and Policy Adjustment

Prompt: "Simulate the financial effect of reallocating high-speed delivery services only to the top 20 percent of high-margin customer segments." *Helps drive tiered service models aligned with value.*

Prompt: "Generate a customer service policy recommendation per segment based on behavioral, financial, and service-level sensitivity data." *Creates data-driven service tiers or policies.*

Questions

1. How can AI-driven micro-segmentation help businesses avoid the pitfalls of one-size-fits-all customer service strategies? What prompts would you use, to have AI help you?
2. When reallocating high-cost services like expedited shipping, what ethical considerations should managers keep in mind beyond just profitability? How would you incorporate loyalty, fairness, and potential brand impact? As a manager, would you override the AI? When?
3. In what ways could incorrect or biased data input into an AI system distort customer segmentation and cost allocation models? Can you suggest a sequence of prompts to avoid this? List the risks of poor data governance.
4. How could a manager use AI simulation tools to test different strategic policy changes before rolling them out in the real world? Explain how you would use scenario planning and digital twins.
5. What tradeoffs should a company consider when deciding whether to shift lower-value segments to slower, lower-cost-service tiers? Explain how that would balance operational savings with customer retention and reputation.

Chapter 7

Impact on Sourcing, or "Buy"

In This Chapter You Will Learn

1. that sourcing or purchasing is about more than just buying; it is a critical strategic function;
2. that AI technologies are critical enablers of today's purchasing function;
3. the details of AI's effect on specific sourcing tasks, from order processing to co-creation;
4. about trade-offs among single versus multiple sourcing;
5. about trade-offs in global versus domestic sourcing; and
6. about outsourcing and supplier risk management.

People think sourcing is just buying stuff, like picking up milk from the store.

In business, sourcing is much more complicated. It is more like planning a community potluck with 3,000 guests, half of whom might not show up on time, and the other half might bring the wrong dish. Take, for example, a small outdoor furniture company trying to source teak from Southeast Asia. It is not just finding a supplier;it is tracking weather disruptions, currency swings, customs delays, and making sure the wood is sustainably harvested to meet environmental regulations. A missed shipment does not just mean empty shelves; it can mean lost customers, delayed payroll, and a real headache for everyone down the line. Sourcing deals with all of this and given the complexity, AI tools are essential.

Consider Artisan Beans Café, a regional coffee shop chain with six locations across the Pacific Northwest. The business prides itself on offering ethically sourced, single-origin coffee, and seasonal baked goods. With rising costs and inconsistent supplier availability, sourcing became a growing challenge for this small, but quality-driven, enterprise. Artisan Beans used to rely on manual processes to manage its relationships with coffee importers and local bakers. Seasonal shortages, delayed shipments, and price volatility—especially in specialty beans—often led to stockouts or overordering. The lack of real-time visibility made it difficult to forecast needs or evaluate supplier performance.

Then the café partnered with a cloud-based AI-sourcing platform tailored to small businesses. Within one year of adoption, Artisan Beans Café reduced sourcing costs by 15 percent, improved order fulfillment rates by 25 percent, and built stronger relationships with high-performing, ethically-certified suppliers. The sourcing function became proactive rather than reactive—without the need for a full-time procurement team.

This is the power of AI in sourcing. Let's dig deeper.

7.1 Sourcing, or "Buy"

Sourcing, or purchasing, is the business function responsible for all activities and processes required to purchase goods and services (Figure 7.1). In business jargon it is often referred to as **Buy**. Sourcing manages the upstream portion of the supply chain, the supply side of the organization.

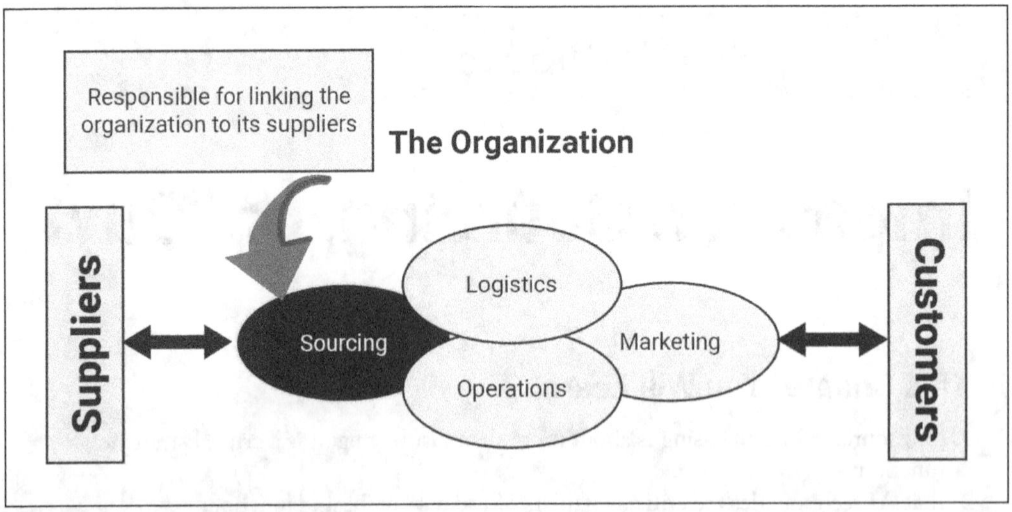

Figure 7.1 The Sourcing Function

For most firms, cost of purchased goods and services accounts for 50 to 80 percent of total revenue. This is a huge amount of money and underscores the strategic importance of effective sourcing. In industries like automotive, more than 75 to 85 percent of a vehicle's total cost comes from externally sourced components, including electronics, drivetrains, and materials. As such, the Buy lever must be managed with precision, leveraging data, supplier collaboration, and strategic sourcing practices to control costs, ensure quality, and reduce risk.

The challenge in managing sourcing is that it is complex. The sourcing function is often referred to by names such as *purchasing*, *procurement*, *sourcing*, *strategic sourcing*, and *supply management*. Although these terms are often used interchangeably, they do not necessarily mean the same thing. These terms also underscore the breadth of decisions for which sourcing is responsible. AI and analytics applications selected along this lever need to consider the differences in focus of what the company needs from this function.

Purchasing is a term that defines the process of buying goods and services. Also called **procurement**, it is a narrow functional activity with duties that include supplier identification and selection, buying, negotiating contracts, and measuring supplier performance. However, it is much more than buying. Over the years, the purchasing function has evolved to include a much broader and more strategic responsibility, which is termed *strategic sourcing* or *supply management*.

Strategic sourcing involves a much more progressive approach to the sourcing function. Also called **supply management**, it goes beyond focusing on just the price of goods, looking at the sourcing function from a strategic and future-oriented perspective. It considers sourcing opportunities that will solve greater problems for the buying firm and give it a competitive advantage. This requires expanding the role of sourcing from mere buying to building close and longer-term working relationships with specially selected suppliers and partners. Figure 7.2 lists various tasks of the sourcing function. In this text we will use the terms interchangeably bearing in mind that the meaning falls on a spectrum from pure buying to strategic sourcing.

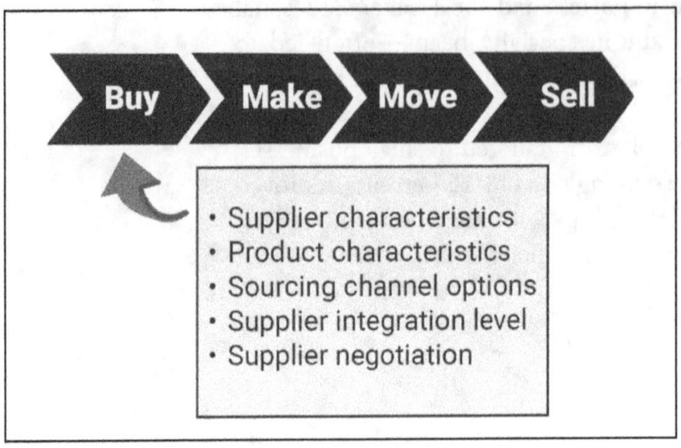

Figure 7.2 Tasks of the Sourcing Function

COMPANY EXAMPLE: APPLE

Apple provides a compelling example of how sourcing can be both broad and strategic. The company's sourcing function spans multiple dimensions, including purchasing and procurement, as well as strategic sourcing and supply management. This reflects the diverse range of decisions Apple must make to maintain its operational efficiency and market leadership.

For example, Apple's purchasing function has evolved to include supplier identification, negotiation, and performance tracking. However, the company's emphasis on strategic sourcing demonstrates a shift toward building long-term partnerships with key suppliers. This approach ensures access to innovative technologies, reliable materials, and consistent quality, factors that contribute to Apple's competitive advantage.

Apple's sourcing strategy is powered by AI, which plays a crucial role in making informed and strategic sourcing decisions. By analyzing supplier performance data, market trends, and demand forecasts, Apple ensures that its sourcing aligns with long-term objectives. This data-driven approach has enabled Apple to reduce procurement costs by 10 percent over the past two years while maintaining its high-quality standards. AI also supports Apple in identifying potential risks within its supply chain, such as disruptions or underperforming suppliers. By proactively addressing these risks, Apple secures the continuity and reliability of its operations. This capability highlights the value of aligning AI applications with strategic sourcing needs.

By embracing strategic sourcing practices and using AI, Apple has transformed the Buy lever into a competitive differentiator. This shift ensures not only cost efficiency but also enhanced supplier collaboration, innovation, and resilience in a dynamic global market.

QUESTIONS

1. How does Apple's approach to strategic sourcing differ from traditional procurement practices, and what advantages does it offer in a highly competitive technology market?
2. In what ways has AI enabled Apple to transform its sourcing function into a source of competitive advantage? Provide examples from the case.
3. What are the potential risks of relying heavily on AI for sourcing decisions, and how might Apple mitigate those risks while still leveraging AI's benefits?
4. Apple emphasizes long-term supplier partnerships. How do these partnerships contribute to innovation and quality, and what challenges might arise in maintaining them globally?
5. If you were advising a mid-sized tech company aiming to replicate Apple's sourcing strategy, what key elements would you prioritize, and how would you scale AI integration effectively?

As AI and analytics have advanced, the sourcing function has become a more strategic and data-driven process, rather than merely buying. New technologies enable companies to improve demand forecasting and make more accurate predictions of what goods they need. As a result, companies are better at managing suppliers and significantly reducing procurement costs. AI capabilities also extend far beyond cost savings. They enhance real-time decision making, strengthen risk-mitigation strategies, and significantly improve overall supply-chain agility and resilience.

Some of the specific areas where AI has affected sourcing include the following:

Predictive Analytics Adoption. Predictive analytics addresses the need to forecast product demands and components with much greater accuracy and anticipate shortages. This is so important that it is estimated that by 2026, more than 80 percent of enterprises will rely on AI-driven predictive analytics to guide sourcing and demand forecasting.

Sustainability Imperative. New technology enables the imperative of sustainability to be baked into the algorithms. With 92 percent of procurement leaders citing environmental sustainability as a top priority, sourcing decisions are increasingly driven by carbon footprint and circular economy goals. Current AI allows automatic inclusion of sustainability constraints into sourcing decisions.

Supplier Collaboration for Innovation. AI tools enable co-creation, discussed later in the chapter, speeding up product innovation and creation. In fact, according to a survey, 70 percent of chief procurement officers believe that closer supplier partnerships are key to driving product innovation, and speed to market.

Digital Transformation. The procurement process itself is increasingly being digitized and automated across all companies and sectors. As such, procurement today looks very different from just a few years ago, and we will discuss this throughout the chapter. Consider that more than 65 percent of global companies are investing in procurement digitization to improve agility, compliance, and cost control.

These improvements in sourcing are occurring for all firms, large and small. Let's look at examples of each. We will start with a large company example, General Motors, and then look at a mid-size and a small firm.

COMPANY EXAMPLE (LARGE FIRM): General Motors

General Motors (GM) exemplifies the power of a technology-driven sourcing function in global manufacturing. Sourcing accounts for approximately 75 percent of GM's total operating costs, reflecting procurement's critical role in cost control, product quality, and supply-chain resilience. Today, more than 80 percent of the cost of a GM vehicle still derives from externally sourced components, reinforcing the strategic necessity of intelligent and proactive supplier management.

GM has an expansive global-supplier network with more than 4,500 suppliers across 70 countries, managing a vast flow of materials to support production across multiple brands and vehicle platforms. Through strategic supplier selection, data-driven contract negotiations, and collaborative relationship management, GM has achieved significant procurement efficiencies, reducing sourcing costs by nearly 18 percent between 2021 and 2024 despite inflationary pressures and global supply disruptions.

GM could not do this without the use of AI. Predictive analytics enables the company to anticipate disruptions such as semiconductor shortages, port delays, and geopolitical risk. AI-based risk models scan supplier financial health, ESG compliance, and geographic exposure in real time, allowing GM to dynamically shift sourcing strategies and maintain production continuity. Also, machine-learning tools identify long-term cost-saving opportunities by analyzing trends in commodity pricing, contract performance, and supplier benchmarks.

By integrating AI into predictive analytics and sourcing practices, GM has elevated its procurement function from a back-office operation to a strategic driver of resilience and innovation in the modern automotive industry.

QUESTIONS

1. How has GM transformed its sourcing function from a transactional process to a strategic capability, and what role has AI played in that transformation? Consider both cost efficiency and risk mitigation in your response.
2. In what ways does GM's use of predictive analytics help manage complex supply-chain disruptions, such as those seen during the semiconductor shortage? Explain the integration of internal and external data sources in proactive sourcing decisions.
3. What are the advantages and potential risks of relying heavily on AI-driven supplier risk scoring in global procurement operations? Explain how such AI systems might misclassify risks or overlook qualitative factors.
4. Identify three specific benefits of AI sourcing on innovation, resilience, and brand reputation.
5. What lessons can other manufacturers learn from GM's sourcing strategy in terms of digital transformation, supplier collaboration, and sustainability? List at least three elements that are most transferable to smaller firms or different industries.

AI procurement platforms are not just for large companies. They are used by mid-size and small companies as well.

COMPANY EXAMPLE (MID-SIZE FIRM): FlexTech Electronics

FlexTech Electronics, a mid-size manufacturer of industrial sensors, recently faced rising production costs due to supplier volatility and raw material price hikes. With more than 78 percent of its product cost tied to sourced components, the company realized that minor inefficiencies in sourcing could drastically impact its profit margins.

What was the result?

FlexTech deployed an AI-powered procurement suite that combined predictive analytics, machine learning, and natural language processing. Here's how they applied it:

Predictive Analytics for Price Forecasting. The system evaluates huge amounts of data such as global commodity price indices, supplier quotes, and historical purchase data to forecast price trends for critical materials such as rare earth metals and semiconductors. This enables the procurement team to time purchases strategically and lock in better contract terms. This process is automated, and the procurement managers only override when needed.

Supplier Risk Monitoring. The AI system continuously monitors supplier performance, financial health, and geopolitical risk using real-time feeds from news, trade data, and ESG compliance sources. When a key supplier in Southeast Asia was flagged for labor violations, the system recommended prequalified alternatives in Vietnam and Mexico, weeks before regulators took formal action.

Smart Supplier Matching. Using machine-learning algorithms, the platform automatically matched sourcing requirements with a global-supplier database, ranking options based on criteria like lead time, quality ratings, sustainability credentials, and total cost of ownership. This cut supplier selection cycle time by more than 40 percent.

Automated RFX (request for X) Processes. FlexTech digitized its request for quote (RFQ) and bid evaluation processes. AI algorithms analyzed proposals not just for price but also for delivery performance and alignment with ESG goals, allowing the team to make data-informed trade-offs.

Procurement Chatbot Assistant. A conversational AI assistant allows sourcing managers at FlexTech to conduct what-if analyses (e.g., "What is the effect of a 15 percent increase in copper prices, on sensor production costs?") and receive real-time responses, enabling faster scenario planning.

What was the result?

Within a year, the company reduced sourcing costs by 12 percent, improved supplier on-time delivery by 18 percent, and mitigated disruption risk by diversifying key suppliers across three regions. This example illustrates that the Buy lever is not just about cost control but also critical in building supply-chain resilience.

QUESTIONS

1. How can small to mid-size manufacturers like FlexTech balance cost, quality, and risk when sourcing accounts for the majority of their production expenses? What role does technology play in enabling this balance?
2. How can FlexTech leverage AI to balance cost efficiency with supply-chain resilience, especially when sourcing critical components from high-risk regions? Explain how predictive analytics and supplier risk scoring might support this.
3. What role should human judgment play alongside AI recommendations in sourcing decisions at FlexTech? When might it be necessary to override or supplement AI-driven insights?

COMPANY EXAMPLE (SMALL FIRM): Olivia's Custom Candles

Background: Olivia runs a small online business from her home, selling custom soy-based candles through Etsy and local markets. She sourced wax, jars, fragrance oils, and packaging materials from multiple online suppliers. As demand grew, she struggled with stockouts during holidays and overordering in slow months, leading to waste and cash flow issues.

The Sourcing Challenge: Olivia managed procurement manually using spreadsheets and vendor websites. She often discovered that jars were out of stock or that prices had changed after she'd accepted large orders. Tracking supplier reliability and planning restocks became increasingly time consuming.

How AI Helped: Olivia began using a simple AI-driven inventory and sourcing assistant designed for small sellers. Here is how it helped her:

- Forecasting Orders: The tool analyzed her sales trends, Etsy traffic, and seasonal data (like Mother's Day and Christmas) to predict which candle types would sell more and when to reorder materials.
- Supplier Monitoring: The AI tool tracked pricing, shipping times, and stock availability across her top three suppliers, alerting her when prices drop or a delay was likely.
- Recommendation Engine: The AI suggested a new eco-friendly fragrance oil supplier with better customer reviews and lower lead times.
- Cash Flow Planning: By forecasting material needs more accurately, Olivia avoided bulk over purchases, improving her available cash for marketing and product development.

Outcome: With AI's help, Olivia reduced sourcing-related delays by 40 percent, cut material waste by 20 percent, and even expanded her product line confidently. More importantly, she saved five to seven hours a week on manual planning, time she now spends on design and customer engagement.

QUESTIONS:

1. How did AI help Olivia transition from reactive to proactive sourcing, and what specific benefits did she gain from this shift? Explain how forecasting and supplier monitoring improved her decision making.
2. What sourcing risks did Olivia face as a small business owner, and how did AI tools help her mitigate those risks? Explain how AI helps with price volatility, stockouts, and overordering.
3. In what ways can AI-driven supplier recommendations affect product quality, customer satisfaction, and brand reputation in a small business like Olivia's? List the trade-offs between cost, quality, and reliability.
4. How might Olivia's use of AI-sourcing tools evolve as her business grows and scales to more markets or platforms? How would multi-supplier coordination, international shipping, or sustainability goals affect her decisions?
5. What challenges might a small business owner like Olivia face in adopting AI tools, and how can she overcome them? Explain the trade-off for both budget and technical know-how.

7.2 AI Technologies in Sourcing

Today's business environment is hyper-competitive with globally connected markets. Purchasing and sourcing is responsible for a huge amount of spending and, as a result, has evolved from a transactional function to a strategic lever. AI is playing a transformative role in this shift, enabling procurement to make faster, data-driven decisions while managing supplier risks and responding to volatile market conditions.

Let's look at specific AI-driven technological capabilities that affect sourcing.

1. Predictive Analytics and Demand Forecasting. Predictive analytics uses historical data, machine-learning models, and real-time market inputs to forecast demand, cost trends, supplier performance, and potential disruptions in supply. In sourcing, this capability helps companies anticipate material shortages, commodity price changes, and supplier risks before they affect the business.

Example: Unilever employs AI-driven predictive models to manage the volatility of raw materials such as palm oil and packaging components. By integrating market data, weather forecasts, and trade flows, the system helps procurement teams negotiate better contracts and hedge against inflationary risks. This has enhanced their sourcing agility and reduced overstock and waste.

2. Autonomous Procurement and Robotic Process Automation (RPA). RPA, often enhanced with AI, automates repetitive procurement tasks such as creating the purchase order, invoice matching, and supplier onboarding. When integrated with natural language processing (NLP), these bots can also interpret supplier documents or contract terms, enabling touchless transactions. This process is now almost entirely automated.

Example: Siemens uses AI-powered bots to manage thousands of purchase requisitions, automating 80 percent of repetitive sourcing tasks. This allows procurement teams to shift focus toward strategic activities such as supplier innovation and risk management.

3. Supplier Risk Monitoring and AI-Based Risk Scoring. AI systems can consider both structured and unstructured data from every source imaginable—news feeds, financial statements, geopolitical events, and social media—to score suppliers on financial health, ESG performance, cyber risk, and operational reliability. NLP enables real-time analysis of qualitative data sources used to evaluate risk.

Example: IBM's Supply Chain Insights platform includes Watson AI which monitors suppliers across 23 risk categories. During the COVID-19 pandemic, this system helped IBM proactively flag suppliers affected by regional lockdowns, enabling the company to reroute orders and prevent production downtime. Obviously human judgment can override the system, but it is mostly automated.

4. Cognitive Sourcing Assistants and Chatbot's AI Assistants, often powered by large language models, act as intelligent sourcing advisers. They are now also called AI Agents. They can suggest preferred vendors, compare specifications, evaluate total cost of ownership, and even negotiate simple contracts or RFQs using historical data and predefined rules.

Example: BASF's procurement team uses an AI assistant that analyses historical procurement data, suggests suppliers based on technical fit and performance, and simulates the effect of pricing and contractual changes. This has shortened sourcing cycle times and improved negotiations.

5. Dynamic Pricing and Market Intelligence Platforms. AI tools also enable dynamic pricing models by analyzing real-time market data, currency fluctuations, supplier capacity, and raw material indices. These tools help sourcing managers optimize contract timing and supplier selection by providing information in real time.

Example: Dell uses AI algorithms to time the purchase of components such as semiconductors and memory chips. These items are very costly and can be in short supply. AI enables Dell to leverage a dynamic market and reduce costs. This has helped the company improve gross margins while navigating global chip shortages.

6. AI-Powered Supplier Discovery and Evaluation. Using AI for supplier discovery involves going through thousands of data points, certifications, production capabilities, and ESG scores across digital platforms to identify and rank qualified suppliers globally. Machine-learning models help match technical requirements to supplier capabilities.

Example: GE uses AI-powered sourcing tools to identify high-potential suppliers across emerging markets, even those without traditional digital footprints. These tools assess production capacity, certifications, geopolitical risks, and ethical sourcing practices, helping GE diversify its supplier base.

7. Contract Analytics and Intelligent Spend Analysis. AI systems can extract, classify, and interpret terms, clauses, and performance metrics from supplier contracts. These insights are cross-referenced with spend data to identify maverick spending (unauthorized spending), savings opportunities, and compliance risks.

Example: Coca-Cola uses AI-based spend analytics to unify procurement data across bottling partners and identify contract leakage or redundant suppliers. This has enabled Coca-Cola to consolidate vendors, renegotiate pricing, and improve procurement efficiency across global operations.

8. Blockchain & AI Integration for Source-to-Contract Integrity. Recall from Chapter 1 that a blockchain is a decentralized, tamper-proof digital ledger that records transactions in a secure, transparent, and immutable manner across a distributed network. Adding AI enhances blockchains.

Blockchains combined with AI provides a tamper-proof, decentralized ledger system that records every sourcing-related transaction, from supplier onboarding and certification to purchase orders and contract execution. AI layers add advanced pattern recognition, anomaly detection, and intelligent automation to these immutable records. This combination ensures that all source-to-contract activities are transparent, verifiable, and fraud resistant.

What else can blockchains bring? *Traceability*: Consider that every contract, bid, invoice, and payment is recorded on the blockchain, offering an end-to-end audit trail that is immutable and time-stamped. *Fraud Detection*: AI algorithms continuously monitor contract behavior and supplier transaction patterns, flagging anomalies such as duplicate invoices, contract tampering, or inconsistent delivery terms. *Credential Verification*: Supplier credentials, certifications, ESG claims, and compliance documents are stored and verified on-chain, ensuring that they are authentic and up to date. *Smart Contracts*: AI-enhanced smart contracts can autonomously trigger procurement actions (e.g., issuing payments or generating reorders) when conditions are met, speeding execution and reducing manual oversight.

Example: Maersk developed a blockchain-based logistics and procurement platform, designed to improve global supply-chain transparency. The platform is now part of the broader ecosystem supporting supplier and contract management. Now every transaction, from the issuance of a purchase order to final delivery, is recorded on a permissioned blockchain.

The application of AI in purchasing and sourcing has accelerated the shift from reactive, cost-centric procurement to strategic, proactive, and data-driven supply management. The business examples showcased here, such as Unilever, IBM, Siemens, and Coca-Cola, demonstrate how AI technologies, ranging from predictive analytics to dynamic pricing algorithms, can deliver significant competitive advantages.

As supply chains face increasing complexity due to geopolitical tensions, climate change, and fluctuating demand, the role of AI in sourcing will only deepen, enabling procurement professionals to manage risk, drive innovation, and create resilient, responsive value networks.

Let's now look at an example of each of these applications in one company.

COMPANY EXAMPLE: Schneider Electric

Schneider Electric is a global leader in energy management and automation and operates a highly complex supply chain across more than 100 countries. With tens of thousands of suppliers and a growing need for sustainability and resilience, Schneider turned to AI to modernize its entire sourcing and procurement process. Let's see how they did this.

1. Predictive Analytics & Demand Forecasting. Schneider begins the sourcing cycle by using AI-powered predictive analytics to forecast demand for circuit breakers and energy control systems across global markets. The system takes in historical sales and production data, global macroeconomic indicators, and regional energy usage trends, as well as lead times and inventory levels. The result is more accurate demand forecasts that allow Schneider to anticipate raw material needs months in advance, reducing both excess inventory and stockouts.

2. Autonomous Procurement & Robotic Process Automation (RPA). Once material needs are forecasted, Schneider's AI-enhanced Robotic Process Automation system takes over.

What does it do?

First, it automatically generates purchase requisitions. Then, it selects preferred suppliers from approved vendor lists and issues RFQs and POs (purchase orders). It then cross-checks invoices for discrepancies.

For routine purchases like copper wiring or plastic enclosures, the entire cycle is completed with zero human touch unless exceptions are flagged.

3. Supplier Risk Monitoring & AI-Based Scoring. Before finalizing high-value contracts, Schneider's AI risk platform continuously evaluates suppliers based on multiple inputs. This includes supplier financial stability (they may look at Dun & Bradstreet reports and filings), ESG metrics (using third-party data and NLP from news sources), cybersecurity risk (via threat intelligence platforms), and geopolitical alerts (e.g., from the Russia-Ukraine conflict).

One key insight: AI flagged a tier-2 supplier in Vietnam with a potential labor violation risk, prompting Schneider to shift a portion of orders to a compliant backup supplier in India.

4. Cognitive Sourcing Assistant. A chatbot-based sourcing adviser assists procurement managers. For example, when a buyer requests options for sustainable aluminum casing suppliers in Europe, the AI tool scans global-supplier databases and internal performance records. Based on this analysis the AI then suggests three vendors with ISO 14001 certification (a certification Schneider requires) and simulates cost and delivery implications for each.

The buyer selects a supplier, and the AI assistant generates a contract with pre-approved clauses pulled from Schneider's legal library. This is a good example of the interaction of the human buyer with the AI assistant providing information.

5. Dynamic Pricing & Market Intelligence. Schneider sources electronic components such as semiconductors, where prices fluctuate weekly. An AI pricing engine tracks spot-market pricing, monitoring currency exchange rates and analyzing global capacity reports.

When the system detects a likely price spike, such as due to a drought affecting hydroelectric plants in Taiwan (where a key supplier operates), it triggers an early buy recommendation. This helps Schneider lock in prices before they escalate.

6. AI-Powered Supplier Discovery. As Schneider expands its green product line, it seeks suppliers offering recycled plastics. The AI system scours global databases and sustainability registries to identify small but certified suppliers in Eastern Europe and Southeast Asia that were previously overlooked.

Machine-learning models match these suppliers' technical capabilities with Schneider's design specifications. Five new vendors are added to the sourcing pool, enhancing resilience and ESG alignment.

7. Contract Analytics & Intelligent Spend Analysis. Finally, the legal and procurement teams use AI-driven contract analytics to extract payment terms, volume discounts, and SLAs (service level agreements) from hundreds of legacy contracts. This data was overlaid with spend analysis and Schneider uncovered $12 million in maverick spending outside approved contracts and 15 percent of contracts that were missing price escalation clauses. This created opportunities to consolidate suppliers for packaging across three business units.

8. Blockchain & AI Integration. Schneider Electric integrates blockchain and AI technologies to enhance supply-chain transparency and efficiency. By combining blockchain's immutable ledger capabilities with AI's analytical prowess, Schneider Electric enables real-time tracking of products from source to destination.

These AI technologies, working in unison for Schneider, have achieved a 15 percent improvement in forecast accuracy, a 40 percent reduction in sourcing cycle time, and 25 percent fewer supply-chain disruptions. This integrated case demonstrates that the real power of AI in purchasing and sourcing comes not from isolated tools but instead from a coordinated ecosystem where predictive, cognitive, and autonomous systems feed insights to each other. As supply chains become more digitized, companies that harness this synergy will be best positioned to manage cost, risk, and strategic value in sourcing.

QUESTIONS

1. How does the integration of predictive analytics and autonomous procurement improve sourcing efficiency and reduce operational risk at Schneider Electric? What is the relationship between accurate demand forecasting and the automation of routine procurement tasks?
2. In what ways does AI-driven supplier risk monitoring enhance resilience and ethical compliance in global sourcing? Explain the use of external data (e.g., ESG reports, geopolitical signals) and how it influences sourcing decisions.
3. What are the strategic advantages and limitations of using cognitive sourcing assistants in supplier selection and negotiation? Give at least one example of when human judgment might still be necessary, despite AI recommendations.
4. How does dynamic pricing supported by AI give companies like Schneider Electric a competitive edge in volatile commodity markets? Discuss real-time decision making and early-purchasing strategies driven by market signals.

5. What challenges might Schneider Electric face in unifying diverse AI systems (e.g., forecasting, risk scoring, contract analytics) into a cohesive sourcing strategy? Evaluate issues like data integration, organizational change, and system interoperability.

7.3 Impact on Specific Sourcing Tasks

Selection of AI applications in sourcing depends in part on whether a company is looking to improve mere purchasing or manage strategic sourcing. AI offers a range of possibilities, from improving current processes of containing spend, to enabling advanced risk scenario planning and automation of the entire procurement process. Regardless of scope, implementation can result in making considerably faster, better, and more accurate sourcing decisions.

7.3.1 Order Processing

One of the most immediate and effective applications of AI in supply-chain management is within order processing, the end-to-end cycle from order placement to fulfillment and payment. Efficient and accurate order processing plays a critical role in shaping customer experience, as it forms the primary interface between the organization and its customers. As of 2025, companies deploying AI in order processing had up to a 35 percent reduction in order-processing time and up to 25 percent fewer fulfillment errors.

These systems enable real-time order tracking, dynamic inventory updates in real time, and proactive issue resolution. They also analyze large volumes of data from multiple sources—online storefronts, ERP systems, warehouse management platforms, and customer service tools—to provide real-time visibility and predictive insights. With advanced analytics, firms are automating repetitive tasks, eliminating manual data entry, and accelerating workflows across the order life cycle. This is all along the procurement cycle from order confirmation to invoicing.

The current order-processing technology uses cloud-based platforms, electronic data interchange, and application programming interfaces (APIs) to transmit orders instantly and accurately between buyers, sellers, and service providers. Automation reduces order cycle time, improves accuracy, and minimizes costly disruptions.

IBM is an example of an enterprise-level order-processing transformation through its use of AI. For example, their order-to-cash (O2C) cycle begins when a customer places an order and moves through manufacturing, logistics, delivery, billing, and post-sales support. AI tools monitor order status, predict disruptions, and automate invoicing and collections, improving both cash flow and customer satisfaction. They also use AI in the procure-to-pay (P2P) cycle that integrates IBM's procurement operations with its accounts payable systems. AI allows visibility into supplier performance (managers can see what is going on), manages risk proactively (flags risks for managers), and streamlines payment workflows (automates this process). In a 2024 interview, IBM's Vice President of Global Supply Chain, Timothy Carroll, emphasized that every aspect of IBM's supplier engagement, from production to travel procurement, is optimized.

7.3.2 Order Standardization

The integration of AI into procurement is also transforming the way organizations standardize and manage the global order cycles. Automated order processing not only accelerates operations but also ensures consistent execution across geographic regions and time zones. For IBM, discussed previously, standardization enabled by AI ensures a uniform level of service and fulfillment, regardless of customer location.

Take, for instance, a scenario where a client in Germany identifies the need for a critical part during off-hours. Rather than waiting for local business operations to resume, the AI-powered global order system automatically routes the request to an available fulfillment center in North

America or Asia. This 24/7 responsiveness is made possible by intelligent order-routing algorithms, multilingual order capture systems, and globally harmonized procurement workflows.

AI also plays a pivotal role in elevating sourcing standardization by transforming how decision makers interact with information. AI tools can now convert complex data sets into intuitive, real-time visualizations, empowering leaders to make faster, data-backed decisions. These systems leverage dynamic dashboards, charts, and simulations to visualize sourcing activity across suppliers, commodities, and geographies.

What is especially useful is that AI-sourcing/procurement platforms integrate internal transactional data (e.g., purchase orders, invoices, SKUs, bill of materials) with external sources like market trends, supplier performance metrics, warranty claims, and risk indices. This creates a 360-degree view of procurement performance.

Consider that these systems have **interactive dashboards** that allow real-time tracking of spend by category, supplier, or region and enable visual what-if simulations to model the effect of price changes, supplier disruptions, or demand shifts. They also enable "drill-down capability" from aggregated spend data to granular part-level metrics and AI-powered alerts for contract compliance, volume thresholds, and supplier risk flags.

According to a recent survey, 78 percent of leading procurement organizations in 2025 use AI-driven analytics platforms to conduct scenario planning and spend optimization, up from just 42 percent in 2020. So, AI enables not just automation but also intelligent standardization and visualization, turning procurement into a predictive, globally coordinated function.

7.3.3 More and Faster Queries

Remember that procurement is responsible for the bulk of a company's spend. The latest generation of AI-powered procurement tools is revolutionizing how companies manage spend and make sourcing decisions. Current systems allow organizations to run millions of queries in real time, enabling procurement teams to test a wide array of sourcing award scenarios under multiple market and operational constraints. With these tools, companies can instantly locate optimal cost-benefit trade-offs across large, multidimensional "spend cubes," interactive data models that segment spend by location, commodity, supplier, or material taxonomy.

As new sourcing opportunities emerge, AI enables rapid creation of customized spend cubes and scenario simulations. The processing speed of advanced analytics platforms, enhanced by cloud computing and in-memory databases, allows near-instantaneous analysis of complex procurement variables.

However, AI doesn't just speed up sourcing; it transforms how procurement teams solve complex problems. Traditional e-sourcing tools have evolved into intelligent systems capable of running large-scale network optimizations for common sourcing events, including less-than-truckload (LTL) and full truckload (FTL) logistics, multi-echelon sourcing, and multi-supplier bundling. Organizations now routinely use AI to model what-if consequences of rare but high-impact scenarios, such as geopolitical disruptions or commodity price spikes.

This ability to simulate, optimize, and adapt sourcing strategies is quickly becoming a strategic differentiator. In fact, as of 2025, 72 percent of top-performing procurement organizations embed AI and scenario analysis in their sourcing workflows.

Walmart is a great example of deploying AI and analytics for global procurement. With more than 10,500 stores in 19 countries and revenues exceeding $638 billion, the scale and complexity of Walmart's supply chain demands world-class spend management capabilities. As discussed earlier, Walmart's proprietary analytics platform enables interactive spend visualization by geography, supplier, SKU, or material group. Imagine having that information in a visual format at your fingertips. These are basically dynamic dashboards that consolidate internal purchasing data with external signals (e.g., inflation trends, FX rates, transportation bottlenecks).

Then there is real-time scenario modeling for supplier selection, cost arbitrage, and freight consolidation. For instance, Walmart regularly runs network optimization simulations for LTL and FTL logistics across its North American distribution network. These simulations help the company identify sourcing efficiencies that reduce freight costs by up to 12 percent, according to its 2023 sustainability and logistics report.

Walmart's visualization tools also allow managers to evaluate supplier risk, contract compliance, and regional sourcing opportunities, all within a single unified platform. This integration of data, visualization, and optimization enables Walmart to make faster, smarter, and more resilient sourcing decisions. This capability is unlike anything before.

7.3.4 Cost Savings

Recall that one of the biggest changes with AI-driven applications in sourcing is that they are fundamentally transforming this function from a reactive buying function into a proactive, strategic capability. Rather than merely responding to cost pressures, systems enable companies to optimize sourcing decisions far in advance of the sourcing event, with a comprehensive view of cost, risk, design specifications, supplier capabilities, and long-term value.

These tools support cross-functional collaboration among design engineers, procurement teams, and suppliers. Using real-time data-sharing platforms, stakeholders can now engage in early supplier involvement, where design decisions are shaped by input from the supply base. AI applications can simulate cost implications of different materials, quality tolerances, or production methods, allowing teams to jointly reconfigure product specifications to improve quality and reduce costs.

COMPANY EXAMPLE: Best Buy

Best Buy, one of North America's largest electronics retailers, exemplifies how AI enables collaborative sourcing at scale. Best Buy partners closely with key suppliers through collaborative planning, forecasting, and replenishment (CPFR) protocols. These involve regular data exchanges, including real-time inventory levels, promotional forecasts, demand projections, and supply constraints, which fuel joint decision-making across the supply chain.

The company's AI-enhanced CPFR platform enables real-time forecast synchronization with suppliers, improving forecast accuracy by more than 25 percent compared to pre-CPFR implementation levels. It also allows shared visibility into inventory and POS data. Visibility means that everyone sees the same data and allows Best Buy and their suppliers to jointly plan allocations during high-demand cycles such as Black Friday or back-to-school. Also, automated scenario modeling helps adjust promotional volumes and distribution strategies within minutes based on updated sales data.

This collaborative approach has led to tangible business benefits. In 2023, Best Buy reported an 18 percent improvement in on-shelf availability during peak promotional periods, and a 12 percent reduction in excess inventory across its national distribution network. In addition, the enhanced supplier coordination has allowed the company to shorten its replenishment cycles and minimize stockouts on high-demand SKUs such as laptops, gaming consoles, and smart home devices.

By enabling early, data-driven collaboration between internal stakeholders and suppliers, companies can design products for optimal cost, performance, and resilience well before the first sourcing bid is ever issued. Best Buy demonstrates how integrated analytics platforms and collaborative models like CPFR can unlock significant competitive advantages in today's volatile global supply chains.

QUESTIONS

1. How does AI-enhanced CPFR improve forecast accuracy and supplier collaboration in Best Buy's supply chain? How does real-time data sharing and joint decision making affect this?
2. What are the key risks and limitations of relying heavily on shared forecasting and real-time data in collaborative sourcing arrangements? How do data integrity, system compatibility, and strategic alignment come into play?

3. How do AI-driven scenario-modeling tools give Best Buy an advantage during promotional periods like Black Friday or back-to-school seasons? Evaluate how speed, agility, and demand responsiveness contribute to performance.
4. What are the trade-offs between minimizing excess inventory and ensuring high product availability? How should Best Buy balance efficiency with service-level goals?
5. In what ways can early supplier collaboration, supported by AI and analytics, influence product design and procurement strategy? How can upstream coordination affect downstream sourcing and logistics decisions?

7.3.5 Predictive Edge

Companies that use AI-powered predictive analytics have given their procurement teams a decisive edge in navigating volatility, mitigating risk, and making smarter sourcing decisions. Traditionally, procurement functions have relied on historical data to explain performance issues, such as unexpected price fluctuations, deteriorating supplier quality, or sudden material shortages. Now, with predictive analytics and scenario modeling, these issues can not only be explained but also anticipated and preempted.

These systems analyze supplier scorecards, commodity market trends, weather disruptions, geopolitical risk, and transportation flows, all in real time. More importantly, they understand when long-standing correlations break down, helping companies plan for the unpredictable. This predictive foresight is transforming procurement from a reactive to a strategic and preventive function. In addition, some of the modern platforms integrate predictive analytics with scenario planning, enabling procurement leaders to run what-if simulations on sourcing strategies. These tools help forecast the potential effect of events such as material cost surges, supplier bankruptcies, transportation bottlenecks, and regulatory changes or ESG noncompliance. Companies can now simulate multiple futures, test contingency plans, and select sourcing strategies based on forecasted cost, risk, and service-level performance.

COMPANY EXAMPLE: Unilever

Unilever is the global consumer goods leader operating in more than 190 countries and has emerged as a benchmark in predictive procurement. Through its procurement intelligence engine, Unilever combines AI-driven market monitoring with advanced risk analytics to proactively manage sourcing activities across its vast global network.

Unilever's AI platform continuously monitors more than 100 external data sources—including commodity markets, geopolitical signals, weather, and social media chatter—to assess supply risk in real time. The platform uses predictive algorithms to forecast raw material price movements up to 12 months in advance, enhancing budget accuracy and contract negotiation timing. It also simulates supplier risk scenarios, enabling proactive adjustments to sourcing strategies when potential disruptions are identified.

In 2023, this system helped Unilever avoid more than $350 million in cost exposure across commodities such as palm oil, aluminum, and packaging materials. It also contributed to a 20 percent improvement in forecast accuracy, allowing for more agile inventory positioning and supplier coordination.

Unilever's success mirrors broader trends among leading firms such as IBM, Schneider Electric, and P&G, all of which have embedded AI-based predictive tools into procurement operations.

As a result, Unilever reported $65.5 billion in 2023 revenue, underscoring the scale at which predictive procurement contributes to operational efficiency. AI and analytics initiatives have also helped Unilever achieve Scope 3 emissions visibility across 80 percent of its supply chain, aligning sustainability with procurement agility. As a result, Unilever exemplifies the future of procurement, where intelligence, agility, and foresight are no longer optional, but essential.

QUESTIONS

1. How does Unilever's use of AI and predictive analytics improve procurement decision making and risk management across its global supply network? How does real-time data monitoring help?
2. What are the benefits and potential challenges of integrating more than 100 external data sources into a centralized procurement intelligence engine? What is the effect on data accuracy and system complexity?
3. How might predictive procurement strategies, like those used by Unilever, support both cost efficiency and sustainability goals such as Scope 3 emissions tracking? Describe the intersection of environmental and financial performance.
4. In what ways could predictive price forecasting influence supplier negotiations and contract timing for companies operating at Unilever's scale? How does better foresight shape sourcing leverage and budget stability?
5. How do Unilever's procurement practices compare to traditional procurement models in terms of agility, responsiveness, and resilience? Does predictive procurement offer a competitive advantage during supply disruptions or market volatility?

7.3.6 Co-Creation

The integration of AI, analytics, and cloud-based platforms is fundamentally reshaping procurement by enabling a new paradigm: co-creation between manufacturers and suppliers. This collaborative approach moves beyond transactional relationships, allowing companies and their supply partners to jointly design, simulate, and optimize new products well before they reach the production stage.

Co-creation is particularly impactful in complex, high-value industries such as automotive, aerospace, electronics, and industrial manufacturing, where products can consist of hundreds of thousands of components sourced globally. In these sectors, early design decisions determine up to 80 percent of total life cycle cost. AI tools enable engineers, procurement professionals, and suppliers to share real-time data, run simulations, and test design alternatives collaboratively, reducing costs, accelerating time-to-market, and improving innovation outcomes.

AI powers co-creation through shared access to digital design environments, digital twins, and generative AI models. Manufacturers and suppliers can simulate product performance under various conditions using AI-powered digital twins. They can run design-for-manufacturing and cost simulations in real time. They can also leverage generative design algorithms to create and evaluate multiple design iterations automatically. In addition, they can use predictive analytics to forecast supply risks or cost implications based on component choices or sourcing locations.

These capabilities transform the early stages of product development, ensuring that designs are optimized not just for function but also for manufacturability, sustainability, and cost-efficiency.

Automotive companies such as Toyota, Nissan, and Stellantis (formerly Fiat Chrysler) have aggressively pursued co-creation strategies, partnering with tier-1 and tier-2 suppliers through integrated digital platforms. Toyota, for instance, uses its Toyota Supplier Support Center in conjunction with AI-powered tools to codevelop components with suppliers. By integrating design feedback loops early in development and using AI simulations to detect flaws, Toyota has been able to reduce new model development time by up to 50 percent, eliminate more than 80 percent of potential defects before building physical prototypes, and achieve 20 percent cost savings on average in collaborative development projects.

Another example is the aerospace sector, where Airbus uses AI-enabled co-creation through its Skywise platform, a cloud-based, open-data ecosystem that connects Airbus with its suppliers, airline customers, and maintenance partners. This shared data environment supports joint simulation of aircraft components, cost optimization of design choices, and proactive identification of engineering and sourcing risks. Airbus uses Skywise in tandem with digital twins and AI algorithms to simulate and refine designs for new aircraft, such as the A321XLR and future hydrogen-powered models. By engaging suppliers early, Airbus has accelerated its design cycles and achieved measurable outcomes.

As we can see, AI has enabled co-creation that engages suppliers at the earliest stages of product design. AI facilitates simulation, testing, and optimization, and firms reduce development costs, increase innovation speed, and deliver superior products to market. In industries where precision, complexity, and innovation are paramount, co-creation, fueled by AI and data transparency, has become a core competency for maintaining competitive advantage.

7.4 Supply-Chain Complexity & AI Needs

How much to invest in AI in sourcing depends on a host of variables, including supply-chain complexity, the number of sourcing options, and the location of suppliers. There are countless applications of AI on the Buy lever, ranging from automating the order-processing cycle to segmenting suppliers and informing supplier negotiation.

Which technologies should companies invest in?

Types of complexities drive how to best utilize AI technologies on the supply side. Just as there are uncertainties on the demand side, there are also uncertainties on the supply side. The supply side of the supply chain can be classified as a *stable* or an *evolving supply process*. A stable supply process is where sources of supply are well established, manufacturing processes used are mature, and the underlying technology is stable. Consider sourcing apples, lumber, or undershirts.

In contrast, an evolving supply process is where sources of supply are rapidly changing, the manufacturing process is in an early stage, and the underlying technology is quickly evolving. Examples include alternative energy sources, certain organic food items, or high-end computer technology. It is challenging to manage supply chains with either demand or supply uncertainty. It is especially challenging to manage supply chains with *both* uncertainties (Figure 7.3).

Supply Uncertainty \ Demand Uncertainty	Low Functional Products	High Innovative Products
Low Stable	**Efficiency Focused SC** (Least Challenging)	**Responsive SC**
High Evolving	**Risk-Hedging SC**	**Agile SC** (Most Challenging)

Figure 7.3 Demand and Supply Uncertainty

Supply chains that face uncertainty on either the demand or supply side are challenging to manage. However, those experiencing uncertainties on both sides demand even greater reliance on big data analytics to enable real-time decision making, scenario planning, and risk mitigation.

COMPANY EXAMPLE: General Electric

GE is a good example of a company that operates across multiple sectors—including aerospace, energy, and healthcare—each with unique supply uncertainties. For stable, high-volume procurement such as standard industrial components, GE applies AI tools to streamline sourcing, reduce cycle times, and lower transactional costs. For evolving processes, such as sourcing advanced materials for jet engines or medical imaging systems, GE integrates predictive analytics to manage uncertainty, ensure supplier reliability, and uphold stringent quality standards.

With a broad and technologically diverse portfolio, GE differentiates its sourcing strategies. For mission-critical components, it employs real-time performance monitoring, AI-driven risk scoring, and supplier collaboration platforms to mitigate disruption. For noncritical goods, GE leverages automation and machine-learning algorithms to conduct dynamic pricing analysis and optimize supplier selection based on cost, lead time, and regional factors. This distinction is very important, one that more companies should emulate.

GE's supply network spans more than 100 countries. To manage this complexity, GE deploys AI-enhanced platforms to monitor supplier performance in real time, conduct scenario-based risk assessments, and optimize global logistics. These tools enable the company to maintain agility and resilience across multiple geographies and economic conditions.

GE's investment in digital procurement technologies has led to measurable improvements. For example, procurement cycle times have been reduced by more than 30 percent through digital sourcing platforms and analytics-driven workflows. Predictive analytics and AI-driven sourcing decisions have delivered up to 15 percent in procurement cost savings across selected business units. In addition, improved visibility and data sharing have enhanced supplier performance, leading to a 20 percent improvement in on-time delivery metrics globally.

GE's case underscores how modern companies can tailor big data analytics investments to the specific demands of their supply-chain functions. Whether optimizing cost in stable categories, or managing risk in technologically complex sourcing, analytics provides the intelligence needed to enhance procurement performance and resilience.

QUESTIONS

1. How does GE differentiate its procurement strategy for stable versus evolving supply chains, and what role does AI play in managing each?
2. What are the specific advantages of using AI and predictive analytics in sourcing critical components, especially in high-tech industries like aerospace and healthcare?
3. In what ways has GE's global-supplier monitoring improved supply-chain resilience, and what technologies enable this capability?
4. What are the trade-offs between cost optimization and risk mitigation in GE's sourcing strategy? How can data analytics help balance these objectives?
5. What are some challenges GE might face in scaling its big data analytics across different business units and geographies?

Let's briefly look at each of these supply-chain types and see what AI technology should focus on for each of these. Their summary is listed in Table 7.1.

7.4.1 Efficiency-Focused Supply Chains

Efficiency-focused supply chains are characterized by low demand uncertainty and stable supply availability, making them relatively straightforward to manage. These supply chains are typically found in industries with high-volume, low-margin products such as commodities, basic consumer goods, and standardized components. Because margins are thin, the competitive advantage lies not in responsiveness or innovation but in relentless cost control, process optimization, and waste elimination.

In such environments, AI plays a central role in enhancing efficiency. Companies can automate routine procurement activities, streamline the O2C cycle, and reduce manual interventions. Tools like electronic auctions, automated reorder systems, and visual dashboards are used to provide transparency, reduce lead times, and minimize administrative overhead. Data-driven standardization allows for tighter inventory control, fewer errors, and improved contract compliance.

The strategic objective in these supply chains is to create a repeatable, scalable, and low-cost operational model. Key performance indicators typically include inventory turnover, fulfillment rate, and cost per transaction, metrics that can be tracked and optimized in real time through integrated analytics platforms.

Georgia-Pacific, a leading manufacturer of paper towels and toilet paper, operates an efficiency-focused supply chain. With demand that is steady and predictable, the company leverages big data to automate order processing, optimize shipping schedules, and reduce procurement cycle time. Through analytics-driven inventory planning and the use of supplier dashboards, Georgia-Pacific has minimized waste and maintained low operational costs in a highly competitive, low-margin market.

Remember that efficiency-focused supply chains have both low demand and low supply uncertainty, and are easiest to manage. These supply chains typically do not have high profit margins. However, their operations are highly predictable, and the gains in these supply chains come from efficiency and elimination of waste.

So what AI capabilities are best to use here?

The efficiency focus means using AI to automate and standardize the order-processing cycle, to rely on electronic auctions, and to use simple visualization and dashboards.

7.4.2 Responsive Supply Chains

Responsive supply chains are used for innovative products that have a stable supply base. The primary challenge here is ensuring quick response to customer demands. Mass customization strategies such as *postponement*, whereby the supply chain is designed to delay product differentiation as late in the supply chain as possible, are effective here. A good example is the case of HP printers that are kept in generic form as long as possible. They are differentiated by country and language specific labels that are added at the last point in the system.

What AI capabilities are most important here?

AI analytics applications can be very effective in designing postponement into the manufacturing and distribution process. Zara is a great example of a responsive supply chain. Fast fashion demands changes rapidly, and Zara thrives by sensing consumer preferences and quickly translating them into new designs that reach stores in just a few weeks. The company uses real-time sales data, store feedback, and RFID tracking to identify what is selling and what is not, then rapidly adjusts production and replenishment. Its vertically integrated supply chain and investment in data analytics allow Zara to reduce markdowns, minimize overstock, and capture emerging fashion trends before competitors. This has resulted in a highly responsive supply chain.

7.4.3 Risk-Hedging Supply Chains

Risk-hedging supply chains are those with high uncertainty on the supply side. These supply chains must do everything possible to minimize risks of supply disruptions and inventory shortages. These supply chains typically rely on higher inventory safety stocks and engage in a practice where resources are pooled and shared between different companies. For example, Toyota and Mazda have an alliance where they share manufacturing facilities and codevelop electric vehicle technology. By pooling R&D resources and plant capacity, they reduce development cost and speed up brining new technology to market.

HP exemplifies the use of risk-hedging supply chains to manage products and components subject to volatility on the supply-side, such as semiconductors, critical metals, and custom assemblies. To maintain resilience HP maintains strategic safety stock buffers of key inventories and leverages multi-sourcing agreements to avoid overreliance on any single supplier or region. HP also engages in inventory pooling through third-party logistics partners and collaborative supplier platforms to dynamically allocate components based on shifting needs across business units.

What AI capabilities should be used here?

For risk-hedging supply chains, scenario planning is a critical tool. To achieve this, HP uses AI-driven scenario planning and risk analytics tools that simulate potential disruptions and assess their impact on lead times, costs, and service levels. These tools support real-time contingency planning, enabling HP to shift orders, reroute logistics, or accelerate alternative sourcing when early-warning signals are triggered. This proactive, analytics-enabled approach helps HP maintain continuity of supply for key components, particularly in the face of global shocks like the COVID-19 pandemic and semiconductor shortages, while minimizing overstock costs.

7.4.4 Agile Supply Chains

Agile supply chains are used in cases of both high demand and supply uncertainty. They are the most difficult to manage as they simultaneously must be responsive to an uncertain demand while using strategies to hedge risks to ensure there are no supply disruptions. These supply chains use mass customization strategies on the demand side, while carrying higher safety stock and engaging in resource pooling on the supply side.

What are the AI capabilities to use here?

Here, both scenario planning and predictive analytics for real-time risk management and supply-chain disruptions are important.

FlexTech Electronics operates an agile supply chain that must accommodate frequent design changes, uncertain customer forecasts, and volatile component availability in the electronics industry. It uses AI-enhanced supply-chain control towers, digital twins, and predictive analytics to simulate disruption scenarios and reconfigure supply flows in real time. During the semiconductor shortages of 2021–2022, FlexTech was able to shift production priorities across its global facilities based on lead-time predictions, customer needs, and real-time supplier data, demonstrating agility in both decision making and execution.

Table 7.1 Enabling Technologies for Each Supply-chain Type

Demand Uncertainty	Supply Uncertainty	Key Objective	Typical Industry Examples	Enabling Technologies
Low	Low	Cost minimization and waste elimination	Commodities, paper products (e.g., Georgia-Pacific)	AI analytics, dashboards, automation
High	Low	Quick response to market changes	Fashion retail (e.g., Zara)	Real-time sales data, RFID, fast-design feedback loops
High	High	Flexibility and adaptability to disruption	High-tech electronics (e.g., FlexTech)	AI control towers, digital twins, predictive simulations
Low to Moderate	High	Risk mitigation and supply continuity	Computers and electronics (e.g., HP)	Scenario planning, AI risk analytics, multi-sourcing

7.5 Single versus Multiple Sourcing

The traditional view of many companies was that multiple sources of supply were best. This would increase cost competition and ensure supply security. This view has been challenged for quite some time, and the notion of single sourcing has become acceptable. Single sourcing focuses on building closer supplier relationships and cooperation between buyers and suppliers. It moves away from arm's length relationships and competitive bidding. Instead, it uses cooperative negotiation and building long-term relationships.

Single sourcing has a number of benefits. Splitting the order among multiple suppliers can be costly as it does not permit consolidating purchase power. Splitting the order also has an effect on quality, as there will be natural variations between different sources, even if minimal. Single sourcing also lowers freight costs and enables easier scheduling of deliveries. Unilever, for example, employs a single sourcing strategy for specific raw materials and packaging components. This approach enables them to build strong, long-term relationships with their suppliers, ensuring supply-chain efficiency and product consistency.

In many cases, however, there may not be a choice but to go with a single source. This may occur, for example, when a supplier is the only source of a particular material. They may be the only one that has the needed process, or they are the sole owner of the patent for the product. Having a single source of supply, however, is risky in practice. When a supplier is small, the buyer's business may utilize most of the supplier's capacity. This makes the supplier highly vulnerable if there is a discontinuity of purchase from the buyer. At the same time, the buyer might not want to be tied to a source that is so highly dependent upon them. Similarly, by relying on one supplier, the buyer becomes vulnerable if there are disruptions in the supplier's production process, such as a fire at a plant, a labor strike, a work stoppage, or if the supplier's supplier has a problem. As the geopolitical landscape has become more volatile, many companies have opted to move away from single sourcing to diversify their risk.

For example, Unilever recognizes this vulnerability and actively works with its suppliers to ensure supply continuity. To mitigate risks, Unilever balances single sourcing with a portfolio of suppliers for other components, allowing the company to maintain flexibility while effectively managing costs and risks.

Choosing between single and multiple sourcing is a critical strategic decision for companies, affecting cost efficiency, supply risk, and operational flexibility. This is where AI can play a transformative role. AI can be critical in guiding this decision by offering data-driven insights, predictive modeling, and scenario analysis that far exceed traditional procurement analysis.

AI systems can process huge amounts of data, from supplier performance records and market trends to geopolitical risk signals and weather forecasts, to assess the stability and reliability of existing or potential suppliers. Using machine-learning algorithms, AI can identify correlations between supplier disruptions and external variables that human analysts might overlook. This allows procurement leaders to evaluate the risk profile of single sourcing more accurately and consider whether the efficiency and volume discounts it offers outweigh the potential costs of disruption.

For multiple sourcing strategies, AI can simulate complex supply-chain scenarios to determine the optimal number of suppliers needed for resilience, without incurring unnecessary overhead. Through predictive analytics, AI forecasts potential risks such as transportation bottlenecks, currency fluctuations, and regional instability that could threaten supply continuity. It can also help model what-if scenarios—for example, "What happens to lead times and costs if Supplier A in Southeast Asia is disrupted?"—thus supporting dynamic sourcing strategy adjustments in real time.

Toyota has historically favored a Lean, just-in-time approach with tightly integrated suppliers, often relying on single source relationships to reduce costs and maintain quality consistency. However, after the 2011 earthquake and tsunami in Japan exposed vulnerabilities in its supply network, Toyota began integrating AI into its procurement planning to better assess supplier risk and diversification strategies.

Today, Toyota uses AI-powered supply-chain control towers to monitor global supply risks and simulate disruptions. For instance, if AI models predict a potential shortage in a critical semiconductor component, Toyota's system evaluates the effect across production lines and recommends switching to a multi-sourcing model, even if temporarily, by identifying alternative qualified suppliers and estimating cost trade-offs, lead-time changes, and risk exposure. This AI-driven flexibility has improved Toyota's resilience while maintaining cost efficiency.

7.6 Global versus Domestic Sourcing

Another challenge to consider is whether to use domestic or global sourcing, also called offshoring. Global sourcing rapidly grew as companies were attracted to cheaper labor costs in other parts of the world. It has been most prominent for sourcing products with easily defined standards, such as in the retail industry. It is also used heavily in the service industry, such as running call centers, processing claim forms, or in software development. Even in medicine, reading of diagnostic tests, such as X-rays, has been offshored.

However, this has been challenged, given the rise in global tariffs and political tensions, uncertainty of fuel prices, and high labor and transportation costs. Recent years have seen a trend of nearshoring or back sourcing. A number of companies are turning toward domestic sources to reduce transportation costs, monitor quality more closely, and have closer buyer-supplier relationships. Also, with an emphasis on sustainability and decarbonization, there has been a push toward local and regional sourcing.

Deciding between global and domestic sourcing is a high-stakes decision that affects cost structure, supply reliability, lead times, and regulatory compliance. AI empowers companies to make these decisions more effectively, by analyzing real-time data and simulating future outcomes under different sourcing strategies. Rather than relying solely on historical data or static-cost comparisons, AI enables dynamic, adaptive decision making based on multiple interdependent variables.

AI systems can evaluate the total landed cost of sourcing globally versus domestically, not just product cost but also tariffs, currency volatility, transportation lead times, carbon footprint, and risk exposure. Machine-learning algorithms detect patterns in demand volatility, logistics bottlenecks, and supplier reliability that can tilt the balance in favor of domestic or global sourcing. Natural language processing (NLP) engines scan news, policy changes, trade agreements, and geopolitical developments to anticipate shifts that could affect global trade—such as new export bans or port strikes—and feed those signals into decision-making dashboards.

Beyond decision support, AI can also manage the sourcing execution process by coordinating supplier onboarding, dynamically rerouting orders, and recommending reallocation of procurement volumes in response to disruptions. It can simulate scenarios where domestic sourcing may mitigate disruption risk but at a higher cost, and help businesses quantify whether that trade-off is justified based on current and predicted conditions. In the case of dual sourcing strategies (both global and domestic), AI optimizes allocation by continuously evaluating supplier performance, market dynamics, and transportation conditions.

GM leverages AI-driven platforms to assess and balance its sourcing footprint between domestic and international suppliers. During the COVID-19 pandemic and subsequent semiconductor shortages, GM faced extensive delays tied to overseas suppliers. In response, GM began using AI-based analytics tools to evaluate critical components across its global supply base and compare the financial and operational trade-offs of relocating certain supply lines to North America.

AI-driven models at GM assessed not only cost and delivery time but also modeled risks such as shutdown probabilities, customs delays, and policy shifts like the US CHIPS Act. The insights helped GM shift some high-risk component sourcing from Asia to US-based and Mexico-based suppliers for improved reliability and government incentive alignment. This strategic shift, guided by AI forecasts and scenario planning, helped GM stabilize supply for its electric vehicle program and mitigate future risks.

7.7 Outsourcing

7.7.1 Why Outsource?

Outsourcing is a practice where companies use external providers to carry out business processes that would otherwise be handled internally. It has become a mega trend as companies focus on their core competencies and shed tasks perceived as noncore. Companies may choose to outsource activities or tasks for many reasons, rather than performing them internally. These include lower costs, access to technical skills, and the ability to free themselves of doing noncore activities. Outsourcing enables companies to tap into specialized skills and capabilities that they currently do not have.

A good example has been the outsourcing growth of transportation of goods to third-party logistics providers (3PLs), such as UPS and FedEx. Transportation and movement of goods requires investment of specialized resources, such as fleets of trucks, aircraft, and state-of-the-art information systems for product tracking. Outsourcing this activity to specialty companies is a much more cost-effective decision for most firms. Good outsourcing decisions can result in lowered costs and a competitive advantage, whereas poorly made decisions can increase costs, disrupt service, and even lead to business failure. Although financial aspects of outsourcing are important, outsourcing has increasingly taken on a broader strategic organizational focus.

It is for these same reasons that we are now seeing the rise of outsourcing in AI and analytics services. We are seeing the rise of the third-party analytics providers (3PAs) as more companies recognize the symbiotic benefits that can arise from working in concert with an external provider. Technological capability is so highly specialized and growing exponentially that companies simply cannot keep up. All indications point toward the continuation of this trend in the foreseeable future.

Google exemplifies the strategic use of outsourcing to enhance its innovation capacity and maintain a competitive edge in the fast-paced technology sector. Rather than viewing outsourcing solely as a cost-reduction tactic, Google leverages it as a key enabler of organizational agility, technical excellence, and global scalability.

Google strategically outsources various business functions, particularly in software development, customer support, content moderation, and IT infrastructure, to access specialized skills and technologies not available in-house. This enables the company to integrate external innovation, accelerate development timelines, and focus internal resources on core competencies such as AI research and advertising technologies. For example, Google contracts with firms in Eastern Europe and India to support parts of its software engineering and testing operations, helping scale product development across markets.

By outsourcing noncore functions, Google ensures its internal teams are concentrated on high-impact areas such as product innovation, machine learning, and cloud services. This targeted allocation of resources has supported the launch of cutting-edge products like Bard, Google Cloud AI, and Tensor Processing Units. Outsourcing also helps the company meet fluctuating operational demands without expanding its permanent workforce unnecessarily.

Outsourcing extends beyond software development to include third-party analytics services, which support Google in understanding global user behavior, forecasting trends, and optimizing product experiences. These insights help Google personalize services such as YouTube, Google Ads, and Maps at scale. According to a recent study, companies like Google that outsource data analytics functions see up to a 25 percent improvement in time-to-insight and a 20 to 30 percent reduction in analytics operating costs.

Google's outsourcing strategy reflects a broader organizational philosophy: *innovation through collaboration.* By engaging with specialized partners, Google enhances its capabilities in areas like cybersecurity, cloud operations, and multilingual customer support. These partnerships allow it to remain flexible in responding to market shifts and emerging technological challenges.

Google's approach demonstrates how outsourcing can be used strategically not just to cut costs but also to gain competitive advantages. By outsourcing noncore functions and partnering with expert providers, companies like Google can reallocate resources to core innovation areas, adapt quickly to changing markets, and enhance long-term resilience. As the digital economy continues to expand, strategic outsourcing decisions will be integral to sustaining innovation and operational excellence across industries.

7.7.2 Outsourcing AI Capability

Analytics has not traditionally been a core competency for most organizations. As businesses increasingly rely on data-driven decision making, many are working urgently to acquire advanced analytical capabilities. Hiring analytical talent is a start, but it's rarely sufficient. Building deep analytical expertise requires highly specialized professionals, individuals who can not only develop sophisticated models but also translate insights into actionable strategies. Equally important are employees who possess hybrid skills: those who can perform analysis and effectively communicate with operations, marketing, and frontline teams. This blend of technical and communication proficiency is in short supply.

A widespread challenge persists across industries: a shortage of talent capable of handling the full spectrum of analytical work. It's not just about mastering one technique; today's environment demands fluency in integrating multiple analytical approaches, from predictive modeling to machine learning and optimization. But even with the best tools and data, companies cannot fully realize the value of analytics without people who know how to use them strategically.

As a result, many companies turn to external analytics providers to close the talent and capability gap. These partners often operate at the leading edge of technology and bring domain-specific expertise, access to proprietary data sets, and advanced software tools. Increasingly, providers such as software vendors, data aggregators, and consulting firms offer end-to-end analytical services, spanning everything from strategic advisory roles to model development and implementation. They enable companies to stay competitive without having to build full-scale analytics capabilities in-house.

An example is Target's collaboration with Palantir Technologies. To improve supply-chain visibility and enhance demand-forecasting accuracy, Target partnered with Palantir to integrate its Foundry platform across merchandising and logistics. While Target has internal data science teams, the retailer relied on Palantir's platform and analytical expertise to build a centralized data ecosystem and apply advanced analytics at scale. This hybrid model of in-house talent, complemented by external analytical capabilities, enabled faster insights, more agile inventory decisions, and reduced stockouts during the 2023 holiday season. The partnership underscores how even data-savvy enterprises leverage outside providers to stay ahead in a competitive landscape.

7.7.3 How Much to Outsource?

Outsourcing provides tremendous benefits, particularly enabling an organization to tap into highly specialized skills it currently does not have. However, outsourcing has inherent risks. Outsourcing to an outside vendor can create a range of risks and dependencies, not the least of which is the leaking of proprietary information. Companies need to give a great deal of thought to the extent of their outsourcing engagement, considering both immediate and long-term benefits. Outsourcing is not a simple "make-buy" decision. It needs to be a strategic decision, carefully considering the skills the company will keep internally versus those it will access externally.

Two key dimensions that help define outsourcing are the **scope** and **criticality** of the outsourced task. **Scope** is the degree of responsibility assigned to the supplier. **Criticality**, on the other hand, is the importance of the outsourced activities or tasks to the organization. The greater the scope of the outsourced task, the larger the relinquishing of control by the buyer. Similarly, the greater the criticality of the outsourced task, the greater the consequences of poor performance to the buying firm and greater the requirement for supplier management. It is important to understand these dimensions when evaluating the outsourcing decision.

Outsourcing can involve just one aspect of analytics. For example, the outsourcing engagement can be small in scope, such as purchasing data or conducting analysis on one data set. It may mean supplementing current staff with either onshore or offshore analytical consultants. On the other hand, it may mean outsourcing most of the company's analytics capabilities. Companies should be especially careful when outsourcing activities with large scope and criticality and should especially monitor such outsourcing engagements to ensure a successful outcome.

The outsourcing trend will continue to increase as AI technology continues to evolve, requiring ever specialized skills, which makes a well-formulated AI outsourcing strategy necessary. This strategy must clearly delineate which AI and analytical capabilities the company wants to build for itself and which it will outsource from partners. It must also specify the long-term plan of building capabilities over time, in addition to meeting short-term goals (Figure 7.4).

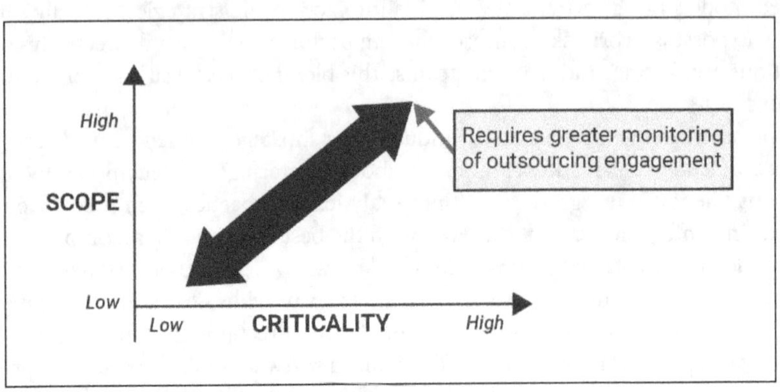

Figure 7.4 Dimensions of Outsourcing

As AI technologies continue to evolve, outsourcing will play an increasingly vital role in providing companies with access to ever specialized skills. A well-formulated outsourcing strategy is essential for long-term success. This strategy should delineate the analytical capabilities a company intends to build internally versus those it will source externally. It should also account for both short-term goals and a long-term plan to develop in-house capabilities over time.

7.8 Risk Management

AI and analytics have a huge application in risk management. They are critical tools to analyze the risk of suppliers, monitor security risks, and manage supply-chain disruptions.

7.8.1 Analyzing Supplier Risk

The success of modern supply-chain operations depends heavily on a resilient and transparent supplier network that includes external vendors, contract manufacturers, and sub-tier suppliers. As global supply chains become more complex and interdependent, evaluating supplier performance and risk routinely is no longer optional; it is essential. Yet, supplier risk analytics remain underdeveloped relative to other supply-chain capabilities. Many companies still rely on static scorecards or basic metrics that fail to capture the dynamic nature of supply-side risks. AI and analytics offer a powerful opportunity to elevate supplier risk management and bring much-needed predictive intelligence to this critical function.

One prominent application is the creation of **dynamic supplier resiliency scorecards**. These scorecards go beyond simple performance tracking by integrating variables such as geographic exposure to natural disasters, geopolitical risk indices, operational capacity buffers, and dependency ratios. By identifying weak links and high-risk nodes in the supply network, companies can take proactive measures such as multi-sourcing, supplier diversification, or jointly developed contingency plans. For example, Samsung Electronics uses a real-time resiliency scorecard modeled after Cisco's to continuously evaluate risks across suppliers, manufacturing sites, and even tooling equipment vendors, all in real time.

AI technologies also allow firms to assess supplier behavior in response to past disruptions. By analyzing third-party data and historical performance, companies can better predict how a supplier may respond to new crises, such as a pandemic, port delay, or raw material shortage. This analysis improves partner selection, contract structuring, and ongoing collaboration.

Also, companies are now incorporating social media and external data feeds into their supplier monitoring systems. These platforms serve as early-warning systems for issues that may not yet appear in formal audits or reports. Signals from workers, activists, or local news outlets can be extracted using NLP to uncover problems at the supplier, or the supplier's supplier, level.

Apple Inc. offers a current, high-profile example of analytics-driven supplier risk management. In 2023, Apple expanded its use of machine learning and geospatial analytics to monitor environmental, labor, and geopolitical risks across its global supply base. The company integrated data from suppliers, third-party monitoring groups, customs records, and social media into a centralized supply-chain risk management platform. Through predictive analytics, Apple flagged suppliers operating near high-risk zones in Southeast Asia prone to climate events and rerouted production before floods caused disruption. In parallel, Apple used AI-driven text analysis to detect potential human rights issues based on social media chatter from factory regions in India and Vietnam. These insights triggered proactive audits and remediation efforts, demonstrating how analytics can serve not only operational resilience but also ethical responsibility.

7.8.2 Security

As supply chains become increasingly data-driven and digitally integrated, they are more capable than ever of managing end-to-end operations with speed and precision. However, this heightened connectivity also introduces new vulnerabilities. Modern supply chains face an expanding array of security threats, from cyberattacks and data breaches to cargo theft, terrorism, and piracy. As a result, supply-chain security has emerged as a strategic priority, encompassing both physical goods and digital infrastructure.

Security concerns today extend far beyond traditional risks like pilferage. They now include advanced persistent threats, ransomware attacks on logistics networks, insider threats, and geopolitical risks that can compromise cargo and data integrity. The financial effect is substantial: in the United States alone, supply-chain related security breaches cost tens of billions of dollars annually.

Leaders are responding with cutting-edge AI-driven technologies. FedEx, for instance, developed SenseAware, a secure, sensor-enabled platform that offers real-time tracking, temperature monitoring, light-exposure detection, and tamper alerts for sensitive shipments. This platform was originally built for healthcare logistics. However, it has since expanded across industries requiring precise and secure delivery, leveraging data from IoT devices and cloud-based platforms to generate actionable insights and prevent loss or damage.

Another example is Maersk, which faces elevated risks due to the scale and geographic reach of its operations. To combat piracy, terrorism, and cybercrime, Maersk has adopted an integrated approach. *First,* it uses predictive analytics to monitor shipping routes and detect unusual activity. *Second*, it uses blockchain technology, enhancing end-to-end visibility and securing digital documentation against tampering. *Third*, collaborative security partnerships with international agencies to ensure compliance and safety in high-risk zones. *Fourth*, it uses cyber resilience measures, implemented after a 2017 ransomware attack disrupted global operations, including enhanced firewalls, endpoint protection, and employee cybersecurity training.

These examples illustrate how technologies such as IoT sensors, predictive analytics, and blockchain are reshaping supply-chain security. They enable real-time surveillance, automated threat detection, and faster response, ultimately enhancing both resilience and transparency.

7.8.3 Supply-chain Disruptions

Supply-chain disruptions are now a constant threat in today's globalized and increasingly volatile environment. Climate change, natural disasters, geopolitical instability, cyberattacks, and pandemics can instantly derail operations, interrupting access to raw materials, stalling production, and delaying delivery to customers. These events reveal underlying fragilities, such as overly long supply lines, dependency on outsourced production, and Lean operations that eliminate inventory buffers. While such practices are efficient under normal conditions, they greatly increase vulnerability during times of stress.

A recent McKinsey Global Institute report reaffirms that globalization has extended supply chains, increasing exposure to disruptions across geographies. Outsourcing, while cost-effective, transfers operational risks to external partners. Lean methodologies, once celebrated for cost cutting, often leave little room to absorb shocks. Compounding these vulnerabilities is the interconnected nature of modern supply chains, where a disruption in one region can ripple across multiple tiers, affecting suppliers, manufacturers, and end customers.

Nestlé provides an example of how environmental challenges can threaten supply continuity. In 2023, the company reported rising volatility in the supply of coffee and cocoa due to prolonged droughts in South America and West Africa, two key sourcing regions. With increasing water scarcity and erratic weather patterns, Nestlé accelerated its investment in regenerative agriculture and supplier education programs to build climate resilience into its supply chain. The company also deployed satellite monitoring and soil moisture analytics to support farmer partners in adapting to climate variability. By addressing environmental risk proactively, Nestlé is working to secure long-term ingredient availability while meeting sustainability goals.

Toyota, long viewed as a model of Lean efficiency, faced substantial disruption during the global semiconductor shortage between 2021 and 2023. Initially less affected than competitors due to its inventory buffers, Toyota eventually had to slash global vehicle output as chip supply constraints worsened. The company's reliance on just-in-time inventory practices and a limited pool of chip suppliers highlighted operational rigidity. In response, Toyota restructured its supply strategy by increasing chip inventory, forging long-term agreements with semiconductor manufacturers, and expanding in-house chip design capabilities for electric vehicles (EVs). This shift marked a strategic move toward supplier diversification and internal capability building.

Tesla's response to COVID-19 disruptions exemplified supply-chain agility. During peak lockdown periods, Tesla encountered delays in battery components and electronic parts, yet it adapted faster than many peers. The company rewrote its software to accommodate alternative microcontrollers when standard chips were unavailable. Tesla also reengineered parts of its production process in near real time and used AI-driven forecasting models to anticipate bottlenecks. By integrating vertical supply capabilities (e.g., acquiring mining rights for lithium) and building localized gigafactories, Tesla reduced geographic risk and increased control over critical materials. These measures helped the company maintain growth while others slowed production.

The experiences of Nestlé, Toyota, and Tesla underscore several critical strategies for mitigating the effect of supply-chain disruptions:

Predictive Analytics & Scenario Planning. Leading firms are deploying advanced analytics to identify weak links, simulate disruption scenarios, and develop contingency plans. Tools such as AI-powered supplier scorecards and climate risk models allow organizations to anticipate issues before they escalate.

Redundancy and Supplier Diversification. Lean models without flexibility increase fragility. Building dual sourcing strategies, nearshoring key components, and holding strategic inventory reserves are essential risk-mitigation tactics.

Operational Agility. The ability to adapt operations quickly, whether by switching inputs, retooling production lines, or realigning delivery networks, ensures continuity amid unexpected change.

Technology & Digital Transformation. AI, digital twins, blockchain, and cloud-based visibility platforms enhance transparency, traceability, and responsiveness. These technologies reduce decision latency and improve collaboration across partners.

In a world where supply-chain disruptions are not a question of "if" but "when" resilience must be embedded into strategy. Predictive intelligence, diversified supplier networks, and technological integration are no longer optional. They are competitive necessities. Companies that learn from disruption, like Tesla, Nestlé, and Toyota, are transforming their supply chains into agile, forward-looking ecosystems prepared for the next shockwave.

Chapter Review

Key Points

1. Purchasing, or sourcing, is about more than just buying, it is a critical strategic function. Purchasing directly affects cost structure, quality, innovation, and speed to market, making it central to competitive advantage. Strategic sourcing involves aligning supplier capabilities with the company's long-term goals, not just negotiating the lowest price. It includes supplier relationship management, capacity planning, ethical compliance, and innovation partnerships.
2. AI is a critical enabler of purchasing. AI enhances visibility, speed, and accuracy in sourcing decisions by analyzing vast data sets on suppliers, markets, and costs. AI automates contract review and compliance monitoring and predicts demand shifts, supplier performance, and risks, allowing proactive sourcing strategies. AI-driven platforms streamline the procurement life cycle from supplier discovery to performance tracking.
3. AI has a direct effect on specific purchasing tasks, from order processing to co-creation.
4. Trade-offs exist between single and multiple sourcing. AI can help quantify these trade-offs by modeling supply risk, lead times, and cost scenarios.
5. There are trade-offs in global versus domestic sourcing. AI enhances global sourcing by tracking regulatory changes, optimizing logistics, and modeling total landed cost in real time.
6. Outsourcing allows companies to focus on core competencies but introduces dependency on external partners. Supplier risk includes operational risk, financial health, geopolitical instability, and ESG violations. AI tools support supplier risk management.

Case Study

IBM's Predictive Supply Chain

Background

IBM, one of the world's leading technology and consulting firms, has long been recognized for its sophisticated, analytics-driven supply chain. Over the past decade, the company transformed from managing fragmented, regional procurement operations into operating a single, integrated, global supply chain. As of 2024, IBM manages more than $25 billion in annual procurement spending across more than 17,000 suppliers in 100-plus countries. The backbone of this transformation is advanced analytics, specifically predictive and prescriptive technologies that help IBM control costs, mitigate risks, and model potential disruptions in real time.

The Challenge

In 2010, the eruption of Iceland's Eyjafjallajökull volcano brought air travel across Europe to a standstill, threatening the continuity of global supply chains, including IBM's. While many companies scrambled to assess the immediate damage, IBM turned to its analytics software to look ahead. Surprisingly, the software flagged Hong Kong, not Europe, as the critical chokepoint. This raised eyebrows among executives, but the prediction proved prescient.

The Response

IBM's predictive analytics had detected a looming risk: once European airspace reopened, a surge of backlogged shipments from Asia, especially from key manufacturing hubs in China, Taiwan, and Malaysia, would flood into Hong Kong. Without early action, IBM would face major delays in moving components to European customers. Acting on this insight, IBM preemptively secured cargo space on both commercial and charter aircraft out of Hong Kong. This strategic move ensured minimal disruption and preserved service levels during a period of intense volatility.

The Outcome

Thanks to prescriptive analytics, IBM avoided costly bottlenecks, maintained supplier performance, and kept customer commitments intact. This event led to a permanent shift in how the company prepares for future disruptions. Today, IBM's supply-chain team uses AI-powered scenario modeling to simulate thousands of disaster scenarios annually, from geopolitical instability to extreme weather. These simulations inform contingency plans and allow IBM to respond proactively rather than reactively.

Strategic Insight

By integrating analytics deeply into its operations, IBM has evolved its supply chain into a cognitive system that not only reacts to global events but also learns from and adapts to them before they strike.

Questions

1. What role did predictive analytics play in IBM's ability to respond effectively to the Iceland volcano crisis? How does this differ from traditional risk management approaches?
2. Why was Hong Kong identified as the critical node in IBM's supply chain during the 2010 disruption? What does this suggest about the importance of analyzing second-order effects in global logistics?
3. How can IBM's approach to prescriptive analytics be applied to other types of supply-chain disruptions, such as cyberattacks or geopolitical conflict?
4. IBM moved from a regionally managed procurement system to an integrated global supply chain. What are the risks and rewards of this centralization strategy?
5. How does AI-powered scenario modeling improve IBM's long-term supply-chain resilience compared with static contingency planning? What are the challenges of maintaining such a system?
6. What organizational changes might be necessary for a traditional manufacturing company to emulate IBM's analytics-driven supply-chain model?
7. Discuss how IBM's supply-chain capabilities align with its broader corporate strategy. In what ways does supply-chain performance influence IBM's competitive advantage in technology and services?

Chapter 8

The AI Roadmap

In This Chapter You Will Learn

1. why an AI roadmap is essential,
2. perils of implementation without a roadmap,
3. steps in the AI roadmap called the GNGN framework
4. the importance of measurement tools to evaluate performance, and
5. details of AI assessment readiness and performance metrics.

Imagine that you are planning a road trip from New York to California.

You begin by defining your destination and selecting your route. You choose your stops, such as key cities or landmarks, based on your time and resources. Then you prepare your vehicle and supplies, making sure your car is ready for the journey. You even bring along people who can help navigate or drive. As you go along, you monitor progress and adjust the plan.

This is analogous to the AI journey of a business. It begins by defining AI goals (e.g., reduce costs, improve customer experience), prioritizing AI use cases, and setting up the data infrastructure and AI tools. On a road trip you may assess the weather, traffic, and road conditions. Similarly, a business assesses internal readiness and market conditions for AI adoption. You might then change plans based on real-time information. Like upskilling employees or partnering with AI experts, you make sure your team has the right skills to keep moving forward. Then revisiting your AI roadmap, you check if you're on track, adjust your timeline, or even explore better routes as new tools or data emerge.

Every company must follow a roadmap in implementing AI. Domino's Pizza is a great example. In the late 2000s they faced declining sales and poor brand perception. They began by acknowledging its product shortcomings and launched a revamped pizza recipe in 2010. This laid the groundwork for deeper changes. Domino's then shifted its focus to technology, building a robust digital ordering platform, launching mobile apps, and integrating with smart devices like voice assistants and smartwatches. The company also introduced real-time tracking for orders, vastly improving the customer experience.

Behind the scenes, Domino's invested in data analytics, AI, and machine learning to optimize operations, personalize marketing, and forecast demand. It built an in-house tech team and even experimented with autonomous delivery vehicles. Its transformation was not an overnight success but a detailed roadmap. It not only turned around its brand but also drove massive growth.

An AI roadmap in business is analogous to a roadmap in everyday life, both serve as structured guides that help you get from where you are now to where you want to be.

8.1 The AI Roadmap

8.1.1. The Need for a Roadmap

What is an AI roadmap?

An AI roadmap is a strategic plan that outlines how an organization will adopt, implement, and scale artificial intelligence technologies over time. It acts as a blueprint that aligns AI initiatives with overall business goals, helping a company move from exploration to execution in a structured and prioritized way.

Whether it is a cross-country trip or an enterprise AI strategy, a roadmap helps avoid aimless wandering and costly mistakes, by guiding choices in a coherent and stepwise fashion. Both types of roadmaps provide direction, milestones, resources, and contingency planning (what happens if something goes wrong?).

As companies increasingly adopt AI to enhance operations, decision making, and customer engagement, having a clear and structured AI roadmap has become essential. An AI roadmap serves as a strategic guide that aligns AI initiatives with business goals, prioritizes use cases, and provides a phased approach to adoption. Without a roadmap, organizations risk fragmented efforts, wasted resources, and failure to realize the true value of AI investments.

An effective AI roadmap ensures that implementation is not driven by hype or isolated experiments but instead by clearly defined objectives and measurable outcomes. It helps companies assess their data readiness, talent needs, infrastructure requirements, and ethical considerations. Just as importantly, it encourages cross-functional collaboration between technical teams and business units, ensuring AI solutions are relevant and scalable. By following a roadmap, organizations can start with high-impact, low-risk projects to build momentum and trust before scaling to more complex applications.

As organizations rush to explore the possibilities of AI, many fall into the trap of focusing on quick wins or isolated pilots without a clear plan. But deploying AI without a strategic framework is like setting off on a journey without a map; initial progress may be exciting, but the risk of misdirection, waste, or failure grows with every step. Here are some specific benefits an AI roadmap brings:

Strategic Alignment. The first reason an AI roadmap matters is alignment. Many companies struggle with fragmented AI efforts—different departments running separate experiments with no coordination or strategic intent. A roadmap ensures that all AI use cases are tied to defined business goals, whether that is increasing revenue, reducing costs, improving customer experience, or enabling new business models. It forces leadership to ask: *Why are we doing this, and what outcome are we targeting?*

Prioritization and Focus. AI opportunities are everywhere, but not all are equally valuable. A roadmap helps businesses prioritize high-impact use cases based on factors like data readiness, technical feasibility, ROI potential, and risk level. Instead of being seduced by the hype, organizations can focus their energy and investment on a carefully chosen portfolio of initiatives, balancing quick wins with long-term innovation.

Risk Mitigation and Governance. AI introduces unique risks: bias, privacy concerns, regulatory challenges, and ethical dilemmas. A roadmap provides the structure to embed governance and oversight into the entire life cycle, from use case selection to post-deployment monitoring. It lays out who is responsible, what principles guide the development process, and how performance and fairness are measured. Without this, organizations face the risk of reputational damage, legal exposure, and employee or customer mistrust.

Resource Planning and Capability Building. AI is resource intensive. It demands skilled talent, reliable infrastructure, clean data, and cross-functional collaboration. A roadmap helps organizations forecast what resources will be needed, when, and why. It guides investments in cloud platforms, data integration, training programs, and hiring. Critically, it helps leaders plan for capability gaps before they become roadblocks.

Scaling. Many AI initiatives fail to scale, not because the pilot did not work, but because the organization lacked a plan for moving from prototype to production. A roadmap outlines how to build repeatable processes, deploy solutions across units or geographies, and evolve the AI platform over time. It transforms AI from isolated innovation into a core enterprise capability.

Cultural Readiness and Change Management. Perhaps the most overlooked benefit of an AI roadmap is that it prepares people for the change. It brings clarity, builds trust, and sets expectations for how AI will change workflows, roles, and decision making. By involving business leaders, frontline employees, and data professionals early in the process, a roadmap promotes adoption and reduces resistance. It becomes not just a technical plan, but a narrative for change.

An AI roadmap is more than a plan. It is a strategic foundation that guides how AI will be used to create value, mitigate risk, and build long-term competitive advantage. In an environment where AI adoption is accelerating across industries, companies without a roadmap may move fast, but companies with a roadmap move smart. The difference is not just in velocity but in direction, resilience, and results.

Let's look at the importance of a roadmap by expanding on the earlier Domino's example.

COMPANY EXAMPLE: Domino's Pizza

Domino's Pizza, once viewed primarily as a traditional fast-food chain, underwent a dramatic digital transformation that turned it into a technology-driven company. By embracing digital tools and customer data, Domino's redefined the pizza delivery experience and outpaced many of its competitors in both sales and customer engagement.

Steps in Domino's Digital Transformation Journey:

1. **Identifying the Problem (2008–2009):** Domino's faced declining sales and poor brand reputation. Customer feedback revealed dissatisfaction with product quality and service. Executives realized that improving product alone was not enough; they needed a total reinvention.
2. **Rebuilding the Product and Brand:** They first addressed the core issue: pizza quality. Domino's launched a bold marketing campaign acknowledging its shortcomings and introduced a completely revamped recipe in 2010. This regained customer trust and served as a foundation for digital efforts.
3. **Building a Digital Foundation:** Domino's invested heavily in technology. They developed a proprietary online ordering platform, launched mobile apps across iOS and Android and even enabled ordering via smartwatches, smart TVs, and even voice assistants like Amazon Alexa.
4. **Integrating Data and Analytics:** Domino's began tracking customer preferences, order habits, and delivery times. They used this data to personalize marketing campaigns, optimize delivery routes and predict inventory needs.
5. **Automating Operations:** They introduced the Domino's Tracker, giving customers real-time updates on their orders. Behind the scenes, Domino's integrated AI and machine learning to improve demand forecasting, automate scheduling and inventory, and enhance location-based delivery precision.
6. **Innovating Through Digital-First Culture:** Domino's rebranded itself as a tech company that sells pizza. They created an in-house tech team (more than 400 developers), invested in self-driving delivery vehicles and robotic pilots. They even launched Domino's AnyWare ordering, allowing customers to order from more than a dozen platforms.

The final outcome is that 70 percent of sales now come from digital channels in key markets. Domino's stock increased more than 30x between 2010 and 2020. The company regularly ranks as a top innovator in the restaurant industry. By 2024, Domino's was delivering more than 3 million pizzas daily across the globe.

Domino's digital transformation succeeded because it began with a clear understanding of customer dissatisfaction, rebuilt trust through product improvement, and then layered digital capabilities across every aspect of the business, from ordering to delivery logistics. It is now a case study in how traditional businesses can thrive by committing to technology and data-driven decision making.

QUESTIONS

1. How did Domino's decision to publicly acknowledge its product flaws contribute to the success of its digital transformation? What can other companies learn about transparency and customer trust from this approach?
2. In what ways did Domino's use of customer data and AI enhance both operational efficiency and customer experience? How does this data-driven strategy compare to traditional restaurant operations?
3. Domino's now considers itself a tech company that sells pizza. What are the implications of this identity shift for internal culture, talent acquisition, and innovation? Should more traditional companies follow this model?
4. How did Domino's integrate digital tools across its entire customer journey, and why was this holistic approach critical to its success? What might have happened if they had digitized only one part of the business (e.g., ordering)?
5. What risks did Domino's face in becoming an early adopter of emerging technologies, like self-driving delivery vehicles and voice-ordering platforms? How can companies balance bold innovation with operational reliability and cost control?

8.1.2 No Map, No Success

The promise of AI is alluring. Businesses are eager to adopt AI tools to gain a technological edge, streamline operations, and unlock new sources of value. However, when companies purchase AI technologies without a clear implementation plan or strategic roadmap, the results are often underwhelming, or even disastrous.

The first, and most immediate risk, is misalignment between technology and business objectives. Without a roadmap, AI investments tend to be reactive, driven by hype rather than strategic necessity. This often leads to deploying tools that solve peripheral problems while core business challenges remain unaddressed. Instead of being integrated into operational workflows, AI becomes an isolated experiment, championed by a few but disconnected from enterprise priorities.

Second, internal capabilities are often overlooked. Companies may underestimate the need for data infrastructure, skilled personnel, and cross-functional collaboration to support AI initiatives. Purchasing advanced AI software without ensuring data quality, governance mechanisms, and talent readiness results in shelfware, sophisticated tools that go unused or fail to scale. Moreover, employees may resist adoption if they have not been engaged in the process, especially if the AI disrupts existing roles or routines.

Third, without a plan, companies fail to measure and capture value. AI projects without defined KPIs (key performance indicators) or success metrics often drift aimlessly. Resources are expended, but there's no clear way to evaluate impact or justify continued investment. This can erode executive and stakeholder confidence, leading to a loss of momentum or abandonment of AI initiatives altogether.

Last, the absence of a roadmap creates fragmentation and duplication. Different departments may independently acquire AI tools, leading to redundant efforts, compatibility issues, and higher costs. Rather than building enterprise-wide intelligence, the company ends up with disconnected pilots and siloed data.

In sum, AI without a roadmap is not innovation; it is improvisation at scale. To succeed, businesses must treat AI not as a plug-and-play solution but as a transformational journey. That requires clear goals, stakeholder alignment, data readiness, and phased execution. Without these, even the most promising AI investments are likely to falter.

What happens when a company tries to implement AI technologies without a plan or roadmap? One notable example of a company that struggled with AI implementation due to a lack of clear planning and strategic alignment is GE, specifically its Predix platform for industrial IoT and AI-based analytics. This case is often cited in business and technology circles as a cautionary tale of how even massive corporations can falter when they pursue AI without a coherent plan, internal alignment, and clear return on investment (ROI) goals. Let's look at what happened.

COMPANY EXAMPLE: GE

In the early 2010s GE's division GE Digital launched Predix, a cloud-based platform to bring AI and predictive analytics to industrial machines. The goal was ambitious: to create a centralized AI-powered ecosystem that could manage data from jet engines, turbines, locomotives, and more, offering predictive maintenance and efficiency optimization.

Despite billions invested, Predix failed to deliver any meaningful revenue or scalability. Customers found the platform too complex and not flexible enough. In 2018, GE announced that it would sell off parts of GE Digital and shift away from its original vision for Predix.

What went wrong?

1. **No Clear Roadmap or Scope Control.** GE lacked a focused AI roadmap. The project tried to do too much too soon—simultaneously building hardware, software, AI tools, and a new business model—without a phased approach or clear prioritization. There was no structured AI maturity model or enterprise-wide governance, leading to misalignment across business units.
2. **Technology Before Business Use Case.** GE focused on building an advanced platform before validating strong use cases with customers. As a result, adoption was low, and the business value was unclear. The platform was developed as a tech-first initiative rather than a solution to real, validated industrial problems.
3. **Lack of Internal Capability Alignment.** Internal teams were not adequately trained or aligned with the AI transformation goals. There was a disconnect between GE Digital (the software/AI arm) and traditional engineering divisions.

The key lesson is that a company needs to have a roadmap, and not technology, lead. Successful implementation means to start with the problem, not the technology. GE built the tech before truly validating product-market fit. AI needs a strategic roadmap aligned with business units, not just a separate digital initiative. Phased implementation with measurable outcomes helps ensure progress and stakeholder buy-in. Cross-functional alignment and talent development are as critical as the algorithms.

QUESTIONS

1. What role should a clear AI roadmap and phased implementation play in the successful scaling of enterprise AI platforms like Predix? How might GE's outcomes have differed with a structured maturity model or governance framework?
2. Why is it risky to adopt a technology-first approach without validated business use cases in AI implementation? Can you think of other examples where this has succeeded or failed?
3. How can companies ensure internal alignment and capability development when introducing transformative AI platforms across traditional engineering or industrial units? What leadership strategies might have bridged the gap between GE Digital and GE's legacy divisions?
4. What does the GE Predix case reveal about the importance of cross-functional collaboration and stakeholder engagement in enterprise AI projects? How could this have been better managed?
5. If you were advising GE in 2012, how would you have designed the launch strategy for Predix to improve chances of success? What would you prioritize first: technical development, customer engagement, or organizational readiness?

8.1.3 The DESIGN Framework

A typical AI roadmap includes seven key steps we call the DESIGN framework and is shown in Figure 8.1. They include the following six steps:

1. **Define.** This means defining **strategic goals and objectives**. Clarify the business problem or strategic goal. Clear articulation of what the company wants to achieve with AI. These must be aligned with overall business values and explicitly stated.
2. **Evaluate.** Next is to **assess the current state**. Analyze current state, data, and capabilities. Evaluate gaps in data, infrastructure, and skills. Identify internal skill gaps. Develop implementation timeline and governance structure.
3. **Select.** Identify and rank **use case** applications based on value, feasibility, and risk. Choose key opportunities or use cases. Focus on high ROI opportunities. Focus on high-impact use cases.
4. **Innovate.** This involves developing *both* AI and talent capability. Design AI-driven solutions and prototypes. This means either build or buy AI tools and talent. Acquire needed platforms, infrastructure, and tools for development and deployment. Plan for training, reskilling, hiring, or partnerships. Develop plans for collecting, cleaning, and managing the data needed to fuel AI models. Ensure quality, accessible and integrated data.
5. **Go-Live.** This step means deploying the AI solution into operation or implementing it. Begin with a pilot. Then, integrate phased milestones that outline short-, medium-, and long-term goals. Successful pilots are scaled across regions, departments, or products. This includes integrating AI with existing IT systems (ERP, CRM), retraining staff, and monitoring models for drift and performance over time. AI models are tested in pilot environments. Here test on small scale, refine based on feedback. Feedback is gathered, performance measured, and models refined. Guidelines for responsible AI use, including fairness, transparency, and compliance. Change management is critical here to ensure end-user adoption.
6. **Navigate.** This means track performance and adapt based on feedback. It means monitoring, measuring, and improving through an iteration process. Post-deployment, companies must monitor performance (KPIs), ethical compliance, and business impact. Continuous learning loops are essential, retraining models with new data, adjusting parameters, and updating strategies as markets evolve.

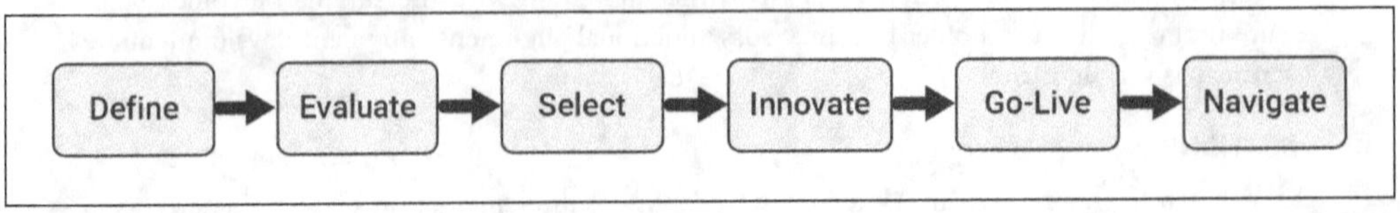

Figure 8.1 Steps in AI Roadmap: DESIGN.

Let's look at an example of Starbucks and their implementation of the DESIGN framework.

COMPANY EXAMPLE: Starbucks: DESIGN In Action

Starbucks began laying the foundation for Project Deep Brew in 2019, publicly announcing the initiative that year during earnings calls. The rollout and development of Deep Brew has been ongoing from 2019 to the present (2026), evolving in phases across digital platforms, store operations, and global markets.

D — Define (2018–2019): In response to competitive pressure from digital-first brands and growing customer expectations, Starbucks' leadership defined a key strategic challenge:

How can we personalize customer experience at scale while improving store efficiency?

This step coincided with CEO Kevin Johnson's push toward a digital flywheel strategy.

E — Explore (2019): The company explored data sources: loyalty rewards programs (90M+ customers), mobile app orders, transaction history, store performance, weather, and more.

They assessed where inefficiencies existed (e.g., long lines, out-of-stock items) and identified where AI could unlock scalable value.

S — Select (2019–2020): Starbucks selected strategic AI use cases aligned with business goals:

- Personalized in-app drink recommendations
- Predictive inventory and staffing models
- Intelligent digital signage (dynamic menus)

These choices reflected a focus on enhancing revenue per customer visit and reducing operational friction.

I — Innovate (2020–2022): Starbucks developed and refined Project Deep Brew, building proprietary AI models and integrating them with customer-facing and back-end systems.

This included:

- Deep learning models for product recommendations
- AI for labor scheduling and inventory forecasting
- Integration into POS and mobile platforms

This period also saw investment in hiring AI talent and building internal capabilities.

G — Go-live (2022 onward): Initial go-lives included AI-powered personalization in the Starbucks mobile app and pilot tests of AI-based store planning. By 2023, elements of Deep Brew were deployed across US and international markets, including personalization features and operational tools in stores.

N — Navigate (2023–2025): Starbucks continues to measure and refine performance metrics:

- Increased digital engagement and mobile order usage
- Improved store productivity and staffing accuracy
- Enhanced customer satisfaction via tailored experiences

Deep Brew is now a core capability, supporting Starbucks' broader growth strategy in the United States, China, and other key markets. Starbucks' multiyear journey (2019–2025) with Project Deep Brew illustrates a well-aligned AI strategy using the DESIGN framework. Each step, from defining strategic needs to navigating post-launch performance, shows how AI can be embedded into the core of a global business to drive sustained, measurable value.

QUESTIONS

1. How did Starbucks ensure alignment between its overall business strategy and its AI strategy through Project Deep Brew? Explain how the company's focus on customer experience and operational efficiency shaped its AI use case selection.
2. What data assets did Starbucks leverage to fuel its AI models, and what challenges might arise in managing such a large and diverse data ecosystem? How does this tie to issues of data privacy, integration, quality, and scalability?
3. Which step of the DESIGN framework do you think is the most critical for ensuring long-term success of an AI strategy? Why?
4. Starbucks chose to build its own AI system (Deep Brew) rather than rely solely on third-party solutions. What are the potential advantages and risks of this approach? What are the trade-offs between customization, cost, talent requirements, and technical debt?
5. If you were advising a smaller retail chain with fewer resources than Starbucks, how would you adapt the DESIGN framework to scale AI efforts responsibly? How different would the prioritization and strategic focus be for small and mid-sized businesses?

8.2. Steps in the Roadmap

AI offers a transformative potential for supply-chain management, along all supply-chain levers, from Purchasing to Marketing. But how do companies collectively leverage these technologies for a competitive advantage?

8.2.1 Define: Clarifying Strategic Goals

The first step in any effective AI strategy is to define clear strategic goals and objectives. This involves articulating the core business problems, opportunities, or performance gaps that AI is expected to address. Without this clarity, AI investments risk becoming unfocused experiments, detached from meaningful outcomes.

Defining strategic objectives ensures that the organization does not begin with the technology and search for a use case afterward, a common mistake in early AI adoption. Instead, this step grounds the AI effort in the broader business strategy, whether the priority is reducing costs, improving customer experience, accelerating innovation, or enabling growth in new markets.

This phase requires cross-functional input. Senior leadership, domain experts, operations managers, and data teams must come together to ask:

- What is the business trying to achieve?
- Where are the inefficiencies or growth opportunities?
- How will success be measured?
- Is the organization ready (culturally and structurally) to act on AI-driven insights?

Importantly, defining goals sets the scope and direction for the entire AI journey. It also enables the organization to prioritize which use cases to pursue and which ones to defer or discard. This focus improves alignment, resource allocation, and long-term ROI.

COMPANY EXAMPLE: Levi Strauss & Co.

In 2019, Levi Strauss began a major digital transformation initiative, but instead of starting with technology, it started with a strategic business objective:

How can we become faster and more responsive to consumer demand across our global supply chain?

Levi's defined goal was to shift from a seasonal production and planning cycle to a demand-driven, responsive model. With this in mind, the company sought to use AI to

- predict shifts in fashion trends more accurately,
- forecast demand for styles and sizes at a store-by-store level, and
- improve speed-to-market and reduce excess inventory.

This clarity guided Levi's toward specific AI solutions such as machine learning for demand forecasting and digital twins to simulate production scenarios. The company did not just implement AI because it was popular; it did so because the technology directly supported its strategic shift to agility and customer centricity.

Defining strategic goals is not a preliminary formality; it is the foundation of AI success. It aligns technological efforts with real business needs and ensures that AI investments drive measurable outcomes. Companies like Levi Strauss demonstrate how clearly defined objectives can transform AI from a buzzword into a strategic asset.

QUESTIONS

1. Why was it important for Levi Strauss to begin its AI journey with a clear business objective rather than a technology-first approach? How did this influence the success of their transformation?
2. How do machine learning and digital-twin technologies specifically support Levi's goal of becoming more demand-driven and responsive? What advantages do these tools provide over traditional forecasting methods?

3. What risks might Levi Strauss have faced if it had implemented AI without a clear strategic focus on agility and customer centricity? How can companies avoid AI for AI's sake?
4. How does Levi Strauss's case illustrate the value of aligning AI investments with measurable business outcomes? What metrics or KPIs might be used to assess success?
5. In what ways can other companies in traditional industries learn from Levi Strauss's approach to AI adoption and supply-chain transformation? What might need to be adapted for different business contexts?

8.2.2 Evaluate: Assessing the Current State

Once strategic goals are clearly defined, the next critical step in an AI roadmap is to evaluate the current state. This phase focuses on determining how prepared the organization is—technically, operationally, and culturally—to achieve the goals using AI. It involves a comprehensive assessment of existing data assets, technology infrastructure, workforce capabilities, and organizational readiness.

Evaluation starts with analyzing the available data: What data do we have? Where does it reside? Is it clean, structured, and accessible? Many organizations underestimate the challenges posed by poor data quality or siloed systems. Without usable data, even the most sophisticated AI models will fail.

Next, companies must evaluate technology infrastructure. Do we have the computing power, cloud platforms, and integration capabilities needed to support AI initiatives? Equally important is assessing internal skills, from data science and engineering to product management and change leadership. Many companies discover they lack in-house expertise in AI, requiring new hiring, training, or external partnerships.

This step also includes identifying organizational gaps and barriers, such as fragmented ownership, lack of cross-functional coordination, or resistance to adopting AI-driven insights. Establishing a clear governance structure, including executive sponsors, ethical oversight, and cross-departmental alignment, is essential to prevent future friction.

Finally, a well-scoped implementation timeline should emerge from this evaluation. Rather than aiming for one large AI deployment, organizations should identify quick wins and phase investments over time. This allows for iteration, learning, and cultural adaptation, making the transformation more sustainable.

COMPANY EXAMPLE: GE—Industrial AI and Digital-Twin Evaluation

When GE began integrating AI into its industrial operations, it started with a bold vision: to become a leader in industrial IoT and predictive analytics across its aviation, energy, and manufacturing divisions. However, during the Evaluate phase, GE discovered serious infrastructure and organizational gaps.

While GE had vast sensor data from equipment, much of it was siloed, inconsistent, or unstructured. The company lacked standardized data governance across business units. Its digital infrastructure varied widely, with some plants using outdated systems. Internal teams had limited AI and machine learning expertise, and business units often resisted centralized initiatives.

To address this, GE conducted a comprehensive audit of its data landscape and digital maturity across divisions. It created a Digital Thread Framework to standardize data flows and integrations. Further GE partnered with cloud providers and AI startups to bridge technical gaps. To this end they launched internal AI academies to upskill employees in data science and AI application. They even set up a central governance team to oversee AI use cases, ethics, and deployment timelines. This evaluation phase helped GE pivot from fragmented pilots to a unified industrial AI strategy, especially moving on from its bad experience with its Predix platform.

The Evaluate phase transforms ambition into realism. By identifying strengths, weaknesses, and readiness across data, infrastructure, and skills, organizations can design AI strategies that are both grounded and scalable. GE's case shows how a rigorous evaluation not only avoids failure but also lays the foundation for long-term success in complex, high-stakes industries. Ignoring this step leads to overpromising and underdelivering. Evaluating wisely builds confidence, capability, and clarity.

QUESTIONS

1. Why is the Evaluate phase critical before scaling AI in complex industrial environments like GE's? What risks can be avoided through a rigorous assessment of infrastructure, data, and capabilities?
2. How did GE's fragmented data systems and lack of governance hinder its initial AI efforts, and how did the Digital Thread Framework address these issues? What does this suggest about the importance of standardization?
3. What lessons can be drawn from GE's approach to upskilling internal teams and building centralized AI governance? How can these efforts help overcome resistance from business units?
4. How did GE's failure with Predix inform the improvements made during this evaluation phase? What does this reveal about the importance of learning from past digital initiatives?
5. In what ways can GE's Evaluate phase serve as a model for other legacy firms attempting to adopt AI at scale? What core elements of their approach are universally applicable?

8.2.3 Select: Prioritizing High-Value Use Cases

After defining strategic goals and evaluating current capabilities, the next step is to select the right AI use cases. This is a pivotal moment in the AI roadmap: choosing where to focus effort, investment, and experimentation. Many companies fail in their AI journey not because the technology underperforms, but because they choose the wrong problems to solve. This phase is about aligning business priorities with feasible, high-impact AI applications. Organizations should filter opportunities based on several criteria:

- **Strategic relevance:** Does the use case support a defined business goal (e.g., cost reduction, revenue growth, risk mitigation)?
- **Data availability:** Is there sufficient clean, historical, and timely data to train models?
- **Technical feasibility:** Can the solution be realistically implemented with current tools and resources?
- **Scalability:** Is the solution replicable across business units or geographies?
- **Time-to-value:** How quickly can the organization see measurable benefits?

It is also wise to prioritize a portfolio of use cases: some quick wins to build momentum, and others that are more transformative and long-term. Involving business leaders, data scientists, and operations stakeholders ensures cross-functional buy-in and prevents misalignment.

Finally, selected use cases should be translated into clear problem statements, with success metrics and expected outcomes. This sets the stage for focused development and rigorous measurement.

COMPANY EXAMPLE: Unilever

Unilever, one of the world's largest consumer goods companies, provides a strong example of disciplined AI use-case selection. After evaluating its global data infrastructure and digital maturity, Unilever faced the challenge of choosing AI projects that would advance its core business goals: improving sustainability, supply-chain agility, and customer-centric product innovation.

Rather than launching dozens of disconnected experiments, Unilever prioritized use cases in three key areas:

1. **Demand Forecasting and Inventory Optimization:** AI was applied to predict product demand by region, factoring in weather, promotions, seasonality, and market signals. This aligned with their strategic goal of reducing waste and avoiding stockouts.

2. **Sustainable Sourcing and Manufacturing:** Machine-learning models helped optimize energy and water use in manufacturing plants, contributing to Unilever's commitment to net-zero emissions and circular production.
3. **Product Innovation through Consumer Insight Mining:** NLP was used to scan reviews, social media, and call center logs to identify emerging consumer trends. This helped R&D teams quickly adapt or create products tailored to changing tastes.

Each of these use cases was selected not just for novelty, but for its alignment with strategic priorities, data readiness, and scalable impact across Unilever's vast operations.

The Select phase is where strategic intent meets operational execution. Choosing the right AI use cases—those with strategic value, technical feasibility, and organizational support—ensures that AI is applied where it can make the most difference. Unilever's approach underscores the importance of focus, discipline, and strategic alignment when selecting AI applications. In a landscape filled with hype, success begins by choosing wisely.

QUESTIONS

1. How did Unilever's focus on strategic alignment influence its choice of AI use cases, and why is this focus critical for scalable impact? What might have happened if they had pursued AI initiatives without this alignment?
2. In what ways do Unilever's AI projects in demand forecasting and sustainable manufacturing reflect its broader corporate values and business goals? How does this help reinforce brand trust and operational efficiency?
3. Why is technical feasibility and data readiness just as important as strategic value when selecting AI use cases? How should companies assess these factors during the Select phase?
4. How did Unilever's use of NLP to mine consumer insights enhance its product innovation process? What competitive advantages can this approach create in fast-moving consumer goods?
5. What lessons can other large, complex organizations learn from Unilever's disciplined approach to selecting and scaling AI use cases? How can they avoid the trap of fragmented or low-impact pilots?

8.2.4 Innovate: Designing, Building & Testing

Once high-value use cases are selected, the next step is to innovate—that is, to design, develop, and test AI-powered solutions that directly address the identified business needs. This step moves from strategy to execution and focuses on building the technical backbone and human-centered workflows that will make AI usable and valuable.

Innovation in this context is not just about technical novelty. It is about problem-solving through experimentation, rapid prototyping, and iterative learning. Organizations must design models that are not only accurate but explainable, ethical, and integrated into business processes. Innovation requires cross-functional collaboration among data scientists, engineers, product managers, operations leaders, and end users.

The Innovate phase typically includes the following activities:

- Building or sourcing AI models (e.g., machine learning, computer vision, etc.)
- Creating data pipelines to clean, structure, and feed data into models in real time or batch
- Prototyping solutions and testing them in sandbox environments or select business units
- Embedding AI into user interfaces or enterprise systems (e.g., ERP, CRM, SCM)
- Validating outcomes through pilot programs and A/B testing

Crucially, this is where companies must also design for trust, fairness, and usability. Even the most accurate model can fail if it's not accepted by users or if it introduces ethical concerns. Therefore, human-centered design, governance, and explainability must be built into the innovation process from the beginning.

COMPANY EXAMPLE: BMW

BMW offers an excellent example of innovation through AI in its automotive manufacturing operations, particularly in automated quality inspection. After defining the goal of improving production efficiency and finding that manual inspection was time-consuming and error-prone, BMW selected computer vision as a high-impact AI use case.

In the Innovate phase, BMW's data science and engineering teams designed an AI-driven visual inspection system for detecting defects in car bodies: scratches, dents, and paint imperfections.

Key steps in the innovation process included gathering and labeling thousands of images of both acceptable and defective parts. They then used deep learning (a form of ML) to train models to identify defects with high accuracy. They integrated cameras and edge devices on the production line for real-time detection. They also developed user interfaces that allowed human inspectors to review and override AI with judgments. Last, they conducted pilots in select factories before scaling across plants worldwide.

BMW's innovation, however, was not limited to technology. The team worked closely with plant workers and line supervisors to ensure the AI solution fit naturally into their workflows. This allowed for buy-in and building a supportive worker culture. This co-creation approach improved adoption and trust.

The results were significant and included

- increased inspection accuracy,
- reduced labor costs and inspection time, and
- faster feedback loops to upstream processes.

Innovation in AI is about building solutions that work—technically, operationally, and culturally. It requires creativity, iteration, and deep collaboration across disciplines. BMW's example illustrates how AI innovation can be grounded in practical needs while pushing the boundaries of what's possible in operational excellence. Successful innovation turns AI from an abstract capability into a concrete engine for business transformation.

QUESTIONS

1. How did BMW's clear problem definition guide its selection of computer vision as the right AI use case for quality inspection? Why is starting with a specific pain point essential in the Innovate phase?
2. What role did cross-functional collaboration among engineers, data scientists, and factory workers play in the success of BMW's AI solution? How does co-creation improve both adoption and long-term sustainability of AI projects?
3. How did BMW balance automation with human oversight in its AI-driven inspection system, and why is this important? What are the benefits of designing human-in-the-loop AI systems?
4. What key steps did BMW take to ensure the AI model was accurate, scalable, and integrated effectively into existing production lines? How can other manufacturers replicate this process for different use cases?
5. How does BMW's AI innovation demonstrate the difference between deploying technology and achieving true operational transformation? What distinguishes successful AI innovation from isolated tech experimentation?

8.2.5 Go-Live: Deploying AI

After designing and validating AI solutions through innovation and pilot testing, the next step is to Go-live—that is, to deploy the AI system into operational environments at scale. This stage marks the transition from experimentation to execution, where AI becomes embedded in the day-to-day workflows of the business.

Going live requires more than flipping a switch. It is a disciplined process of scaling, operationalizing, and supporting AI in production. Key elements of a successful Go-live phase include the following activities:

- **System Integration:** Embed AI models into enterprise systems such as ERP, CRM, or custom platforms.
- **User Training and Onboarding:** Ensure that employees understand how to use AI tools, interpret outputs, and trust automated decisions.
- **Process Redesign:** Adapt or streamline business processes so they are optimized for AI-enabled decision making.
- **Monitoring and Support:** Establish dashboards, alerts, and feedback loops to monitor model performance, flag anomalies, and escalate human intervention when needed.
- **Change Management:** Actively engage stakeholders across departments to manage expectations, encourage adoption, and address resistance.

Without careful planning, Go-live can trigger bottlenecks, errors, or employee pushback. However, when well managed, this step marks the moment when AI begins to deliver tangible business value: increased speed, better accuracy, improved customer service, and smarter resource allocation.

COMPANY EXAMPLE: FedEx

FedEx provides a compelling example of the Go-live process with its use of AI for predictive maintenance across its massive fleet of aircraft and delivery vehicles. After developing and piloting machine learning models that predict equipment failures based on sensor data, FedEx reached the Go-live stage. The company did not just install software—it executed a comprehensive deployment strategy:

- **System Integration:** AI models were integrated with FedEx's maintenance management system, enabling automatic flagging of at-risk vehicles or parts before a breakdown occurred.
- **Process Embedding:** Maintenance workflows were redesigned so that predictive alerts generated scheduled inspections or preemptive repairs, replacing reactive maintenance with proactive intervention.
- **Workforce Enablement:** Technicians received training to interpret AI-generated maintenance recommendations and validate them through physical inspection.
- **Governance and Monitoring:** A central control team monitored model performance in real time, adjusted thresholds for alerts, and escalated unresolved issues to human supervisors.

As a result, FedEx saw significant improvements. They included reduced downtime and improved fleet availability, lower repair costs through early intervention, and enhanced reliability in delivery timelines, supporting its brand promise. FedEx did not treat Go-live as a technical event. It treated it as a business transformation milestone, ensuring AI success through operational alignment, employee readiness, and robust support structures.

QUESTIONS

1. How did FedEx's approach to integrating AI into existing maintenance systems demonstrate that Go-live is more than just launching technology? What made this phase a true business transformation milestone?
2. What role did workforce training and enablement play in the successful deployment of predictive maintenance AI at FedEx? How can neglecting the human element undermine Go-live efforts?
3. Why is embedding AI outputs into day-to-day workflows (like maintenance scheduling) critical for realizing operational benefits? What are the risks of failing to operationalize AI insights?
4. How did FedEx's real-time monitoring and governance structure contribute to the reliability and adaptability of its AI systems post-launch? Why is ongoing model supervision essential after deployment?

5. What lessons can other logistics or asset-intensive companies learn from FedEx's Go-live process in terms of aligning AI with business impact? How can these lessons help avoid stalled or low-impact deployments?

The Go-live phase is where AI becomes real. It is not about launching software but about launching new ways of working. Effective deployment means aligning people, systems, and processes so that AI is not just accessible, but usable and trusted. FedEx's example shows how disciplined, well-integrated deployment allows AI to deliver measurable results at enterprise scale.

8.2.6 Navigate: Monitoring, Adapting & Scaling

The final step in a successful AI strategy is to Navigate—the ongoing process of measuring, managing, and evolving AI solutions after they have been deployed. This phase ensures that AI continues to deliver value as business conditions, data, and organizational priorities change. Rather than viewing implementation as the end point, Navigate treats it as the beginning of continuous learning and optimization.

The Navigate step includes the following activities:

- Monitoring model performance in real time by tracking accuracy, precision, latency, and business outcomes.
- Updating models as new data becomes available or as patterns shift (e.g., seasonality, consumer behavior, supply-chain shocks).
- Recalibrating KPIs and goals to reflect changing strategic contexts.
- Collecting feedback from users to assess usability and trust in the system.
- Identifying new opportunities to expand the AI solution or deploy new use cases across departments or geographies.

Governance is essential here. Organizations need well-defined protocols for retraining models, managing drift, and escalating exceptions. They also need mechanisms for human oversight, ensuring that AI decisions are explainable and accountable, especially in high-risk domains.

Ultimately, Navigate turns AI from a one-time initiative into a sustainable, evolving capability, what some call "AI at scale."

COMPANY EXAMPLE: Microsoft

Microsoft offers a clear example of the Navigate phase through its use of AI in customer support services. Several years ago, the company launched AI-driven virtual agents (chatbots and virtual assistants) to reduce wait times, resolve routine queries, and support global users 24/7.

After the initial Go-live, Microsoft did not walk away from the system; it actively navigated its evolution through the following activities:

- **Performance Monitoring:** Microsoft tracked customer satisfaction, resolution rates, escalation volumes, and AI response accuracy. When the AI missed intent or misunderstood a query, it triggered learning feedback loops.
- **Model Retraining and Tuning:** Based on new customer inputs and language patterns, Microsoft continuously retrained its NLP models to improve relevance and response quality.
- **Human-in-the-Loop Feedback:** Live support agents were encouraged to review AI interactions, tag problems, and contribute to system improvement.
- **Scaling Use Cases:** Once performance stabilized, Microsoft scaled the AI system across new languages, regions, and product lines, extending the benefits to other business units.
- **Governance and Ethics:** Microsoft's Responsible AI team ensured that the AI models adhered to fairness, privacy, and transparency standards as they evolved.

The Navigate process helped Microsoft's AI solution move from a narrow tool to a mature, trusted enterprise platform, improving both customer satisfaction and internal efficiency.

QUESTIONS

1. How did continuous monitoring and feedback loops help Microsoft improve the performance and reliability of its AI customer support agents? What does this say about the importance of post-launch adaptation?
2. Why is human-in-the-loop feedback essential for refining AI systems over time, especially in customer-facing applications? How does it balance automation with human judgment?
3. In what ways did Microsoft's Navigate strategy enable the scaling of AI across languages, regions, and product lines? What prerequisites must be in place to successfully scale an AI solution?
4. What role did governance and ethical oversight play in Microsoft's ongoing AI management? How can Responsible AI practices impact trust and long-term success?
5. How does Microsoft's approach to the Navigate phase illustrate the shift from deploying an AI tool to maturing it into an enterprise-wide capability? What lessons can other companies take from this to avoid stagnation post-Go-live?

Navigate is about stewardship, not maintenance. AI systems are dynamic; they interact with the world, learn from new data, and affect real people. Organizations that commit to navigating AI after deployment, position themselves for long-term strategic advantage. Microsoft's example shows how thoughtful iteration, governance, and feedback loops keep AI aligned with both business objectives and ethical responsibilities.

8.3 Implemention Across Supply Chains

Implementing AI within supply chains presents a distinct set of challenges and opportunities that set it apart from other business functions. Unlike isolated departments, supply chains span the entire enterprise and extend outward to suppliers, partners, logistics providers, and customers. This interconnectedness means that AI implementation cannot occur in silos; instead, it must be embedded across the end-to-end value network. The complexity of these linkages, ranging from procurement and production to distribution and customer service, demands a highly coordinated approach that accounts for variability, dependencies, and dynamic demand signals.

One of the core differentiators in supply-chain AI implementation is the need to work across functions and organizations, matching supply and demand, and aligning efforts. To effectively deploy AI, functions such as IT, operations, procurement, and finance must align not only on data standards but also on shared objectives and success metrics. Without this collaboration, AI initiatives risk delivering fragmented insights or suboptimal outcomes that do not scale across the chain. Let's look at these next.

8.3.1. Cross-Functional Collaboration

One of the most persistent myths about AI is that its success hinges solely on algorithms and computing power. In reality, the most successful AI initiatives are not just technical; they are organizational. They thrive at the intersection of domain expertise, data fluency, operational knowledge, and executive alignment. That's why cross-functional collaboration is not optional in an AI roadmap. It is fundamental.

As organizations move beyond pilots and begin scaling AI, they must break down silos between departments. Data scientists alone cannot build meaningful AI solutions without input from business leaders who understand customer needs, operational managers who control implementation, and legal and compliance teams who guard against unintended consequences. The roadmap must therefore embed collaboration from day one.

Why does cross-functional collaboration matter?

1. **Translating Business Problems into AI Solutions.** Business units understand the pain points—inventory inefficiencies, churn risk, and customer delays—but often do not know how to frame them as data problems. Data science teams, meanwhile, may build sophisticated models that do not solve practical problems. Cross-functional teams bridge this gap by translating business questions into well-structured AI use cases.
2. **Accelerating Adoption and Usability.** AI systems must be embedded into real workflows. Operations teams, frontline employees, and end users help design AI systems that are intuitive, trustworthy, and practically useful. Without their input, AI tools often go unused, regardless of how accurate the model is.
3. **Building Trust and Shared Ownership.** AI initiatives require trust. Business leaders want confidence in the results, legal teams want accountability, and employees want transparency. Cross-functional collaboration ensures that multiple perspectives are built into the system, increasing transparency and reducing fear or resistance.
4. **Navigating Risk and Ethics.** Collaboration between data, legal, risk, and compliance functions ensures that AI systems meet ethical standards and legal requirements. This is especially important in regulated sectors like finance, healthcare, or public services, where unintended consequences carry real-world harm.

How can companies enable effective cross-functional collaboration?

1. **Create an AI Governance Structure.** Form an AI steering committee or task force with representatives from data science, IT, operations, marketing, legal, HR, and finance. This body should guide strategic decisions, approve use cases, and coordinate initiatives across silos.
2. **Establish Common Language and Metrics.** Miscommunication often stems from language. Train business teams in AI fundamentals and data teams in business strategy to build mutual understanding. Align on shared success metrics (e.g., time saved, customer satisfaction, accuracy uplift) to focus on outcomes.
3. **Appoint Translators and Integrators.** AI translators", professionals with both technical fluency and business acumen, who can connect domain experts and data scientists. These roles are critical for aligning scope, refining models, and guiding implementation.
4. **Pilot Cross-Functional Teams on Specific Use Cases.** Start with one or two AI projects that are visible, low-risk, and strategically important. Use these as learning laboratories for how different functions can work together, then scale that structure.

COMPANY EXAMPLE: P&G

Procter & Gamble's AI success stems not only from advanced analytics but also from deep collaboration between business and technical teams. When developing AI-powered demand forecasting tools, P&G brought together

- supply-chain planners who understood production constraints,
- retail teams who contributed insights on promotions and seasonality,
- data scientists who built predictive models using historical data and external signals,
- IT teams who integrated the tools into enterprise platforms, and
- executives who ensured alignment with broader business priorities.

By co-designing the tools, P&G not only improved forecast accuracy but also enhanced trust in the system, leading to faster adoption and measurable improvements in service levels and inventory optimization.

Conclusion: AI is not just a technology; it's an organizational capability. To build that capability, companies must foster collaboration across functions, ensuring that AI initiatives are relevant, adoptable, and responsible. Without it, even the most advanced models can fail to deliver. With it, AI becomes a powerful, integrated force for business transformation.

QUESTIONS

1. Why was cross-functional collaboration essential to the success of P&G's AI-powered demand forecasting initiative? How did each team's input strengthen the final outcome?
2. How does co-designing AI tools with business users improve adoption, trust, and long-term effect? What risks exist when AI is developed in isolation from end users?
3. What role did executive alignment play in ensuring that P&G's AI initiatives supported broader business priorities? How can leadership drive cross-functional accountability in AI projects?
4. In what ways does P&G's example reinforce the idea that AI is an organizational capability, not just a technical solution? What changes must organizations make to treat AI this way?
5. How can companies replicate P&G's collaborative model to ensure that their AI initiatives are not only technically sound but also strategically relevant and operationally sustainable?

8.3.2 Supply-Chain Segmentation

Customer segmentation has long been a foundational strategy in marketing, relying on attributes such as demographics, purchasing behavior, and lifestyle preferences. This same principle is now critical in supply-chain management. By segmenting customers into groups with similar needs and expectations, companies can tailor supply-chain operations to better serve each segment. This means offering different products, through different channels, supported by distinct supply-chain structures, all aligned with the unique characteristics of each customer group.

What has changed is the scale and granularity made possible by big data. Companies no longer need to settle for broad segments; instead, they can create microsegments based on detailed, real-time data. These microsegments reveal hidden patterns, emerging trends, and behavioral outliers, enabling a more dynamic and responsive supply-chain design. With this capability, businesses can identify and act on opportunities for differentiation, cost savings, and customer satisfaction like never before.

A critical element of segmentation is defining competitive priorities for each segment. These priorities, whether cost-efficiency, speed, quality, flexibility, innovation, or service, guide the operational strategy behind each supply chain. For instance, a segment prioritizing speed may require expedited logistics and minimal handling time, while another focused on cost-efficiency may emphasize bulk purchasing and lower-cost suppliers. This alignment allows organizations to deploy distinct supply-chain models within a unified business framework, optimizing performance across diverse customer needs.

A compelling example of advanced supply-chain segmentation is Amazon. The company uses AI to continuously analyze customer behavior, preferences, and purchasing patterns. Through this analysis, Amazon dynamically segments customers and adjusts its fulfillment strategies accordingly. Prime Members, for example, are part of a high-priority segment where speed and service are paramount. For these customers, Amazon optimizes its network of fulfillment centers, uses predictive shipping, and ensures same-day or next-day delivery.

Meanwhile, customers who opt for value over speed may receive slower, consolidated shipments, often fulfilled from centralized warehouses with lower operating costs. This differentiation allows Amazon to reduce supply-chain costs while maintaining high service levels where it matters most. Additionally, Amazon's regional segmentation supports localization strategies such as stocking seasonal items or regionally preferred products in nearby fulfillment centers, further improving delivery speed and relevance.

This segmented approach has helped Amazon remain both efficient and customer centric. The company's ability to align supply-chain resources with customer expectations has contributed to industry-leading fulfillment metrics, customer satisfaction, and profitability. It exemplifies how modern supply-chain segmentation, powered by data and tailored execution, supports strategic agility and long-term growth.

In short, supply-chain segmentation enables businesses to move beyond a one-size-fits-all approach. By understanding the distinct needs of different customer segments and aligning operational models accordingly, companies can improve responsiveness, reduce costs, and enhance customer value, all while strengthening their competitive position in an increasingly data-driven world.

8.3.3 AI Alignment Across the Supply Chain

AI and advanced analytics can only deliver real value when aligned with an organization's strategic goals and supply-chain priorities. Without this alignment, even sizable investments in AI, machine learning, or predictive tools risk producing fragmented insights that fail to support meaningful business outcomes. Strategic alignment avoids siloed decision-making, ensures cross-functional coherence, and allows analytics to drive competitive advantage across the enterprise.

True alignment integrates processes across departments and external partners, from customer-facing CRM systems to internal logistics, procurement, and production planning. Leading firms are leveraging AI to dissolve these traditional boundaries, syncing demand signals with supply capabilities in real time.

Apple, for instance, exemplifies integrated alignment in action. Its tightly coordinated Sales and Operations Planning (S&OP) process uses AI-powered demand sensing, supplier analytics, and real-time sales data to align supply decisions with global market shifts. By sharing forecasts and operational plans with suppliers, Apple minimizes inventory costs, reduces lead times, and enhances responsiveness to product demand, critical for product launches and global delivery.

Unilever offers another example. The company employs AI-driven tools across its supply chain to align marketing campaigns with production planning and procurement. Its cloud-based control towers provide end-to-end visibility, enabling teams to coordinate decisions from raw materials to shelf availability. This strategic alignment not only improves service levels but also supports sustainability targets by reducing waste and emissions.

Companies like Merck and Hershey's have also advanced their alignment efforts by embedding predictive analytics into their S&OP frameworks. By harmonizing production schedules with real-time demand signals and risk forecasts, they reduce variability, improve responsiveness, and safeguard margins.

In today's interconnected environment, alignment is not a one-time exercise; it is a continuous process, reinforced by predictive analytics, cross-functional collaboration, and transparent data sharing.

8.3.4 Matching Supply & Demand

One of the most enduring challenges in supply-chain management is achieving the delicate balance between supply and demand. Overproduction leads to excess inventory and markdowns; underproduction results in stockouts and missed opportunities. AI and big data analytics are transforming how companies navigate this balance, by enabling real-time demand sensing, intelligent forecasting, and responsive supply-side decision-making.

AI-powered demand sensing uses data from diverse sources—such as point-of-sale systems, social media trends, weather forecasts, and customer behavior—to detect changes in demand patterns early. On the supply side, predictive analytics helps manufacturers and distributors adjust production plans, replenishment cycles, and inventory strategies dynamically.

Target, for example, has invested heavily in machine learning models to forecast demand at the SKU and store level. By integrating customer behavior data, historical trends, and local events,

Target's system enables automatic replenishment and optimal inventory positioning. This reduces stockouts and avoids overstocking, particularly during seasonal peaks or promotional campaigns.

In the automotive industry, Tesla exemplifies how AI can tightly align supply with customer demand. Tesla uses real-time order data and customer configurations to guide production. Its vertically integrated model, coupled with in-house software platforms, allows Tesla to synchronize manufacturing with demand signals directly from customers. This build-to-order approach minimizes unsold inventory and increases responsiveness to market shifts.

Walmart uses demand forecasting powered by AI to synchronize decisions across its vast supply-chain network. Its algorithms consider a broad range of inputs—from local weather forecasts to macroeconomic trends—to adjust procurement, distribution, and replenishment. During events like hurricanes or pandemics, Walmart can predict spikes in demand for specific products and preposition inventory in affected regions, ensuring product availability without overstocking.

Another leader that exemplifies responsive supply chains is Shein, the fast-growing fashion retailer. Shein collects real-time data on customer browsing, purchasing, and return behavior to drive ultra-fast product design and replenishment. Using AI to analyze microtrends, the company produces small batches of new items, gauges demand and scales up only for bestsellers. This demand-driven model has helped Shein achieve rapid growth with minimal inventory waste.

These examples highlight how the integration of AI enables the match between supply and demand. The benefits include the following:

- SKU-level accuracy in forecasting and replenishment
- cross-functional alignment between merchandising, procurement, and logistics
- faster reaction times to market volatility and consumer behavior
- reduced inventory costs with improved service levels

Aligning supply with demand is no longer a forecasting function alone; it is a real-time, AI-enabled orchestration of insights and actions across the entire supply chain. Companies that succeed in this synchronization are not just reducing costs; they're improving customer experience and gaining a competitive edge.

8.4. Measurement Tools

Implementing AI is a strategic endeavor that extends far beyond simply deploying algorithms or software. For AI to deliver measurable business value, organizations must establish measurement tools that assess both the conditions under which AI can succeed and the outcomes it delivers. Two critical instruments in this process are the **AI-Readiness assessment** and **AI Performance Measurement** frameworks. Together, they enable organizations to navigate the complexities of AI integration with clarity, accountability, and strategic alignment.

AI-Readiness Assessments are essential before any meaningful implementation begins. These tools help organizations evaluate their current infrastructure, data quality, governance, skill sets, culture, and leadership alignment to determine whether the necessary foundations are in place for AI to thrive. Without such an assessment, companies risk misaligning AI investments with business capabilities, leading to delays, cost overruns, or outright failure. Readiness tools often include diagnostic surveys, capability maturity models, and heatmaps to visualize organizational gaps and guide prioritization. For instance, a business may discover that while it has ample data, it lacks skilled personnel to interpret model outputs, prompting early investment in talent development.

AI Performance Measurement, on the other hand, focuses on tracking and quantifying the effectiveness of AI systems once deployed. These tools ensure that AI solutions are not only technically sound but also deliver value aligned with business goals, such as improved forecast accuracy, reduced operational costs, faster decision making, or increased customer satisfaction. KPIs might include precision and recall rates for predictive models, ROI, adoption rates among users, or impact on customer experience metrics. Ongoing measurement is vital for adjusting models, optimizing workflows, and ensuring continued relevance as business conditions evolve.

In essence, readiness assessments ensure the organization is prepared for AI, while performance measurement confirms the AI is performing as intended. The absence of either increases the risk of AI becoming a costly experiment rather than a strategic asset. Organizations that integrate these tools into their AI roadmap foster a culture of continuous improvement, accountability, and long-term success.

Let's look at these in more detail.

8.4.1 AI-Readiness Assessment

Before any organization embarks on an AI transformation, it must answer a critical question:

Are we ready?

AI implementation is not just about algorithms or infrastructure; it touches data governance, talent strategy, leadership mind-set, business processes, and cultural adaptability. Readiness assessment tools offer a structured way to evaluate an organization's current state, identify gaps, and chart a realistic path forward. They transform ambition into actionable strategy by ensuring that companies do not just pursue AI but are prepared for it.

What is an AI-readiness assessment tool?

An AI-readiness assessment tool is a diagnostic framework used to evaluate an organization's maturity across several critical dimensions that influence successful AI adoption. These tools typically assess the following:

- **Data Readiness:** availability, accessibility, quality, and integration of data assets
- **Technology Infrastructure:** cloud capabilities, computing power, data platforms, and scalability
- **Talent and Skills:** in-house AI expertise, training programs, leadership support
- **Organizational Culture:** willingness to experiment, openness to AI-driven decisions, collaboration across functions
- **Use Case Maturity:** clarity on business priorities, alignment of AI efforts with strategy
- **Governance and Ethics:** responsible AI policies, regulatory compliance, model oversight processes

These assessments may be quantitative (e.g., maturity scores) or qualitative (e.g., interviews, workshops) and can be developed in-house or sourced from consulting firms and technology providers.

Why readiness matters?

AI implementation without readiness often leads to fragmented or stalled projects, as well as poor data quality and low model performance. In addition, without the right readiness a business may experience resistance from employees or decision makers. Also, there can easily be misalignment between technical capabilities and business needs.

Readiness tools help organizations prioritize foundational investments—such as data cleanup, talent development, or infrastructure upgrades—before deploying complex AI systems. They also support risk mitigation by identifying gaps that could derail deployment.

COMPANY EXAMPLE: Shell

Shell, the global energy company, used an internal AI-readiness framework to guide its digital transformation. Before scaling AI solutions across its exploration, production, and retail operations, Shell conducted a comprehensive evaluation of

- data maturity across business units,
- cloud infrastructure compatibility,
- AI literacy among engineers and field operators,and
- governance standards for high-risk applications like predictive maintenance and safety alerts.

This readiness assessment informed Shell's roadmap in several key ways. It led to the creation of Shell.ai, a centralized center of excellence for scaling AI tools and standards. It prompted targeted hiring of data scientists and retraining of domain experts in AI and analytics.

It prioritized pilot use cases in areas with high data quality and low regulatory risk, such as equipment diagnostics.

Because Shell took time to assess and address readiness gaps, they were able to scale AI responsibly and strategically, deploying more than 100 AI projects across its portfolio, from autonomous drilling systems to energy efficiency modeling.

There are many popular readiness assessment frameworks. Some are developed internally and others by organizations or consulting firms that offer structured tools. Here are four of the most popular readiness assessment frameworks and their specific focus:

1. McKinsey's AI-Readiness Index (measures across five domains)
2. Deloitte's AI-Readiness Canvas (focus on strategy, data, ethics, and talent)
3. PwC's Responsible AI Toolkit (focus on governance and trust)
4. World Economic Forum's AI Procurement in a Box (geared toward public sector)

Of course, organizations may also develop custom readiness frameworks tailored to their industry, size, or strategic focus.

An AI-Readiness assessment is not a one-time audit; it's a strategic checkpoint that ensures an organization is equipped to pursue meaningful, scalable, and responsible AI adoption. It enables companies to make informed decisions, avoid costly missteps, and build confidence across the organization. Shell's example demonstrates that asking *Are we ready?* is not a sign of hesitation; it is the first step toward executing an AI strategy that delivers real-world results.

Let's look at an example of these tools in action.

COMPANY EXAMPLE: Readiness Tools Applications

Here are some examples of each of the AI-readiness assessment tools developed by leading organizations, and how they can be applied in practice. These examples illustrate how a sample business might use each framework to evaluate their preparedness and guide AI strategy. Consider the weaknesses identified by the tools and the actions each company took.

1. MCKINSEY'S AI-READINESS INDEX

McKinsey's framework evaluates readiness across five dimensions: strategy, talent and leadership, data and technology, adoption, and ecosystems. Each area is scored from nascent to advanced.

One company that uses this tool is Carrefour, a multinational retail chain. The company used this readiness approach to benchmark its maturity across business units. This is what the assessment tool found in terms of scores and assessment:

Strategy: High — AI is embedded in digital transformation goals.
Talent: Medium — Strong in IT but limited in data science talent.
Data & Tech: Low — Data was siloed and inconsistent across markets.
Adoption: Medium — Some successful pilots in pricing and inventory, but low scale.
Ecosystem: Medium — Active partnerships with Microsoft and startups.

Action Taken: Carrefour identified its weaknesses. As a result, they created a centralized data platform, hired a chief data officer, and prioritized AI use cases like demand forecasting and dynamic pricing in well-prepared units.

2. DELOITTE'S AI-READINESS CANVAS

Deloitte's canvas spans eight categories: vision, use cases, talent, governance, data, technology, operations, and risk. It emphasizes alignment with business value and responsible deployment.

An example of the use of this tool is in the Cleveland Clinic. They used Deloitte's canvas to evaluate readiness for applying AI in clinical decision support. This is what they found:

Vision: Clear commitment to AI in healthcare.
Use Cases: Identified diagnostics and patient flow optimization.
Talent: Strong in medical expertise; growing in data science.

Governance: Lacking ethical oversight for AI.
Data: High-quality Electronic Health Record data but concerns over privacy.
Technology: Investment in cloud and AI tools underway.

Action Taken: The clinic formed an AI ethics board, developed HIPAA-compliant data pipelines, and launched pilot projects in radiology interpretation and patient scheduling.

3. PWC'S RESPONSIBLE AI TOOLKIT

PwC's toolkit emphasizes governance, transparency, accountability, fairness, interpretability, and robustness. It is designed to ensure AI is trustworthy and compliant. An example of use of this tool is the fintech company Klarna, a payments and credit platform. They used PwC's toolkit to assess the ethical implications of AI-driven credit decisions. This is what they found:

Governance: Strong model development pipeline but no formal AI ethics oversight.
Fairness: Concern about demographic bias in credit scoring algorithms.
Transparency: Limited explainability of automated loan decisions to users.
Robustness: Good internal testing but lacking external audit.

Action Taken: Klarna built a model audit function, incorporated explainable AI tools, and began publishing algorithmic accountability reports to build consumer trust and prepare for EU regulation.

4. WORLD ECONOMIC FORUM'S (WEF) AI PROCUREMENT IN A BOX (PUBLIC SECTOR FOCUS)

This tool is designed for government agencies, and the framework provides guidelines for defining AI needs, evaluating vendors, managing risks, and ensuring accountability during public procurement. One example of where this was used is Singapore's Smart Nation Initiative, which is a government department. Singapore used this tool to guide AI procurement for its Smart Nation sensor network, which supports urban planning, traffic management, and public safety. This is what they found:

Needs Definition: Clarified objectives (e.g., congestion detection).
Vendor Assessment: Evaluated providers on accuracy, privacy, and local compliance.
Risk Management: Required vendors to provide bias testing and audit trails.
Accountability: Embedded citizen feedback loops into deployment.

Action Taken: The initiative selected vendors that met WEF-aligned criteria, implemented real-time dashboards for public oversight, and ensured data remained under sovereign control.

QUESTIONS

1. How do the different AI-Readiness frameworks (McKinsey, Deloitte, PwC, WEF) reflect the unique priorities of private versus public sector organizations? How does this relate to the differences in goals, accountability structures, and stakeholder expectations?
2. Carrefour used McKinsey's AI-Readiness Index to uncover key weaknesses in data and talent. Why is benchmarking across business units critical before scaling AI initiatives?
3. Deloitte's Canvas emphasizes responsible deployment of AI. In what ways did Cleveland Clinic's response (e.g., creating an ethics board) illustrate the importance of governance in healthcare AI?
4. PwC's Responsible AI Toolkit focuses heavily on fairness and transparency. Why are these principles especially important in industries like fintech, as demonstrated by Klarna's case?
5. What role does cross-functional collaboration play in addressing gaps uncovered by readiness assessments, such as data privacy, ethical governance, or talent shortages?
6. Singapore's use of WEF's Procurement in a Box highlights the importance of public accountability in AI projects. How might citizen feedback loops influence the design and adoption of AI systems in public policy?
7. Across all four examples, readiness tools led to targeted action plans (e.g., hiring a chief data officer, forming an ethics board, or launching audit functions). How do these follow-up actions determine the long-term success of AI strategies?
8. If you were advising a mid-size company with limited AI expertise, which of the four readiness assessment tools would you recommend they start with, and why?

Before any organization embarks on an AI transformation, it must answer the question: Are we ready? Remember that AI implementation is not just about algorithms or infrastructure; it touches data governance, talent strategy, leadership mind-set, business processes, and cultural adaptability. Readiness assessment tools offer a structured way to evaluate an organization's current state, identify gaps, and chart a realistic path forward. They transform ambition into actionable strategy by ensuring that companies do not just pursue AI but are prepared for it as well.

8.4.2 Performance Measurement Metrics

A popular business adage commonly attributed to Peter Drucker is *"If you can't measure it, you can't manage it."*

This principle is especially relevant in the implementation of AI, where the stakes for measurable impact are high. To move beyond experimentation and hype, organizations must develop clear, actionable, and strategically aligned KPIs. These metrics serve as both a compass and a feedback loop, clarifying direction, enabling continuous improvement, and ensuring that AI initiatives deliver value aligned with core business objectives.

AI has revolutionized how organizations define and use metrics. No longer limited to retrospective KPIs, looking backward, companies now use dynamic, predictive, and behavior-driven metrics that more closely reflect real-time performance.

A useful analogy comes from the film *Moneyball*, where baseball decision makers abandoned outdated metrics, like batting averages, in favor of more insightful indicators like on-base percentage, unlocking a new way to assess value. Similarly, organizations today are reevaluating what they measure to better align data with outcomes.

One example is Amazon, which uses AI-enhanced metrics to continuously optimize both its supply chain and customer experience. For instance, Amazon measures forecast accuracy at the SKU and fulfillment-center level, tying predictions directly to inventory placement and last-mile delivery performance. It uses a Delivery Promise Accuracy metric that is tightly coupled with customer satisfaction and used to fine-tune operations in near-real time. Amazon also tracks metrics related to AI model impact, such as revenue uplift from personalization algorithms and reduced stockouts from predictive inventory management. These KPIs are tied to both financial results and customer-centric objectives.

Another case is Unilever, which leverages AI across its product innovation pipeline. KPIs include time-to-market for new products, consumer sentiment analysis from social media, and sustainability impact metrics like emissions per product unit. Unilever's AI success is not evaluated by model sophistication alone but by how it accelerates innovation while aligning with broader ESG goals.

Remember that metrics must be understandable and actionable at every level of the organization. If workers and managers do not understand the metrics, they will not be used. A common pitfall is producing overly complex dashboards that overwhelm rather than inform. Simplicity and clarity are essential.

Consider this: Rather than targeting a 50 percent reduction in defect rates, a more intuitive goal might be fewer than two product complaints per 1,000 units shipped. When employees understand the goal, see how their actions influence it, and know what success looks like, they are far more likely to engage and contribute meaningfully.

Metrics serve several essential roles within the AI roadmap:

1. **Clarity and Focus.** Metrics define what success means. A machine learning model that achieves 95 percent accuracy is meaningless unless it solves a business problem, such as reducing churn or improving demand forecasts. KPIs must bridge the gap between technical performance and business impact.
2. **Accountability Across Teams.** Clearly defined KPIs foster alignment and ownership. Data scientists, business leaders, and IT teams must be measured against shared

outcomes, not siloed activities. For instance, a supply-chain AI deployment might be jointly assessed on both model precision and actual inventory turnover improvement.

3. **Learning, Adaptation, and Scaling.** In AI, failure is often part of the process. Metrics help identify where models underperform, why certain pilots stall, or where assumptions fail. These insights enable rapid iteration, refinement, and eventually, scaling of successful solutions.

Complementing performance metrics, AI-Readiness Assessments play a foundational role. These assessments evaluate whether an organization is equipped with the right data infrastructure, skills, governance, and cultural alignment to support AI adoption. Without a clear view of readiness, organizations risk launching initiatives that are misaligned, under-supported, or unscalable.

Remember that metrics and readiness assessments are twin pillars of AI implementation success. Metrics ensure that organizations measure what matters, track progress, and generate insight. Readiness assessments ensure that the environment is fertile for AI to take root and grow. Together, they move organizations from vision to execution, turning the potential of AI into performance that matters.

COMPANY EXAMPLE: Netflix

Netflix provides a compelling case study for how strategically aligned metrics and milestones can drive AI success. At the heart of its platform is an advanced recommendation engine—responsible for curating personalized content for more than 260 million global subscribers. But what sets Netflix apart is how it evaluates the performance of its AI systems not just technically, but holistically, aligning algorithm performance with core business outcomes.

From a technical standpoint, Netflix uses metrics such as click-through rate (CTR), engagement lift in A/B tests, and algorithmic precision to evaluate improvements in its recommendation models. However, these metrics are only part of the story. The company closely monitors business-critical indicators such as minutes watched per user, user retention rates, churn reduction, and "time to first play" after login, a key signal of how effectively the platform connects users with engaging content.

This dual focus—on technical efficacy and business value—ensures that AI serves strategic goals rather than becoming a siloed innovation. For example, a new model may outperform a previous one in CTR, but if it leads to lower total watch time or increased user drop-off, it is reevaluated. Every AI experiment is grounded in cross-functional metrics that link data science, product design, and revenue goals.

Netflix also adheres to clearly defined milestones in its AI development process. These include successful offline evaluation, live A/B testing across diverse user segments, and gradual deployment at scale, each gated by performance thresholds. This structured approach helps Netflix avoid overhyping experimental features and ensures that only high-impact models are deployed widely.

The lesson is clear: AI without metrics is guesswork. AI without milestones invites chaos. Netflix's disciplined use of performance indicators and staged evaluation enables it to continuously learn, iterate, and improve. In a landscape where personalization drives competitive advantage, Netflix demonstrates how organizations can move from experimentation to excellence, by measuring what matters and staying tightly aligned with their users and strategy.

QUESTIONS

1. Why is it important for companies like Netflix to use both technical and business metrics when evaluating AI performance? How might an overemphasis on technical indicators lead to missed strategic goals?
2. How do milestones contribute to the success of AI projects at Netflix, and what risks might arise if milestones are not clearly defined?
3. What are the advantages and challenges of using A/B testing to evaluate AI-driven personalization features like recommendation engines?

4. In what ways can cross-functional collaboration (e.g., between data science, product, and business teams) enhance the effectiveness of AI metrics?
5. Netflix tracks "time to first play" after login as a key performance metric. What does this reveal about the company's approach to user engagement, and how might other businesses adopt similar metrics?
6. If an AI model increases click-through rates but leads to reduced viewing time or higher churn, should it still be considered successful? Why or why not?
7. How can the Netflix case guide other companies in developing a framework for scaling AI responsibly and effectively? What principles are transferable across industries?

Chapter Reveiw

Key Points

1. An AI roadmap is a critical strategic tool that provides structure, direction, and alignment for AI initiatives. It ensures AI investments support business objectives and are integrated across the enterprise, rather than operating in silos.
2. Without a roadmap, AI implementation often fails. Common pitfalls include disconnected pilots, unmet expectations, poor adoption, lack of governance, and wasted resources. A roadmap mitigates these risks by aligning AI development with organizational readiness and capacity.
3. The DESIGN framework outlines the key phases of AI deployment:
   D — Define strategic goals that AI should support.
   E — Evaluate the current state of data, skills, infrastructure, and processes.
   S — Select opportunities for AI application across functions.
   I — Innovate development of use cases with proper governance and timelines.
   G — Go-Live implementation by scaling pilots and managing change.
   N — Navigate outcomes through performance tracking and refinement.
4. Measurement is vital to AI success. Metrics and evaluation tools allow organizations to track progress, demonstrate value, and make informed decisions. These include both implementation metrics (e.g., project milestones, adoption rates) and impact metrics (e.g., cost savings, accuracy, customer satisfaction).
5. AI transformation is not just technical; it is strategic and cultural. The roadmap integrates planning, talent, technology, and governance to ensure AI delivers meaningful and sustainable value.

Case Studies

Schneider Electric & The DESIGN Framework

Schneider Electric, a global leader in energy management and automation, offers a model example of how organizations can undertake large-scale digital transformation of their supply chains using a strategic, structured approach. The company's journey through its "Tailored Sustainable Connected Supply-Chain 4.0" initiative aligns closely with the DESIGN framework: Define, Evaluate, Select, Innovate, Go-Live, and Navigate. This structured methodology allowed Schneider to build resilience, reduce environmental impact, and respond to increasing global complexity.

Step 1: Define Strategic Goals and Objectives

Facing intensifying global disruptions, Schneider Electric recognized the urgent need to rethink its supply chain. Operating in more than 100 countries with 200-plus manufacturing sites, the company defined clear strategic priorities:

- Enhance supply-chain resilience against volatility and disruptions
- Improve customer responsiveness through real-time demand visibility
- Advance sustainability by reducing emissions and energy usage
- Drive operational efficiency while enabling greater agility

These goals formed the backbone of the company's Supply-Chain 4.0 vision, transforming a traditional, linear supply chain into a dynamic, interconnected ecosystem.

Step 2: Evaluate Current State and Capabilities

Before launching its transformation, Schneider conducted a rigorous readiness assessment of its existing infrastructure, technologies, processes, and skills. Key findings included the following weaknesses:

- Limited real-time visibility across global operations
- Fragmented data silos across procurement, production, and logistics
- Underutilization of emerging technologies like IoT, AI, and digital twins

- Operational bottlenecks caused by a lack of integrated planning systems

This readiness assessment revealed that while Schneider had scale and expertise, it lacked the digital cohesion necessary for next-generation performance.

Step 3: Select Opportunities

The evaluation enabled Schneider to pinpoint specific gaps:

- **Technological gap:** Absence of end-to-end digital connectivity
- **Data gap:** Fragmented data streams limiting predictive insights
- **Skill gap:** Insufficient AI and analytics talent for advanced decision making
- **Sustainability gap:** High carbon emissions due to inefficient operations

Concurrently, the company identified strategic opportunities in areas such as predictive maintenance, demand sensing, and scenario planning. These informed the design of its future-state supply chain.

Step 4: Innovate

Schneider developed a transformation roadmap centered on its proprietary EcoStruxure platform. This architecture integrated IoT sensors, edge computing, cloud-based analytics, and AI-driven control towers. Key initiatives included

- smart factories equipped with robotics, AI, and energy optimization systems;
- digital twins to simulate and optimize production flows;
- predictive analytics to anticipate equipment failures and supply disruptions; and
- control towers for real-time monitoring and decision support.

These initiatives were guided by a governance structure with cross-functional collaboration across IT, operations, procurement, and sustainability teams.

Step 5: Go-Live

Execution occurred in phases across global regions. Schneider transformed its flagship facilities into "Lighthouse" factories, deploying digital solutions and agile workflows. Critical to success was a strong change management plan:

- Upskilling employees in AI, analytics, and digital operations
- Establishing global centers of excellence to support scaling
- Embedding sustainability KPIs into business performance metrics

By fostering a digital-first culture and aligning incentives with innovation, Schneider accelerated adoption and minimized resistance.

Step 6: Navigate

The outcomes were measurable and significant:

- On-time delivery improved by more than 20 percent.
- Carbon emissions were reduced by 25 percent, supporting ESG goals.
- Operational costs dropped due to smarter energy use and inventory optimization.

Supply-chain resilience was tested, and validated, during the COVID-19 pandemic. Real-time scenario planning helped reroute logistics and maintain continuity. In addition, several Schneider factories were designated as World Economic Forum "Lighthouse" sites, serving as benchmarks for Industry 4.0 transformation.

Schneider Electric's digital supply-chain transformation illustrates the power of a structured, goal-driven approach. By applying the DESIGN framework, the company was able to align vision with execution, implement advanced technologies at scale, and achieve measurable gains in efficiency, resilience, and sustainability. For other global organizations navigating complexity, Schneider's journey offers a blueprint for turning digital ambition into operational reality.

Questions

1. What were Schneider Electric's core strategic objectives for undertaking its supply-chain transformation? How did these goals reflect both internal priorities and external market pressures?
2. How did Schneider's evaluation of its existing supply-chain capabilities shape its transformation strategy? What specific weaknesses or gaps did they uncover, and why was this step critical?
3. What types of gaps—technological, data, skills, or sustainability—were most significant for Schneider Electric? How can identifying such gaps early improve transformation outcomes?
4. How did the EcoStruxure platform enable Schneider's digital strategy? What role did emerging technologies like AI, IoT, and digital twins play in shaping a connected supply chain?
5. Why is cultural change just as important as technology implementation in a digital transformation? How did Schneider support workforce adoption of new tools and workflows?
6. What metrics did Schneider use to evaluate the success of its transformation? How did real-time visibility and scenario planning help during the COVID-19 pandemic?
7. How did Schneider incorporate sustainability goals into its supply-chain transformation? What are the benefits of embedding environmental KPIs into supply-chain operations?
8. What departments or teams needed to collaborate for Schneider's transformation to succeed? How might silos between IT, operations, and procurement have affected progress?

9. In what ways did Schneider Electric improve its ability to respond to disruption? What lessons can other global manufacturers take from Schneider's approach to building resilience?
10. How does the DESIGN framework help structure a successful digital transformation? Could this framework be applied to smaller companies or other industries? Why or why not?

Siemens — AI Success with Readiness Assessment and Performance Metrics

Overview

Siemens AG, a global technology leader in industrial automation and digitalization, recognized early that AI could significantly improve its operations and product offerings. However, rather than jumping straight into implementation, Siemens took a methodical approach: it began with an AI-Readiness assessment across business units, followed by the establishment of rigorous performance metrics to measure AI's impact. This two-step approach helped Siemens avoid costly missteps, align AI with strategic goals, and scale successful initiatives across its global enterprise.

Step 1: AI-Readiness Assessment

In 2018, Siemens launched a corporate-wide initiative called AI@Siemens aimed at embedding AI across manufacturing, product development, and service functions. Before rolling out AI technologies, the company partnered with internal strategy teams and external experts to conduct AI-Readiness Assessments at over 50 facilities worldwide.

The readiness assessment focused on six critical domains:

1. **Data Infrastructure.** Were systems in place to collect, store, and process large volumes of quality data?
2. **Talent and Skills.** Did the workforce include enough AI specialists, data engineers, and business translators?
3. **Leadership Alignment.** Was executive support strong enough to prioritize AI investments?
4. **Process Maturity.** Were operations stable, standardized, and ready for algorithmic optimization?
5. **Governance and Ethics.** Were there frameworks for data privacy, algorithm transparency, and responsible AI?
6. **Cultural Openness.** Were employees receptive to adopting AI-based decision support tools?

This readiness assessment revealed critical insights. For example, while many plants had high levels of automation, few had standardized or labeled data sets suitable for training machine learning models. Others had strong data capabilities but lacked cross-functional teams to integrate AI into existing workflows. These findings led Siemens to prioritize certain plants for early AI pilots and direct targeted investment to build data pipelines and retrain staff.

Step 2: Implementing AI with Performance Metrics

After establishing which sites and processes were AI ready, Siemens launched several pilot projects, particularly in predictive maintenance, energy optimization, and automated quality control. To measure success, the company developed KPIs across three levels. One was technical KPIs that addressed issues of model accuracy and data quality. A second was operational KPIs that looked at operational measures such as increase in production throughout or reduction in unplanned downtime. The third category was business impact KPIs that looked at financial metrics, such as ROI on AI investments.

For example, in one flagship project at a Siemens gas turbine manufacturing plant in Berlin, AI models were used to predict equipment failure and optimize maintenance schedules. The result was a 20 percent reduction in unplanned downtime, a 15 percent improvement in asset utilization, and a return on investment within 18 months. These results, verified through dashboards and monthly performance reviews, led to the expansion of the project across other facilities.

Continuous Monitoring and Scaling

Success metrics are not static. So, Siemens implemented feedback loops to refine AI models and retrain them as new data became available. The performance of AI solutions was tracked quarterly through an internal AI performance scorecard, which was reviewed by both business unit heads and the central AI steering committee. When a solution met both operational and business KPIs for three consecutive quarters, it was flagged for global scaling. This structured, metrics-driven, approach helped Siemens avoid the common pitfall of stalled pilot projects and ensured that AI added measurable value.

Siemens' disciplined approach, starting with a comprehensive AI-Readiness assessment followed by implementation anchored in performance metrics, illustrates how large enterprises can maximize the impact of AI. By ensuring that teams, infrastructure, and governance were aligned before deployment, Siemens reduced risk and increased adoption. And by tracking both technical and business outcomes, the company translated AI from a promising idea into a proven competitive advantage. This case underscores a broader lesson:

AI success is not just about algorithms; it is about preparation, measurement, and continuous learning.

Questions

1. Why did Siemens begin its AI transformation with a readiness assessment rather than immediate implementation? What risks might they have faced by skipping this step?
2. Which of the six readiness domains do you think is most critical for AI success? Why?
3. How did Siemens use the results of its readiness assessment to prioritize AI investments and pilots? What lessons can be drawn for other companies?
4. What role did cross-functional collaboration play in preparing Siemens' plants for AI deployment? How can silos hinder AI readiness?

5. How did Siemens balance technical metrics (like model accuracy) with business metrics (like ROI and downtime reduction) to evaluate AI success?
6. What are the advantages of using a tiered KPI system (technical, operational, business) when measuring the performance of AI projects?
7. Siemens required three consecutive quarters of successful KPIs before scaling an AI solution globally. Why might this milestone-based approach be effective?
8. How did Siemens ensure continuous improvement in its AI initiatives after initial deployment? What is the role of feedback loops in AI performance management?
9. In what ways did the readiness assessment contribute to long-term scalability and sustainability of AI across Siemens' global operations?
10. Based on Siemens' approach, what recommendations would you offer to a mid-size company beginning its own AI journey with limited resources?

CHAPTER 9

Implementing AI & Scaling

In This Chapter You Will Learn

1. why implementation and scaling are dual challenges for enterprise AI;
2. about the three pillars of implementation: people, processes, and technology;
3. how the three pillars must be managed simultaneously for AI success;
4. about the AI implementation maturity curve;
5. how to effectively scale AI implementation; and
6. about sales & pperations planning (S&OP) as a practical entry point for AI implementation.

There is an old saying, "Ideas are easy, implementation is hard."

Think about our personal life. Most of us have had the experience of deciding to get in shape. We buy new workout gear and download fitness apps, but after a few days, we begin skipping workouts. It may be because we did not make a realistic schedule or plan for how tired we would be after work. The point is that the idea was easy, implementing it every day was the hard part.

Turning ideas into reality requires discipline, coordination, and persistence. It is true in our personal life and in business, and it is true for AI implementation.

A well-documented example of AI-implementation challenges is Marks & Spencer (M&S), the British multinational retailer. In 2021 the company pushed to modernize operations. M&S launched an ambitious digital transformation strategy that included AI-driven supply-chain optimization. However, the company encountered major challenges during implementation, particularly around integrating systems and aligning cross-functional teams. Departments such as merchandising, logistics, and IT operated in silos, which hindered data sharing and delayed decision making. Legacy systems were difficult to integrate with the new system. Additionally, budgetary constraints and a lack of internal digital expertise slowed adoption, forcing M&S to rely heavily on external consultants. This resulted in large budget overruns.

Despite significant investment, progress was uneven, with limited effect in early phases due to fragmented technology systems and unclear accountability. Expected benefits did not materialize, and M&S learned the need for strong cross-functional governance and clear ownership of AI outcomes.

9.1 AI Implementation & Scaling

AI is transforming how organizations operate, compete, and innovate. Yet unlocking this value requires more than experimentation; it demands successful implementation and thoughtful scaling.

Implementation is the bridge between AI strategy and tangible results. It marks the moment when AI moves from concept to application, transforming processes, augmenting decisions, and automating tasks. But AI implementation is not straightforward. Unlike traditional IT systems, AI systems are probabilistic, require continuous learning, and depend heavily on high-quality

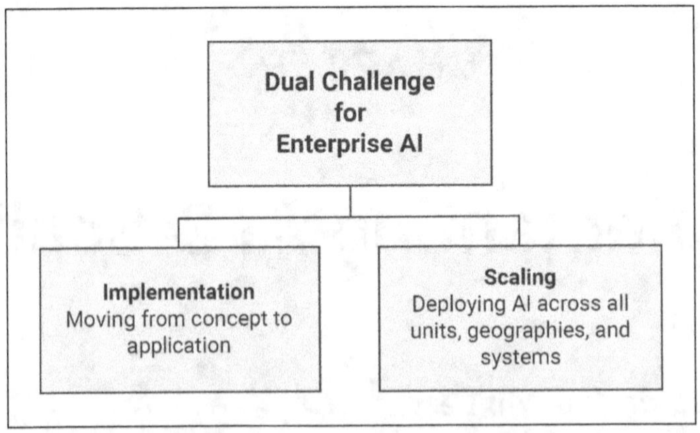

Figure 9.1 Dual AI Challenge: Implementation & Scaling

data. This makes implementation more iterative, technical, and organizationally complex. Poor implementation often leads to what is called "pilot purgatory," where AI initiatives remain stuck in experimentation or pilot stage, without ever delivering enterprise value.

Even when implementation succeeds at the pilot level, many companies fall short in scaling. Scaling refers to deploying AI across multiple business units, geographies, or systems and embedding it into the core of business operations. This is where AI's full return on investment can be realized. However, scaling is not a matter of simple replication of what was done at the pilot stage. As we will see, it requires rethinking infrastructure, processes, talent models, and governance. Organizations must move beyond data silos, legacy systems, and cultural resistance to build an ecosystem where AI can be used at scale.

Both implementation and scaling present distinct but interconnected challenges, shown in Figure 9.1. They are a dual challenge. Implementation requires technical readiness, leadership alignment, and a solid business case. Scaling requires operational maturity, standardization, cross-functional collaboration, and change management. Organizations that ignore either stage risk underutilizing their AI investments.

The following case study of Schneider Electric demonstrates how a global company navigated these two critical phases to achieve enterprise-wide AI transformation.

COMPANY EXAMPLE: Schneider Electric

Schneider Electric, a global leader in energy management and automation, operates in more than 100 countries with a complex supply chain of more than 200 manufacturing sites. Faced with rising operational complexity, increasing customer demands, and sustainability goals, Schneider Electric launched a major AI-driven digital transformation initiative: the Tailored Sustainable Connected Supply Chain 4.0.

PHASE 1: IMPLEMENTATION

Schneider began by piloting AI in demand forecasting and supply planning. Using machine-learning models, they predicted product demand by integrating historical sales data, market signals, and external factors like weather and geopolitical events.

Key implementation steps included deploying IoT sensors across production lines to collect real-time operational data. They built predictive maintenance models to reduce downtime. They also established data pipelines with Microsoft Azure to consolidate information across systems.

Initial results were impressive. They showed a 20 percent reduction in forecasting errors, a 30 percent improvement in asset utilization, improved service levels, and on-time delivery in pilot regions. However, despite this early success, the company recognized that pilot projects alone would not yield enterprise-wide value. This is where scaling comes in, and Schneider moved to the next phase.

PHASE 2: SCALING

Schneider took deliberate steps to scale AI. First, they focused on their platform infrastructure. The company scaled through a central digital architecture, its EcoStruxure platform. This enabled integration of AI across all business functions, from production to logistics to customer service. Next, Schneider created a global AI Center of Excellence (CoE) responsible for AI governance, model management, and knowledge sharing. This ensured consistency, reusability, and responsible AI deployment across units.

At the same time, they worked on talent, culture, and governance to transform their entire organization. Schneider trained more than 50,000 employees in digital and AI literacy. It also promoted cross-functional teams, pairing data scientists with supply-chain planners and factory managers to ensure alignment between AI models and real-world operations. In addition, a model risk governance framework was put in place to ensure explainability, bias detection, and auditability. Every model was reviewed and monitored for drift and regulatory compliance.

The outcome was excellent. By 2024, Schneider Electric had scaled AI to more than 70 percent of its supply-chain operations. Notable outcomes included a 30 percent reduction in inventory holding costs and a 40 percent decrease in unplanned downtime across smart factories. In addition, Schneider had enhanced their sustainability with optimized energy use and reduced waste.

Lessons learned for Schneider were that implementation is a proving ground: start small, learn fast, and iterate. Scaling requires a systemic approach: build platforms, governance, and skills in parallel. Schneider discovered that AI is not just a technology challenge, it is a cultural and operational transformation.

QUESTIONS

1. What factors made Schneider Electric's initial AI pilot projects successful? Why were these pilots insufficient on their own to drive enterprise-wide transformation?
2. How did Schneider use their centralized platform? How did the global AI CoE contribute to its ability to scale AI across diverse operations and geographies?
3. In what ways did Schneider address the human and cultural dimensions of AI transformation? Why are these elements critical to successful scaling?
4. What role did AI governance, including model risk management and auditability, play in Schneider's ability to deploy AI responsibly at scale?
5. Based on Schneider Electric's experience, what lessons can other global supply-chain organizations draw about aligning AI implementation with long-term strategic goals like sustainability and resilience?

9.2 Strategy Sets the Direction

Companies that succeed with AI do not implement it haphazardly. They strategically deploy these capabilities in areas that promise the greatest competitive advantage, areas defined by their business strategy. The core mistake many organizations make is to implement AI tools simply because competitors are doing it. This follow-the-leader mentality has long plagued business trends, from total quality management to Lean, Six Sigma, and ESG initiatives. But adopting AI without strategic alignment leads to underwhelming returns and wasted resources.

A successful AI initiative begins with a clear understanding of what the company is trying to achieve. Not every firm needs predictive analytics for every function or the most advanced AI for every decision. What it needs is technology that directly supports its unique goals, capabilities, and competitive positioning.

Recall from Chapter 2 that business strategy outlines a company's long-term vision, competitive positioning, and the roadmap for achieving growth and differentiation. It answers foundational questions like

- What markets do we serve?
- How do we compete?
- What makes us different?

For instance, luxury fashion brands like Moët Hennessy Louis Vuitton compete on exclusivity and brand experience, while Walmart focuses on operational efficiency and low prices. These choices ripple throughout the organization, from marketing and supply chain to how data is collected and used.

Strategic decisions set the course for tactical ones. While strategy is long-term and broad, tactical decisions are short-term and operational, concerning day-to-day execution, resource allocation, and process efficiency. The two must remain tightly aligned. Tactical decisions both implement and inform strategy. Without a clear strategic anchor, AI efforts may become fragmented, duplicative, or misaligned with what truly drives value. These strategies then translate into tactical decisions regarding three critical pillars of implementation: people, processes, and technology.

Successful implementation of AI must involve each of these pillars simultaneously, as seen with the example of Schneider Electric. This is shown in Figure 9.2.

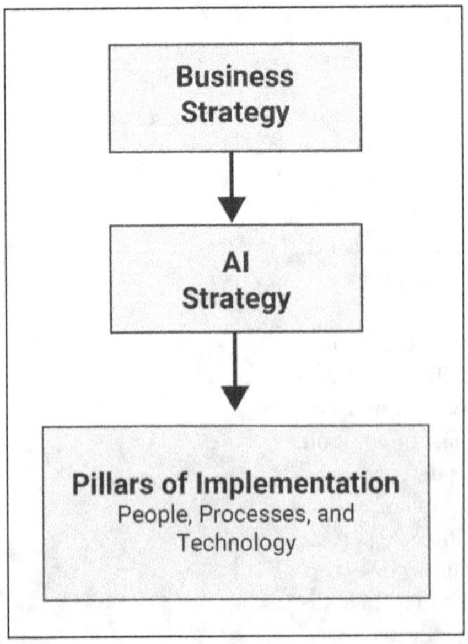

Figure 9.2 Strategy Drives Implementation

9.2.1 Translating Strategy into Tactics

AI implementation goes well beyond following a technology roadmap. It must include every aspect of the enterprise and its supply chain. It should ensure that data architectures, governance, and use cases are not siloed but instead unified under a shared purpose. Often companies make the mistake of having fragmentation of data across functions such as marketing, sales, service, and finance. Then each function or department uses its own tools, definitions, and metrics, hindering implementation efforts.

There are five requirements for AI strategy to translate into tactical decisions. *First,* AI and business strategy must be aligned. In 2025, according to IDC, 74 percent of digital transformation failures are due to poor alignment between technology initiatives and core business objectives, which are outlined in the business strategy discussed earlier. *Second,* AI strategy must have cross-functional alignment. This means that AI initiatives must span departments, touching finance, operations, marketing, and IT. *Third,* there must be governance and a process for managing data quality. According to Gartner, poor data quality costs businesses an average of $12.9 million annually. *Fourth,* there needs to be seamless cloud integration. Consider that the global cloud AI services market is expected to surpass $200 billion by 2027. Cloud-native AI is essential for scale, cost-efficiency, and agility. *Fifth,* there must be assurance of privacy and regulatory compliance. With AI regulations tightening globally, firms must embed governance frameworks for explainability, fairness, and accountability from the start.

Let's look at two companies and how they have managed translating strategy into tactics.

COMPANY EXAMPLE: Amazon versus Unilever

Let's compare Amazon and Unilever with regard to their AI strategies, alignment, and integration.

Amazon has a business strategy focused on offering low prices, convenience, and operational efficiency. This has translated into an AI strategy where machine-learning algorithms drive personalized product recommendations, fraud detection, demand forecasting, warehouse automation, and Alexa's natural language processing. The AI systems are integrated across all levels. For example, AI powers Amazon's anticipatory shipping system, smart fulfillment centers, and real-time pricing, deeply entwining analytics with its business model. Amazon's AI efforts are not isolated IT projects; they are core to delivering the strategic promise of speed, selection, and personalization.

Unilever, on the other hand, has a business strategy of sustainable growth, consumer responsiveness, and global brand management. This has translated into an AI strategy of using predictive analytics for supply-chain agility, consumer sentiment analysis for brand decisions, and AI-powered R&D for product formulation. By aligning AI with its Compass strategy, Unilever reduced supply-chain planning time by 30 percent and improved demand forecast accuracy by 15 percent.

How do they compare and contrast?

Amazon is driven by a customer-obsessed, efficiency-focused strategy. It aims to deliver low prices, wide selection, and fast delivery by optimizing every part of its operations. Unilever, on the other hand, is oriented around sustainable growth, consumer responsiveness, and global brand stewardship. Its strategy balances economic performance with environmental and social goals. Amazon emphasizes speed and scale, while Unilever emphasizes sustainability and adaptability. Both aim for excellence in customer experience, but the underlying strategic values differ—efficiency versus responsibility.

Amazon uses AI to amplify its core strengths; personalization (e.g., recommendation engines), logistics optimization (e.g., anticipatory shipping), and real-time decision making (e.g., dynamic pricing). Unilever leverages AI to enable strategic agility in product innovation, marketing, and supply-chain planning, especially through predictive analytics and consumer insights. AI at Amazon is deeply operationalized and productized, fueling real-time interactions and logistics. At Unilever, AI is more strategic and insight driven, helping inform long-cycle brand and supply-chain decisions.

When it comes to integration, Amazon has embedded AI across nearly every function, including fulfillment, pricing, fraud detection, and voice-enabled services. AI is not a support function; it is a core operating model. Unilever integrates AI through targeted programs, such as aligning with its Compass strategy, improving forecasting, and enhancing R&D capabilities. Amazon has industrialized AI at scale, while Unilever applies AI in a more programmatic, business-aligned fashion, focused on specific strategic goals.

As a result of these differences, their outcomes and focus areas differ. For example, Amazon achieves outcomes like hyper-personalization, rapid delivery, and operational scale, driving continued customer growth and market dominance. Unilever sees improvements in forecast accuracy, supply agility, and product innovation, supporting its goals of resilience and sustainability. Amazon's AI outcomes are tied to market performance and operational edge, while Unilever's are aimed at resilience, consumer alignment, and responsible growth.

Table 9.1 lists these comparisons.

Table 9.1 Comparing Amazon and Unilever

	AMAZON	UNILEVER
Business Strategy	Speed, scale, customer obsession	Sustainability, brand responsiveness
AI Focus	Personalization, logistics, automation	Predictive analytics, consumer insights, R&D
Integration Approach	Fully embedded in operations	Aligned to strategic programs
Key Outcomes	Efficiency, personalization, real-time decision making	Agility, innovation, sustainability
Cultural AI Orientation	Tech-centric, execution-driven	Purpose-driven, insight-led

QUESTIONS

1. How do Amazon and Unilever's differing business strategies shape their approach to AI adoption and integration?
2. In what ways is Amazon's AI implementation more deeply embedded than Unilever's, and what are the advantages and trade-offs of this approach?
3. How do Amazon and Unilever differ in their use of AI for customer engagement and product delivery, and what can each company learn from the other?

4. What does Unilever's experience tell us about aligning AI with long-term strategic goals like sustainability and resilience?
5. Given Amazon's real-time, data-heavy AI use and Unilever's insight-driven AI approach, which model is better suited for future business resilience and why?
6. If you were leading AI strategy for a new global company, would you model your integration approach more after Amazon or Unilever? Justify your answer.

9.2.2 Data Sharing Across Supply Chains

AI implementation across supply chains means data sharing with partners. Companies cannot treat data sharing in their AI-enabled supply chains as an afterthought. To unlock predictive demand sensing, resilience modeling, and real-time exception management, they must fuse partner feeds (e.g., logistics IoT, retailer point-of-sale, weather, and social signals) with their own transaction records inside a single analytics environment. That requires data aggregation, records tracking, and software protocols so that each stakeholder sees the "same version of the truth" in real time.

Transparency, however, does not mean openly sharing everything with every node in the supply-chain network. Privacy and security are critical and require selective exposure. Leaders now apply what is called zero-trust data design. Here they map which partners genuinely improve forecast accuracy or service-level performance, grant them the minimum viable data set, and continuously monitor usage. This selective exposure balances the coordination upside with the real downside of expanded attack surfaces.

A cautionary tale is the MOVEit supply-chain breach. In May 2023 the notorious Cl0p ransomware gang exploited a MOVEit file-transfer tool, an application used by thousands of organizations for automated data exchanges with suppliers and service providers. Within five months more than 2,500 companies and roughly 66 million individuals were confirmed victims. Analysts estimated the cumulative financial toll at approximately US $9.9 billion in response costs, litigation, and operational disruption. High-profile casualties ranged from British Airways and the BBC to multiple US state agencies. The episode underlined how a single integration point can cascade across an entire digital supply chain, exposing sensitive HR, payroll, and customer records that had been entrusted to "trusted" intermediaries.

This illustrates the risk associated with data sharing and the importance of governing the process. According to Gartner's 2024 survey, organizations that embed a formal data-governance framework report 66 percent fewer security incidents tied to data misuse and 52 percent fewer compliance breaches, hard numbers that have put governance on the chief operating officer (COO) agenda rather than relegating it to IT alone. There are three traits shared by most data-governance frameworks:

1. Clear ownership and stewardship for every critical data set.
2. Policy-driven access controls that are audited.
3. Continuous monitoring and automated alerting for abnormal data flows.

While data-sharing powers AI-driven supply chains, without well-planned governance data sharing can quickly bring down companies. It is important for companies to understand this and treat data as a strategic asset and a regulated resource and protect it so that collaboration never undermines resilience.

9.2.3 Balancing Short- and Long-Term Capabilities

Partnering across supply chains requires more than just aligning AI strategies. Short-term collaborations may quickly solve immediate needs—like improved forecasting or route optimization. However, these partnerships need to also be aligned with long-term strategic goals.

AI innovation is occurring at a very rapid pace. This means that investments made without a future-oriented roadmap can quickly become obsolete. As such, organizations need to think beyond cost-efficiency or convenience and focus on building partnerships that help them develop lasting capabilities and be able to evolve with technological shifts.

Short-term solutions address current friction points, but long-term partnerships build the foundation for sustained competitive advantage. Organizations that prioritize future-proof capability today will be tomorrow's digital leaders. Let's look at an example to see how this works.

COMPANY EXAMPLE: Unilever (Part I)

Unilever offers a compelling example of how to pursue long-term capability through strategic collaboration. In 2023, Unilever deepened its multiyear partnership with Microsoft to embed Azure AI, IoT, and digital twin technologies across its global supply chain. The objective was not merely short-term performance boosts but also a comprehensive digital transformation that would support long-term agility and sustainability.

The partnership included digital twins for end-to-end supply-chain visibility. Unilever used Azure to create digital replicas of entire supply-chain nodes—from factories to distribution centers. This allowed them to simulate disruptions and proactively adjust plans, improving both risk management and responsiveness.

Another aspect was AI-driven demand forecasting. Rather than relying solely on legacy statistical models, Unilever adopted machine-learning models to dynamically forecast demand at the SKU and regional level, improving both accuracy and inventory efficiency. In addition, the collaboration enabled sustainability integration. With the advanced analytics, Unilever was now able to track energy consumption, waste, and emissions in real time. This enabled it to embed sustainability metrics into supply-chain decisions, rather than treating them as separate targets.

This collaborative investment helped Unilever cut costs, reduce waste, and improve service levels. More importantly, however, it laid the groundwork for a digitally intelligent supply chain that can adapt as AI capabilities grow more sophisticated.

This case illustrates why businesses must resist the lure of short-term AI "bolt-ons" that deliver immediate but isolated gains. These fixes often lead to brittle systems and technical debt. Instead, companies should look for partners who not only offer current expertise but also co-invest in the transfer of skills, governance models, and scalable infrastructure. These long-term partnerships can evolve in step with the company's strategic growth.

9.3 Pillars of Implementation

Three pillars, or building blocks, of AI implementation are people, processes, and technology (Figure 9.3). They are the touchpoints where implementation happens in a simultaneous and coordinated way. Certainly, the characteristics of these pillars will vary based on the company's business strategy and AI strategy, but together they need to support the AI strategy. The following sections look at these pillars and how implementation should occur along their dimensions.

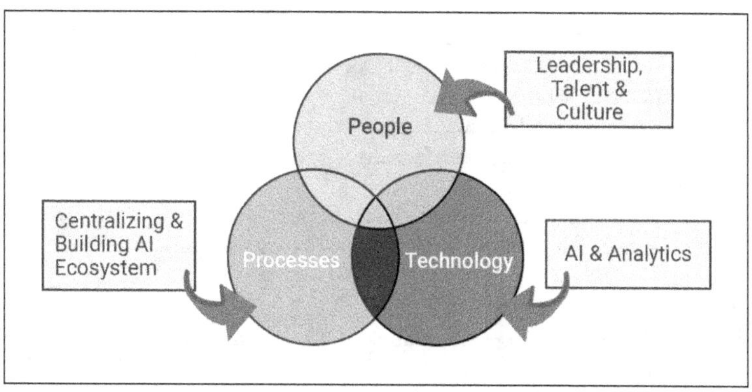

Figure 9.3 The Pillars: People, Processes, and Technology

9.3.1 People

Here we discuss the role of people, the first pillar of implementation. People include leadership, talent management, and the overall culture of the organization. It is often overlooked in AI implementation, yet it is often the most important. Let's look at each in detail.

Leadership

AI does not reduce the need for leadership; it magnifies it. Far from being a purely technical transformation, AI implementation is a strategic and cultural challenge. It is leadership that sets the vision, defines the strategic priorities, secures the resources, and cultivates the culture required to unlock value from AI. Without committed, informed, and forward-thinking leaders, even the most advanced technologies fail to deliver impact.

Successful AI-driven organizations are led by individuals who see the transformative potential of AI to redefine business models, operations, and customer experiences.

Why does leadership matter?

Leaders are the ones that define the "why" and the "what." They articulate a compelling case for AI that aligns with the company's mission and long-term business goals. Leaders also drive cultural change. They foster a data-literate, experimentation-driven culture in which decisions are informed by insights, not intuition alone.

AI implementation requires collaboration along silos, and unifying teams. It is leaders that enable collaboration across IT, data science, business units, and customer-facing teams. They are the ones in charge of budgets and make the investment in capability building. They allocate budget for talent, training, and infrastructure and ensure that these investments are aligned with strategic objectives. They also model the change. They champion data-driven thinking in their own decisions, setting the tone for the organization.

As we will see in this chapter, there are numerous barriers to implementation. It is leadership that must find ways to overcome key organizational barriers (legacy systems, skills shortages, risk aversion, and fragmented data strategies) by guiding the company through a structured and patient transformation process.

Technology alone does not drive transformation; leadership does. It is leadership that inspires the shift in mind-set, priorities, and processes needed for AI to scale and deliver value. It is interesting to note that according to McKinsey in 2024, companies with strong analytics leadership are 1.5x more likely to achieve above-average revenue growth, and 2x more likely to achieve operational efficiency targets.

Next, let's look at leadership of three companies and how each drove AI implementation.

COMPANY EXAMPLES: Comparing Leadership at Capital One, Microsoft, and Tesco

1. CAPITAL ONE

Founder Richard Fairbank had a strategic vision and recognized early on the power of data to redefine banking. Under his leadership, Capital One became one of the first financial institutions to integrate AI into credit scoring, fraud detection, and customer personalization. Capital One invests heavily in data science talent, internal training programs, and AI-ethics initiatives. Analytics is seen not as a support function, but as a strategic enabler embedded in product design and risk management.

The result: A 20 percent reduction in credit losses through predictive risk models and a 15 percent increase in customer retention via hyper-personalized marketing campaigns.

2. MICROSOFT

Under CEO Satya Nadella, Microsoft has embraced AI as a foundational part of its growth strategy, across Azure cloud services, Office 365, GitHub Copilot, and enterprise tools. But Nadella also made AI ethics and responsible AI a leadership priority.

Microsoft established its AETHER (AI, Ethics, and Effects in Engineering and Research) committee to guide responsible AI practices. Nadella also created a cultural shift. His promotion of a growth mind-set changed Microsoft from risk averse to learning driven and experimental.

The result: This leadership has positioned Microsoft as a global AI leader, with $119 billion in intelligent cloud revenue in 2023, and Copilot is now integrated across its ecosystem.

3. TESCO

Tesco's leadership has made data central to retail strategy, supply-chain management, and customer loyalty. By leveraging data from their loyalty program, Clubcard, and real-time analytics, Tesco customizes offers, forecasts demand, and optimizes pricing dynamically.

Senior executives sponsor data projects and ensure that analytics are actionable at every level, from headquarters to store managers.

The result: Tesco reports a 7 percent increase in average basket size and reduced food waste through improved demand forecasting.

COMPARATIVE ANALYSIS

At Capital One, leadership began with a clear strategic vision from founder Richard Fairbank, who foresaw the transformative power of data in banking. His early investment in AI for credit scoring, fraud detection, and personalization set the tone for a data-first culture that treats analytics as a core business driver, rather than a back-office function. This vision was operationalized through talent development and AI-ethics programs.

Microsoft, under CEO Satya Nadella, took a more sweeping enterprise-wide approach. Nadella made AI central not just to products but to organizational philosophy. His emphasis on a growth mind-set helped shift Microsoft from a defensive, bureaucratic culture to one that embraces experimentation and learning. Importantly, Microsoft also took early steps in AI governance, forming the AETHER committee to ensure ethical AI development and deployment. This blend of cultural shift and institutional oversight allowed Microsoft to scale AI across cloud, productivity, and developer ecosystems.

At Tesco, the leadership initiative is more operational and retail focused, but no less strategic. Tesco's senior executives have made it clear that data is not confined to analysts or IT; it's a leadership responsibility. By empowering store managers and frontline employees with actionable analytics, Tesco has localized decision making and improved both customer experience and operational efficiency. Its leadership is evident in tangible outcomes, like reduced food waste and increased basket size.

What do these companies have in common?

Despite differences in industry and scale, several key themes unite these leadership initiatives:

1. **Leadership.** In all three cases, AI adoption was not left to technical teams alone. Senior leadership played a direct role in making AI a strategic priority. Fairbank at Capital One, Nadella at Microsoft, and Tesco's executive team all championed analytics at the top.
2. **Cultural Transformation.** Each company recognized that technology alone is not enough. They invested in shifting culture, whether through Microsoft's growth mind-set, Capital One's talent strategy and AI ethics, or Tesco's frontline data empowerment.
3. **Responsible and Ethical AI.** All three have emphasized some form of governance and accountability. Microsoft formalized this with AETHER, Capital One through ethical data practices, and Tesco through structured sponsorship and operational discipline.
4. **Business Impact.** These are not theoretical AI strategies. Each organization achieved measurable results. Capital One reduced credit losses; Microsoft grew cloud revenue and platform stickiness; Tesco boosted sales and reduced waste. AI is tightly linked to business performance.

These examples underscore a critical leadership lesson: AI success depends as much on people, culture, and vision as it does on algorithms and infrastructure. Companies that embed AI into strategy, empower their workforce, and govern it responsibly are best positioned to unlock its full value.

QUESTIONS

1. How did leadership vision shape the adoption of AI at Capital One, Microsoft, and Tesco? What can other companies learn from their approaches?
2. In what ways did each company's leadership influence organizational culture to support AI innovation and adoption?
3. How do the companies differ in their approach to AI governance and ethics? Why is this important for long-term sustainability?
4. What role does talent development and internal capability-building play in each company's AI strategy?
5. What evidence shows that AI initiatives in these companies are generating measurable business value? How is this value communicated internally and externally?
6. If you were leading a company just beginning its AI journey, which of these leadership models would you most want to emulate, and why?

Talent Management

While AI and analytics are driven by algorithms and data, their success ultimately depends on people. No matter how advanced a tool may be, its potential remains untapped without the right talent to develop, deploy, and interpret it, and without a culture that embraces data-driven thinking across the organization.

Effective AI implementation requires a dual focus: attracting and nurturing top analytical talent; and building a culture that rewards evidence-based decision making. These two pillars, talent management and organizational culture, form the backbone of any data-driven transformation.

A critical effort is recruiting and developing analytical talent. To compete by using AI, companies must build a critical mass of skilled data scientists, machine-learning engineers, and analytics translators. But deep technical expertise alone is not enough. Organizations must also develop managers and frontline employees who can understand and act on analytical insights.

There are numerous recruitment strategies. Top-performing companies are proactive in recruiting analytics talent by sourcing globally. For example, companies like Apple and Meta tap international talent pools and partner with universities and bootcamps. Another strategy is building internal pipelines. Firms such as Procter & Gamble and Siemens offer rotational programs that expose early-career hires to multiple analytics functions. Yet another is partnering with academia. Walmart, for example, collaborates with the University of Arkansas to fuel its data science talent pipeline and research in retail analytics.

AI adoption, however, is not just about hiring new talent. It is also about upgrading existing capabilities. This means investing in reskilling current talent. Many organizations are now investing in enterprise-wide upskilling programs to help staff better understand AI tools and decision making. For example, Capital One's internal learning platform, Capital One University, offers advanced training in data storytelling, experiment design, and AI ethics. These programs ensure that both analysts and business leaders have the tools to use analytics effectively. The result is a shared fluency across the organization that accelerates AI adoption and value delivery.

Building analytics capabilities at scale requires integration with business strategy and leadership. Effective talent management involves aligning analytical roles with business functions. This means job descriptions and tasks have to support the business functions. To keep great talent, companies need to ensure clear career paths for technical experts (e.g., data scientists and ML engineers). Last, there have to be incentives for collaborative problem solving and experimentation. Without the right incentives talent will not participate or be supportive.

For example, Google has long been a leader in both attracting top talent and fostering lifelong learning. Its internal training programs cover everything from advanced machine learning to ethical AI practices. Google also pioneered the concept of T-shaped talent, encouraging employees to develop both deep expertise and broad business context.

Culture

Talent flourishes in the right environment. That is why AI implementation must be supported by a company-wide culture that values evidence over intuition, experimentation over hierarchy, and continuous improvement over perfection.

A data-driven culture is characterized by decision making grounded in data. From strategy to operations, decisions are supported by measurable evidence. There has to be transparency and accountability. Performance metrics need to be cleared, shared, and performance tracked across roles. Last, curiosity and experimentation need to be encouraged. Employees need to be encouraged to test, learn, and iterate. This is the only way growth and learning will take place.

Progressive Insurance, for example, has cultivated a strong analytics culture over the past two decades. Every process and product change is piloted, tested, and validated before full deployment. Compensation and career progression are tied to quantifiable results, including customer outcomes and financial KPIs. Whether it is usage-based insurance models or chatbot deployment, innovation at Progressive is validated through data analysis. This culture has enabled Progressive to maintain pricing accuracy, reduce underwriting risk, and outperform competitors in customer satisfaction and claims processing.

Navigating Culture—Talent Tensions

While data-driven culture is a strategic asset, it can sometimes create friction with creative or entrepreneurial instincts. For instance, R&D teams often thrive in environments where ideas can be pursued without immediate validation. The challenge is balancing structure with creativity.

Leading organizations learn to adapt their cultural expectations across functions, allowing flexibility in areas like R&D while maintaining discipline in customer-facing or operational decisions. For example, at Amazon, experimentation is tightly linked to measurable outcomes, which promotes rapid iteration. At 3M, the culture accommodates blue sky innovation, which famously gave birth to the Post-it Note after years of informal development.

In fact, one of the keys of successful implementation is the ability to attract exceptional talent, empower them through education and tools, and create an environment in which insight leads to action. As we can see, AI transformation is as much about people as it is about platforms. Without the right talent and a supportive culture, even the most sophisticated AI systems will not deliver results. But when people, process, and purpose align, AI becomes a force multiplier.

9.3.2 Processes

While talent and culture provide the human foundation for AI, it is processes that enable scale, consistency, and impact. Processes determine how AI is embedded in the flow of work—how models are deployed, insights are shared, and decisions are made across business units. Three key dimensions of process enablement are centralizing AI, building an AI ecosystem that connects internal capabilities with external partnerships, and deciding which AI capabilities to keep in-house and which to outsource.

Centralizing AI

Historically, analytics functions have been siloed within departments; marketing ran its dashboards, supply chain had its forecasting tools, and finance managed its models. This fragmented approach led to duplication, inefficiencies, and missed opportunities. As AI becomes a strategic capability, leading organizations typically have centralized AI and analytics functions to ensure coherence, governance, and strategic alignment.

Why does centralization matter?

The first issue is that centralization eliminates redundancy. It gives what is called "one version of the truth"—single data sources, unified platforms. It also enables cross-functional insight. For example, marketing, logistics, HR, and finance can all act on shared intelligence. It also accelerates deployment. It enables streamlined model development and rollout across business units. It also allows for enhanced governance. It is much easier to enforce data quality, compliance, and model risk controls.

Walmart has moved toward a hybrid centralized-decentralized analytics model. While business units maintain autonomy, centralized AI hubs ensure data consistency, platform coordination, and enterprise-wide analytics strategy. The Data Ventures division also monetizes Walmart's analytics capabilities externally.

Similarly, Target has created a central Advanced Analytics and AI team responsible for enterprise-level model development, testing, and ethics reviews. Localized teams execute with support from this central unit, allowing for scalability without sacrificing agility.

Lowe's has established an Enterprise Data and Analytics Office to align data governance, platform strategy, and analytics talent across departments, helping the company move from descriptive analytics to predictive and prescriptive capabilities.

These are all examples of centralized functions that serve as internal consultants, helping local teams define use cases, deploy models, and interpret results. This structure maintains proximity to decision-makers while embedding AI in core processes.

Building an AI Ecosystem

No organization can succeed with AI alone. The pace of technological advancement, the complexity of use cases, and the breadth of data sources demand a networked approach to analytics. This is where the concept of the analytical ecosystem becomes critical.

What is an AI ecosystem? An AI ecosystem includes the following:

- External data providers (e.g., Nielsen, Dun & Bradstreet, Weather Company)
- AI and analytics vendors (e.g., Snowflake, Databricks, SAS, DataRobot)
- Cloud infrastructure partners (e.g., AWS, Microsoft Azure, Google Cloud)
- Consulting and implementation partners (e.g., Accenture, Deloitte, Cognizant)
- Academic and research alliances (e.g., MIT-IBM Watson AI Lab)

These partners supplement internal capabilities, accelerate innovation, and enable access to talent, tools, and best practices. Consider the example of P&G. P&G has one of the most sophisticated analytical ecosystems in consumer goods. For example, P&G is known for its collaborative forecasting, where P&G shares data with suppliers and retailers to synchronize demand and inventory across the supply chain. P&G has a strategic partnership with Microsoft. It works with Microsoft to use Azure AI for demand planning and quality prediction in manufacturing. P&G also has an extensive talent management system. Through partnerships with analytics firms and universities, P&G supplements its internal talent pool with external experts in machine-learning, AI ethics, and supply-chain optimization.

This ecosystem model has enabled P&G to improve forecast accuracy, reduce supply-chain waste, and shorten product innovation cycles.

What AI capabilities should be outsourced?

As companies develop their AI strategy, a critical process decision involves capability sourcing: which activities should remain in-house, and which should be outsourced. This is basically a make-or-buy decision. Table 9.2 lists guidelines for deciding what to keep in-house and what to outsource.

Table 9.2 AI Capabilities to Keep In-House vs. Outsource

Keep In-House	Outsource
• Core strategic models (e.g., pricing, customer segmentation)	• Specialized AI model development (e.g., NLP, computer vision)
• Proprietary data assets	• Cloud infrastructure and platforms
• AI governance and compliance	• Short-term data science projects or pilots
• Enterprise data architecture	• Training and upskilling programs

This means that a company has to understand its core capabilities and core strategic models. It also means deciding what companies to partner with. Companies like Unilever, Coca-Cola, and Ford frequently work with vendors like Palantir, Accenture, and Databricks to access the latest AI tools while maintaining control over strategic direction and IP.

AI processes are what transform experimentation into execution. Without robust processes (centralized coordination, cross-functional alignment, partner integration, and governance frameworks) AI remains stuck in the lab. But, with the right processes, organizations can scale AI, drive efficiency, and create competitive advantage. Whether it is synchronizing global inventory or delivering personalized customer experiences, AI must be operationalized through smart processes that connect strategy with execution.

9.3.3 Technology

AI technology is the engine behind the analytics revolution. While strategy, leadership, and culture shape direction, it is technology, especially AI, that powers the execution of data-driven decisions at scale. Over the past five years, dramatic advances in data infrastructure, cloud computing, machine learning, and real-time analytics have transformed the way organizations collect, manage, and act on data. In addition to analyzing what happened, with AI companies are predicting what will happen and automating responses in real time.

It is important to remember that AI is not a single algorithm but an ecosystem of foundation models. These are huge neural networks pretrained on internet-scale multimodal data and then adapted to niche tasks. These models (e.g., transformers such as OpenAI's family) reason step-by-step, write and execute code, interpret images, and even call external tools. Their "smaller siblings" now run privately on mobile devices, providing models for summarization, extraction, and dialogue without sending data to the cloud.

Current technology extends the base model with a longer deliberate-thought window and more reliable execution where accuracy overrides speed. But these technologies are also ultra fast. Consider Google's Gemini 2.5 Flash–Lite (announced 17 June 2025), which compresses a million-token context into real-time answers, making video frames, sensor streams, and enterprise resource planning (ERP) data searchable by the same prompt. Combined, these trends enable AI to be a continuous, multimodal decision agent. In supply chains, they can read a purchase order, watch a conveyor-belt camera feed, write Python to test scenarios, and then offer justification for every step.

AI has now moved beyond novelty. It is now embedded in product development, marketing, customer service, and content strategy. For example, Coca-Cola uses OpenAI's DALL·E and GPT models to co-create ad campaigns. Morgan Stanley uses GPT-powered advisers to assist wealth managers with tailored recommendations. Adobe Firefly allows designers to create visuals from text prompts, accelerating creativity cycles.

What does this mean for supply chains?

AI technology implications for supply chains cannot be overstated—from real-time visibility and communication to autonomous processes and dark warehouses. These models are also at the heart of digital twin control rooms. For example, Maersk now streams IoT data into twin replicas of ports and vessels, runs what-if simulations in minutes, and pre-positions containers before bottlenecks occur. On the factory floor, Siemens' Industrial Copilot uses AI, so line operators get autonomous statistical analysis within seconds, without sending IP-sensitive data off-site.

Here are some especially significant impacts of AI for supply chains:

AVs and Logistics Optimization. AI-driven autonomous vehicles are reshaping logistics by enabling self-driving trucks, delivery drones, and robotic last-mile solutions. These systems use computer vision and LiDAR (Light Detection and Ranging), a remote sensing method for distances and creating detailed 3-D replicas. They combine this with deep learning algorithms to perceive their environment, navigate complex routes, and optimize delivery times in real time. Companies like Einride and Aurora are deploying AI-powered electric freight vehicles that reduce driver dependency and improve route efficiency. AI also integrates with telematics and traffic data to dynamically reroute deliveries, predict maintenance needs, and minimize fuel consumption. This significantly enhances supply-chain performance, improves resilience, and reduces environmental impact.

AI in Assembly Lines and Smart Manufacturing. On the factory floor, AI is embedded into smart assembly lines through machine vision, predictive maintenance, and robotics coordination. Cameras and sensors powered by AI inspect products in real time, identifying defects with higher accuracy than human inspectors. Predictive analytics use sensor data to forecast machine failures before they occur, drastically reducing downtime and unplanned maintenance costs. Collaborative robots (Cobots) learn from human operators, using reinforcement learning and computer vision to assist in tasks like welding, packaging, and precision assembly, thereby improving throughput and worker safety.

Production Process Optimization. AI is also central to production planning and process optimization. Algorithms can simulate entire production workflows using digital twins (virtual replicas of physical manufacturing environments) enabling planners to test configurations, material mixes, and machine setups in a simulation before deployment. This allows companies to model supply disruptions, optimize production sequencing, and minimize energy use. In process industries like chemicals or food and beverage, AI uses multivariate analysis to maintain consistent quality and adjust variables. It can simultaneously control temperature, pressure, and ingredient flow in real time, ensuring efficient and sustainable operations.

Together, these AI applications form the foundation of what is increasingly referred to as Industry 4.0, a data-driven, highly automated, and intelligent manufacturing and logistics environment.

Implementation Challenges for Supply Chains

While AI technologies hold transformative potential for supply chains, from autonomous vehicles and intelligent assembly lines to predictive production analytics, their implementation in real-world business environments is often far more difficult than anticipated. These challenges span technical integration, organizational readiness, cultural resistance, regulatory constraints, and the need for new governance models. Understanding these barriers is important for companies to move beyond pilot projects and into scalable, enterprise-wide adoption.

Here are some of the most significant of these barriers:

1. **Complex Integration with Legacy Systems.** Most large firms operate with deeply entrenched legacy infrastructure (ERP systems, AI-driven platforms, and custom-built logistics software) that were never designed for AI or real-time data processing. Integrating new AI technologies with these systems is technically complex and often requires substantial middleware development, custom application program interfaces (APIs), and data transformation pipelines. For example, deploying AI-enabled autonomous vehicles within existing logistics systems demands seamless data exchange among fleet management, route planning, warehouse scheduling, and inventory databases. Without this integration, AI tools cannot function as intended or deliver actionable value.
2. **Data Fragmentation and Quality Issues.** AI depends on vast quantities of high-quality, labeled data, something many organizations lack. Supply-chain data is frequently siloed across departments, inconsistent in format, or riddled with inaccuracies. In manufacturing environments, sensor data may not be reliably collected or stored in a usable format. For autonomous vehicles or AI vision systems to work effectively, they must be trained on diverse, accurate data sets, ranging from traffic patterns and road conditions to object detection in complex industrial environments. Poor data quality leads to poor AI performance, undermining confidence and delaying deploy-

ment.

3. **Organizational Resistance and Workforce Misalignment.** One of the most under-appreciated challenges is cultural and organizational resistance. AI often changes workflows, displaces traditional job roles, and requires new skills. Plant operators may distrust predictive maintenance tools that override manual inspections; logistics staff may fear that automation will eliminate jobs; and senior managers may not fully grasp the strategic implications of AI. Without strong leadership, reskilling initiatives, and communication, these changes breed resistance and stall progress. In addition, cross-functional collaboration among IT, operations, supply chain, and data science is rarely the norm, making AI deployment feel fragmented and slow.
4. **Cost and ROI Concerns.** AI implementation is capital-intensive, particularly in physical domains like autonomous vehicles and robotics. Companies must invest in hardware, infrastructure, integration, testing, and ongoing support. Even when pilot projects show promise, scaling them across multiple sites or geographies involves significant expenditure. Many firms struggle to justify these investments without a clear line of sight to measurable ROI. Also remember that AI outcomes are probabilistic, not deterministic. That can make performance harder to predict and success harder to quantify in traditional business metrics.
5. **Safety, Regulation, and Accountability.** AI systems, especially autonomous vehicles and industrial robots, must meet stringent safety and compliance requirements. They are efficient but can be very hazardous. In sectors like automotive, food production, and pharmaceuticals, regulatory oversight is intense. Missteps can lead to operational disruptions, fines, or reputational damage.

Also, AI introduces new questions around accountability: If an autonomous forklift damages goods or a machine-learning model halts production in error, who is responsible? The software vendor, the AI developer, or the company's IT department? Businesses need clear governance structures to handle these risks, yet few have mature AI governance practices in place.

As we can see, the path to AI implementation across supply chains has many pitfalls. While the technologies themselves are maturing rapidly, the human, organizational, and infrastructural elements of implementation often lag behind. That is why implementation has to address these pillars simultaneously. To overcome these challenges, companies must adopt a comprehensive change-management approach, investing not just in tools but also in talent, cross-functional collaboration, and ethical governance.

9.4 AI-Implementation Maturity Curve

Like any large endeavor, implementation occurs in stages. Even established companies struggle to implement AI effectively and have to monitor their progression through the stages of what is known as a maturity curve. AI maturity curves are widely used to assess where an organization stands in its AI journey, and to guide the progression from early exploration to enterprise-wide adoption.

Organizations follow a journey as they build their AI capabilities over time. In Chapter 8 we described an AI roadmap that companies typically follow. They evolve from early experimentation to full AI transformation. These stages are listed in Table 9.3 and described next.

1. **Ad Hoc Experimental AI.** At this stage, AI efforts are informal, isolated, and largely experimental. Organizations are testing AI technologies without a clear enterprise strategy or governance model. Projects are typically driven by enthusiastic individuals or small innovation teams, often without executive sponsorship or defined KPIs. Common use cases include pilot chatbots, basic dashboards, or AI proof-of-concept models. However, these rarely make it into production. For example, a regional retail

chain might launch a pilot to use computer vision in one store to test shelf monitoring. While technically promising, the initiative lacks scalability due to unclear data integration plans and little cross-functional buy-in. Results then become localized and hard to generalize.

2. **Foundational.** In this phase, organizations begin to recognize the strategic potential of AI and invest in the foundational elements required for scaling. This includes setting up data infrastructure (e.g., data lakes and cloud platforms), hiring data scientists, and establishing preliminary governance structures. AI projects may still be siloed but are now better funded and aligned with specific business needs, such as fraud detection or demand forecasting. A good example here is Delta Airlines, which moved into the foundational stage when it began using predictive analytics for flight delays and maintenance scheduling. The company invested in IoT sensors and a centralized data platform to support basic machine-learning models, reducing flight disruptions and improving efficiency.
3. **Integrated.** At this stage, AI is no longer confined to an IT department or data teams. It becomes embedded in core business processes across departments such as marketing, supply chain, finance, and HR. Cross-functional collaboration is standard, and AI tools support decisions at both operational and strategic levels. The organization adopts MLOps (Machine-Learning Operations), a set of practices that automate machine-learning models to manage the AI life cycle. For example, Unilever integrates AI into product development, supply-chain planning, and sustainability initiatives. Its AI-driven demand forecasting system works seamlessly with their ERP systems, and AI tools assist R&D teams in formulating eco-friendly products faster. AI is viewed as a lever for both efficiency and innovation.
4. **Scaled.** In the scaled stage, AI becomes a competitive differentiator. Models are deployed at enterprise scale, often using cloud-native infrastructure and APIs. AI applications span the customer journey, from personalization to real-time logistics, and are monitored for performance, bias, and drift. Governance is now mature, data literacy is widespread, and executive leadership champions AI as a core pillar of business growth. Amazon exemplifies this stage. At Amazon, AI powers nearly every aspect of its operations: personalized recommendations, warehouse automation, dynamic pricing, fraud prevention, and supply-chain optimization. Machine learning is embedded in both customer-facing services and backend logistics at scale.
5. **Transformational.** Organizations at this level do not just use AI; they are redefined by it. AI is woven into every aspect of the business, enabling entirely new business models, digital products, and customer experiences. AI is not just a support tool. It is a strategic engine. These organizations continually reimagine processes using AI and influence entire ecosystems (e.g., suppliers, partners, and regulators). Ethical, explainable AI practices are deeply embedded, and human–AI collaboration is standard. Tesla operates at the transformational stage. AI is not just supporting its autonomous driving features, it is also the foundation of its entire vehicle operating system. Tesla's self-learning fleet and OTA updates represent a new model for vehicle development, customer interaction, and global mobility innovation. Tesla's continuous data feedback loop allows it to evolve its product post-sale, reshaping what a car company can be.

However, the AI-maturity curve is not linear, as you can see in Figure 9.4. Companies tend to go through a nonlinear progression, where some stages (like moving from integrated to scaled) take significantly more time, investment, and organizational change than others. For example, moving from ad hoc to foundational may be relatively fast. However, achieving scaled and transformational stages requires deep integration, cultural alignment, and governance, thus taking considerably longer.

Table 9.3 AI-Implementation Maturity Stages

Stage	Name	Description
STAGE 1	AD HOC/ EXPERIMENTAL AI	• AI used in isolated projects • Pilot projects in silos • No AI governance • Limited data infrastructure • Low executive awareness
STAGE 2	FOUNDATIONAL	• Organization recognizes value of AI • Begins building infrastructure and strategy • Data initiatives under way • AI team formation • Executive level buy-in • Some governance policies emerging
STAGE 3	INTEGRATED	• AI aligned with business goals • AI integrated into select processes • Cross-functional collaboration • AI use case tied to KPIs • AI embedded in workflows • Basic ML operations in place
STAGE 4	SCALED	• AI is scaled across functions with reusable assets • AI uses standardized platforms • AI CoE established • Enterprise data lake or mesh • Models reused across units • Strong governance and compliance
STAGE 5	TRANSFORMATIONAL	• AI is a core strategic asset • AI drives continuous innovation and autonomous systems • AI drives automation • Self-learning systems • Culture of experimentation • ESG and ethical AI embedded

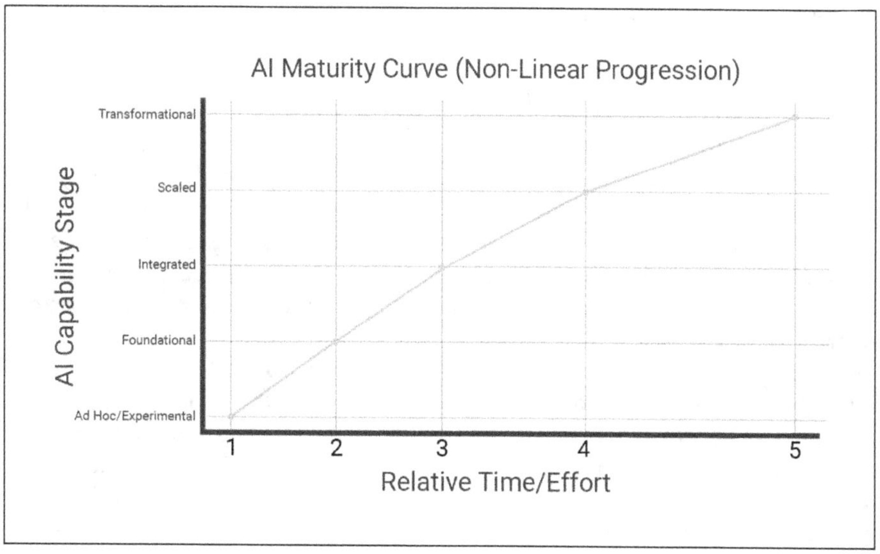

Figure 9.4 Organizational AI Maturity Curve

9.5 Scaling AI Implementation

The successful implementation of AI in one part of an organization marks an important milestone, but it is not the finish line. True value from AI emerges when pilot projects are scaled across the enterprise to drive widespread business transformation. Scaling AI involves moving beyond isolated use cases to embed intelligent systems into core business functions, creating a network of capabilities that can learn, adapt, and evolve with the organization. This process, while essential for unlocking ROI, presents many challenges that must be addressed with strategic planning, cross-functional coordination, and strong governance.

9.5.1 What Is Scaling?

Scaling AI refers to the process of expanding successful AI solutions from limited trials or departmental pilots to broader, enterprise-wide deployment. This includes replication, or applying proven AI models across multiple business units or geographies. It also requires integration, or embedding AI into everyday business processes and enterprise systems (e.g., ERP, CRM, SCM). Next, it requires operationalizing AI with an infrastructure, standardized workflows, monitoring performance, and life cycle management. It also requires equipping the workforce with the tools, training, and mind-sets necessary to use and trust AI outputs.

In essence, scaling transforms AI from an experimental tool to a core capability that is systemic, reliable, and sustainable.

Why is scaling AI challenging?

Despite initial success, many organizations struggle to scale AI effectively. Scaling an AI pilot into an enterprise-wide capability is rarely a straight technical upgrade. It is a systemic change that exposes weak links across data, technology, people, and governance.

Most pilots are fed with carefully curated data sets. Once the use case has to draw on production data from finance, supply chain, marketing, and customer service systems, the reality of fragmented data ecosystems surfaces. Siloed databases, mismatched formats, and conflicting master-data definitions turn simple integration into a multi-million-dollar plumbing exercise, and poor data quality quickly erodes model accuracy. At the same time, many proofs of concept are put together on legacy platforms and cannot deliver performance at full scale enterprise level.

Even with the right architecture, scaling stalls if the organization lacks the talent and literacy to run it. AI at scale is a team sport. Most firms struggle to assemble and retain the right mix of people. Also, the scarcity of experienced AI product managers or analytics translators leaves initiatives stranded between technical promise and business reality.

Then there are cultural issues that compound the problem. Employees who were comfortable with a confined pilot may resist when AI recommendations begin to alter day-to-day decisions, compensation, or job scope. This requires cultural investments to go hand in hand with formal governance. As models touch more customers and critical processes, executives must answer regulators, boards, and the public on issues of bias, privacy, accountability, and auditability. Without explicit ownership, version control, and fairness checks, every additional model rollout multiplies the organization's risk surface.

Finally, senior leaders typically do not approve funding unless the business case for scaling is convincing. This depends on the outcomes of pilots. However, performance metrics used in pilots are often limited in what they measure or deliver narrow cost-savings. During pilot projects it is important to have clear value roadmaps that link AI outcomes to metrics important to senior leaders: revenue growth, working-capital reduction, or customer experience. These will all be important to get sustained financial investment for AI.

Here is a summary of these AI challenges:

1. **Fragmented Data Ecosystems.** Scaling requires unified, high-quality data. Many companies operate in silos, making data integration complex and expensive.

2. **Technical Debt.** Pilot projects often rely on bespoke setups or legacy systems that are not built for scale or real-time performance.
3. **Lack of Talent and AI Literacy.** Scaling requires cross-functional teams, including domain experts, engineers, data scientists, and change agents.
4. **Organizational Resistance.** Cultural pushback and lack of trust in AI can hinder adoption beyond the initial use case. This is why culture, mentioned earlier, is critical.
5. **Weak Governance.** Without clear accountability, auditability, and fairness measures, risks such as bias or unintended consequences increase with scale.
6. **Unclear ROI.** Senior leaders may be reluctant to invest further without clear metrics or a roadmap showing enterprise value.

If any one of these elements is weak, attempts to push AI beyond the lab will stall—making scaling not just a technical challenge but a strategic one.

9.5.2 Scaling AI Effectively

Pilots serve an essential purpose: they validate feasibility, demonstrate value, and build internal confidence in AI capabilities. However, pilots are not the end goal. The true value of AI is realized when it is integrated into core processes, workflows, and decision-making systems at scale. Enterprise-wide implementation unlocks network effects, standardization, and cumulative learning across business units.

To scale AI, companies must treat it as a capability-building journey rather than a one-time deployment. Scaling requires some of the same deployment issues as implementation, prioritizing high-impact, low-complexity use cases and also establishing CoEs to support and measure AI deployments.

The first of these is establishing a clear vision for AI that aligns with business goals. However, not just for the pilot project, but for the entire enterprise and its supply chain. This requires getting top-down commitment and cross-functional ownership. As we discussed earlier, leadership sets the tone and direction for the organization.

So, what does leadership need to do for scaling?

1. **Develop Infrastructure.** Scaling requires putting in place "connective tissue" for the entire enterprise. It means data infrastructure and connective platforms. An important option is to invest in cloud-native platforms and unified data lakes. Also critical is standardizing data pipelines and ensuring governance and security. Without such a broad infrastructure in place, scaling AI initiatives cannot take place.
2. **Operational Integration.** This means ensuring continuous integration, delivery, and monitoring of AI models across systems.
3. **Culture & Governance.** This involves upskilling employees through AI literacy programs; engaging business teams early to co-design use cases; building a governance framework for model validation, explainability, and compliance; and defining performance KPIs for outcomes.

COMPANY EXAMPLE: Unilever (Part II)

Once again Unilever offers a compelling example of scaling AI effectively. After early success with machine learning for demand forecasting, the company developed a global strategy to scale AI across its supply chain, marketing, and sustainability initiatives. Unilever followed the exact steps we discussed in this chapter:

Phase 1 Pilot Projects: Unilever initially deployed AI in local markets to improve forecasting accuracy using machine-learning models that analyzed historical sales, promotions, and weather data. These early pilots demonstrated improved service levels and inventory reduction.

Phase 2 Building Infrastructure: To scale, Unilever invested in a cloud-based data lake and partnered with Microsoft Azure. This allowed the integration of more than 1000 data sources and enabled real-time visibility across global operations.

Phase 3 Operational Integration: Unilever developed an internal platform called Digital Factory that allowed business units to access and deploy AI models on demand. These models were embedded in ERP and supply-chain systems, enabling intelligent demand sensing, automated procurement, and optimized production planning.

Phase 4 Cultural and Governance Shift: Recognizing that people are central to AI success, Unilever launched AI upskilling programs for more than 30,000 employees. The company also established an AI Governance Board to oversee ethical use, data compliance, and model risk management.

The results of the scaling were hugely significant. Forecasting accuracy improved by 15 to 20 percent while inventory costs were reduced by $1 billion globally. Customer satisfaction improved through more agile supply response. AI was also applied in marketing (e.g., dynamic ad targeting) and sustainability (e.g., optimizing energy use in factories).

Scaling AI is not simply a matter of duplicating successful projects. It requires a concerted effort to transform infrastructure, processes, governance, and culture. The real competitive advantage lies in moving from scattered wins to a systemic capability that enhances decision-making, resilience, and innovation. Companies like Unilever demonstrate that, with the right strategy and leadership, AI can be scaled to deliver transformative enterprise-wide impact.

9.6 S&OP

We end this chapter with a discussion of S&OP as it is one of the most practical entry points for implementing AI across the enterprise and supply network. Although we discussed S&OP in earlier chapters, it was in broad terms. Let's look at S&OP in more detail, as this is an important bridge to cross-functional AI implementation.

Unlike isolated AI experiments, S&OP naturally connects multiple functions (sales, marketing, finance, supply chain, and executive leadership) around a shared process, shared data, and shared outcomes. This makes it ideal for piloting and embedding AI technologies in a controlled, governance-rich environment.

Because it follows a recurring monthly cadence, with defined inputs, decision points, and KPIs, S&OP provides a closed feedback loop. It is an ideal setup for ML models to learn, iterate, and improve over time. Companies can deploy predictive demand algorithms, test supply-constrained optimizers, and use AI to narrate executive summaries, all within an existing organizational rhythm.

In this way, S&OP does not just benefit from AI, it actively matures an organization's AI readiness by exposing gaps in data, collaboration, and decision governance that must be addressed before broader implementation.

9.6.1 What Is S&OP

Sales and operations planning is the management process that aligns demand, supply, and financial targets into one executable game plan, typically on a monthly cadence. It was conceived in the mid-1980s as a way to synchronize production and demand. It has since evolved into today's integrated business planning (IBP) discipline, yet the core idea remains the same: cross-functional leaders gather around a single, trusted set of numbers to steer the business ahead of the market.

S&OP is unique in that it reconciles market demand (sales and marketing forecasts) with operational capability (manufacturing, procurement, logistics). By nature, it is cross-functional and converts a group consensus into a supply plan and financial outlook. S&OP creates a forward-looking lens (usually 12–24 months) so leadership can see imbalances in sales and production early enough to act—before customers, cash, or capacity are at risk. This rhythm repeats every month, with a complementary weekly S&OE (sales and operations execution) meeting to manage near-term exceptions. As such, S&OP delivers many hard and soft benefits as listed in Table 9.4.

Table 9.4 S&OP Delivers Hard and Soft Benefits

Hard Benefits	Soft Benefits
Improved customer service	Improved team work
On-time deliveries	Tight linkage between strategic plans and day-to-day activities
Reduced inventory levels	More effective communications
Improved inventory turns	Better informed decision making
Improved plant efficiency	More advanced financial plans
Less plant downtime	Greater accountability
Reduced logistics costs	Larger amount of control

Technology is essential for this process, from advanced forecasting engines to AI-driven digital twins. It allows everyone to see the data and focus on trade-offs. PepsiCo, for example, is now sharing real-time checkout data with major retailers and using AI to sharpen demand forecasts at pack-size level, an illustration of where modern S&OP is heading.

9.6.2 The S&OP Process

The S&OP process is conceptually simple to understand. A properly working S&OP process aligns the company's supply and decision making with the customer's product demand and desired service levels, on a monthly basis, while looking at a 12- to 24-month planning horizon. This process is composed of five distinct steps: generation of month-end reports, demand planning, supply planning, alignment meeting, and executive S&OP meeting. It is important to remember that these steps make up a cycle that is continuous in nature. Step 5, which occurs at the end of the cycle, leads to step 1, which is the beginning of the new cycle. Mistakes from the previous cycle serve as lessons for the next cycle following continuous improvement principles (Figure 9.5).

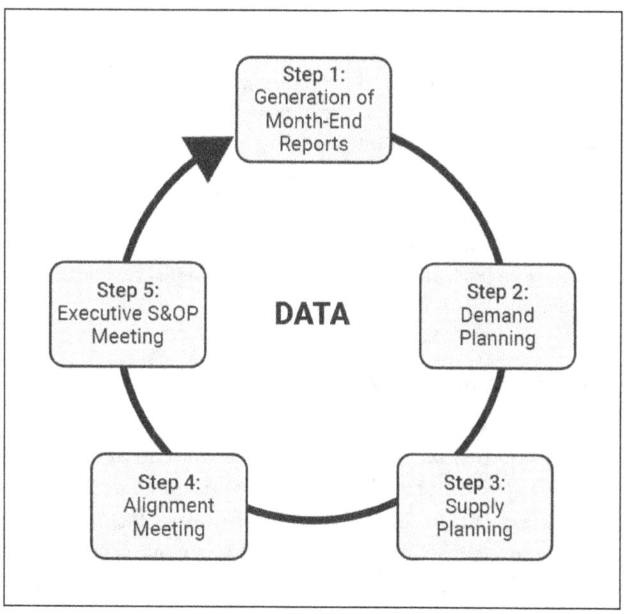

Figure 9.5 Monthly S&OP cycle

There are five steps to the monthly S&OP cycle:

Step 1: The monthly S&OP process begins with the generation of month-end reports. The data is formatted and prepared for the demand planning team. In addition, production and inventory data is sent to operations to be used as part of their evaluation of manufacturing capacity. Having clean, reliable, and trustworthy data is essential, as it will drive decisions.

Step 2: The next step is demand planning, which is the responsibility of sales and marketing with support from product development. The goal is to generate a quantitative demand forecast indicating expected demand, sales, and the resulting expected revenue. Statistical analytics in conjunction with an assessment of current and future business conditions aids in developing this forecast. This single demand plan is passed on to the supply planning process team.

Step 3: The supply planning process is the responsibility of operations, including purchasing, materials planning, and manufacturing. This determines whether the company has the necessary inventory and manufacturing capacity to meet the forecasted demand plan. In addition, financials are evaluated for material and human resources necessary to meet the required capacity. The supply planning team proposes adjustments to the demand plan where it is necessary or practical.

Step 4: This is a joint pre-executive management meeting to review the adjusted demand forecast and resolve any outstanding resource or capacity issues. This meeting must include sales, marketing, product development, operations, sourcing, finance, and logistics.

Step 5: The fifth and final step is the executive S&OP meeting. This is a high-level, fact-driven review with the goal of finalizing the monthly demand forecast and the associated operations plan. In addition, it gives executive management the opportunity to modify these two monthly plans to fill identified gaps between future projections and the company's strategic objectives as a result of changes in the internal and external business environments.

COMPANY EXAMPLE: Unilever (Part III)

Unilever embarked on a global S&OP overhaul in the early 2020s, rebranding it as IBP but retaining the five-step backbone. Key design points are

1. a single global template hosted on a cloud planning platform, so that every country runs the same cycle and KPIs;
2. a rolling 24-month horizon linked directly to the corporate P&L, allowing finance to lift and drop the approved S&OP numbers into quarterly financial forecasts;
3. AI-assisted forecasting that blends statistical models with external demand signals (weather, social sentiment) for short-life products; and
4. tightly defined escalation rules mean that for any demand-supply gap, €3 million is routed to the executive S&OP within 48 hours.

Results after two full cycles included a 20 percent reduction in supply-chain waste. There was a 6 percent increase in return on invested capital (ROIC) driven by leaner inventory and higher on-shelf availability. Beyond the metrics, Unilever's leadership highlights a softer win: "finance, marketing, and supply chain now speak the same language," shortening decision loops from weeks to days when markets shift.

9.6.3 Data: The Foundation for S&OP

S&OP is not just a technical process. It is a structured conversation with members of different departments. And like any good conversation, it depends on a shared language. In S&OP, that language is data. Without trustworthy, timely, and clearly presented data, the process breaks down into misalignment, miscommunication, and missed opportunities. Remember that S&OP requires different departments to come together on agreement. When done correctly, S&OP becomes less about defending departmental agendas and more about making integrated, data-informed trade-offs that align with enterprise goals.

S&OP requires aggregating inputs from multiple systems and silos, such as demand forecasts, supply constraints, inventory levels, financial targets, and customer signals. However, more data does not always mean better decisions. The goal is to deliver the right data, in the right format, at the right time. Too much irrelevant or low-quality data can paralyze the process.

Two key decisions are critical at the outset:

What data matters most?

It is important to identify the few critical data elements that drive most of the variance in forecasts, supply plans, and profitability.

How should data be aggregated?

One answer is to align data granularity with the planning time horizons: weekly for demand sensing, monthly for consensus planning, quarterly for financial alignment.

AI can assist here by identifying data patterns, filling gaps through statistical estimation, and even recommending which data streams have the highest predictive power. The reliability of the S&OP process is only as strong as the consistency of the data that supports it. That means data must be continuously cleansed and validated. Version control must be enforced, especially when multiple functions access the same data sets. Ownership must be clear—every data set used in S&OP should have a defined steward responsible for accuracy and timeliness.

Last –something you may not have considered - effective S&OP requires intelligent visualization. This relates to how the data is visually presented. For executive teams to make fast, high-quality decisions, the data must be presented in an intuitive visual format. This may involve AI-powered dashboards and digital twins that do more than report facts—they can simulate and visually present scenarios (e.g. demand surges, supplier disruptions). They can highlight key exceptions and root causes so that, instead of overwhelming leaders with data, these tools focus their attention on what matters most.

Chapter Review

Key Points

1. Implementation & scaling are dual challenges for enterprise AI.
    - Implementing AI is complex due to technical, organizational, and cultural barriers.
    - Many companies succeed in pilot projects but struggle to move beyond proof of concept.
    - Scaling AI involves not just replication but also integration into core workflows, decision-making processes, and enterprise systems.
    - Dual challenges arise because implementation demands depth (customization, alignment, change management), while scaling demands breadth (repeatability, infrastructure, governance).
2. Three pillars of implementation are: People, Processes, and Technology.
    - People: Refers to leadership, cross-functional teams, training, and change readiness.
    - Processes: Involves aligning AI with business workflows, data governance, and operating models.
    - Technology: Includes the AI models, data infrastructure, cloud platforms, and integration capabilities.
3. The three pillars must be managed simultaneously for AI success.
    - AI initiatives fail when one pillar is prioritized over others (e.g., focusing on tech but ignoring users or workflows).
    - Coordination is essential. Tech must enable processes, and people must adopt and trust the system.
    - Cross-functional collaboration, executive sponsorship, and feedback loops are critical for synchronized progress.
    - Success depends on managing organizational change along with technical development.
4. AI implementation has a maturity curve.
    - The curve outlines stages from experimental pilots to enterprise-wide AI integration.
    - Key phases:
        1. Explore: isolated experiments, low coordination.
        2. Pilot: proof of concept with defined metrics.
        3. Adopt: scaling to specific functions with change management.
        4. Integrate: AI embedded into enterprise systems.
        5. Transform: AI shapes strategy, innovation, and business models.
    - Maturity reflects both technical capabilities and organizational readiness.
5. Effectively scaling AI implementation requires the following:
    - Identify high-impact, repeatable use cases that align with business strategy.
    - Build reusable data assets, model libraries, and platform capabilities.

- Establish centralized governance (e.g., AI CoE) with local deployment flexibility.
- Invest in training and culture to ensure adoption and trust at scale.
- Measure ROI and continuously monitor performance to justify expansion.

6. S&OP is a practical entry point for AI implementation.
    - S&OP is data-driven, cross-functional, and operationally critical, making it ideal for early AI applications.
    - AI enhances demand forecasting, inventory planning, and scenario analysis within S&OP.
    - It offers visible, measurable results that can build trust in AI capabilities.
    - Success in S&OP creates a foundation for broader AI adoption in supply chain and enterprise functions.

Discussion Questions

1. Why is it important for companies to align their AI implementation with business strategy rather than simply following a technology roadmap? Consider how the contrasting approaches of Amazon and Unilever reflect different strategic priorities and long-term outcomes.
2. What are the main challenges companies face when trying to share data across supply chains, and how can these be overcome? Explore technical, organizational, and trust-related barriers, particularly in cross-company collaboration.
3. In building AI capabilities, how should companies balance investments in short-term tools versus long-term organizational learning and infrastructure? Reflect on Unilever's phased approach and how it has contributed to scalable success.
4. How do leadership, process integration, and the right technology platform interact to create a strong foundation for AI implementation? Use the examples of Capital One, Microsoft, and Tesco to examine how different firms prioritize these pillars.
5. Where do most organizations stall on the AI-implementation maturity curve, and what strategies can help them move toward a transformational stage? Discuss how maturity models can guide self-assessment and long-term planning.
6. What makes scaling AI fundamentally different from simply deploying more AI pilots, and what factors most commonly determine success or failure at scale? Evaluate how Unilever's approach in Part II demonstrates effective scaling beyond experimentation.
7. Why is S&OP such a critical use case for AI implementation, and how does data quality shape the success of this process? Discuss Unilever's S&OP journey and the importance of integrating AI into cross-functional decision making.

Case Study

Nike's AI-Driven S&OP Transformation

Nike offers a powerful example of how artificial intelligence can transform S&OP into a dynamic, intelligence-driven process that enables resilience and customer-centricity. As part of its broader Consumer Direct Acceleration (CDA) strategy launched in 2020, Nike undertook a multiyear initiative to integrate predictive analytics, ML, and real-time data into its end-to-end supply chain, reshaping traditional planning processes to respond faster to global market volatility.

Strategic Context—From Forecasting to Insight-Driven Operations: Faced with fluctuating demand, digital sales growth, and supply disruptions during the COVID-19 pandemic, Nike recognized the limitations of traditional forecasting methods. The company responded by expanding its use of AI in the S&OP process to enhance demand sensing, scenario planning, and inventory optimization. By incorporating external signals such as macroeconomic trends, digital engagement metrics, weather patterns, and regional COVID-related closures, Nike moved beyond historical sales data to anticipate changes in demand with greater precision.

AI-enhanced S&OP—Human-Machine Collaboration in Action: At the core of Nike's S&OP transformation is a closed-loop planning ecosystem where AI-driven tools continuously feed updated insights into decision-making cycles. Machine-learning algorithms analyze massive volumes of structured and unstructured data, identifying patterns that influence product preference, delivery timelines, and market-specific shifts. This real-time intelligence is then synthesized by cross-functional teams—including marketing, finance, supply chain, and merchandising—who use the outputs to make strategic trade-offs regarding production volumes, inventory placement, and channel allocation.

Rather than replacing human planners, AI augments them. Nike's planning teams act as orchestrators, validating machine-generated forecasts, setting priorities, and navigating trade-offs. For example, when a supply constraint emerges, planners can simulate various fulfillment scenarios using AI tools and assess their effect on customer satisfaction, margin, and market share.

Tangible Results and Metrics of Success: Nike's results underscore the business value of AI-augmented S&OP:

Inventory Efficiency: Nike reported a reduction of more than 10 percent in excess inventory by aligning supply with predicted demand more accurately.

Forecast Accuracy: Enhanced models improved demand forecast accuracy significantly, particularly for seasonal and fast-moving product lines.

Lead Time Reduction: Digitally enabled planning cut lead times, allowing faster replenishment and adaptation to shifts in e-commerce demand.

Improved Margins: By minimizing stockouts and markdowns, Nike improved gross margins while simultaneously elevating customer experience.

In fiscal year 2023, CEO John Donahoe emphasized that Nike's AI and analytics investments were "mission-critical" to building a direct-to-consumer engine that responds in real time. He credited the digital supply-chain transformation as instrumental in supporting Nike's strong growth in digital sales, which now account for more than 25 percent of total revenue.

Governance and Change Management: Behind the scenes, Nike also invested in governance structures, data integration platforms, and talent development. The company centralized its data architecture through its Enterprise Data & Analytics platform, ensuring that AI models operate on clean, unified, and reliable data. At the same time, Nike reskilled hundreds of employees in data literacy and agile planning techniques, fostering a culture of experimentation and digital collaboration across regions.

Lessons Learned

Nike's experience offers valuable lessons for companies seeking to embed AI into their S&OP processes:

AI is not a plug-and-play tool. It requires synchronized investments in data infrastructure, process redesign, and cultural readiness.

Cross-functional integration is essential. S&OP is where marketing meets operations, and human judgment is vital to interpret AI insights.

Speed and scalability matter. Nike's ability to replan rapidly helped it navigate disruptions in Vietnam (a key sourcing country), port delays, and shifting consumer demand during the pandemic recovery.

Questions

1. What role did artificial intelligence play in enhancing Nike's S&OP process, and how did it change the nature of decision making within the organization? How did this shift affect the balance between automation and human judgment?
2. Nike invested heavily in both technology and workforce upskilling. Why is it critical to align talent development with AI implementation, and what risks arise if this balance is ignored?
3. In the case of Nike, AI supported demand forecasting and inventory optimization. What additional areas of supply-chain planning could benefit from AI integration, and what challenges might arise in those areas?
4. How does Nike's approach to AI-enhanced S&OP reflect the concept of "human-in-the-loop" systems, and why is this hybrid model valuable in dynamic business environments?
5. Nike emphasized centralized data governance and platform integration as key enablers of AI success. How important is data governance in AI implementation, and what might go wrong without it?
6. Discuss how Nike's AI-driven planning helped mitigate disruption during the COVID-19 pandemic. What can other companies learn from Nike's agility and use of scenario modeling in times of crisis?
7. Nike's CDA strategy relies on responsiveness and personalization. How does AI-enabled S&OP support broader business strategy, and how can firms ensure that AI initiatives stay aligned with strategic goals?

Chapter 10

AI Governance, Culture & Leadership

In This Chapter You Will Learn

1. why AI governance is necessary for AI success,
2. the role of organizational culture,
3. steps in the change management process to implement AI,
4. the role of leadership in AI, and
5. what leaders should do to ensure AI success.

It is not the algorithms that transform organizations but rather the decisions and culture that guide them. Just like in our personal lives, it is not what tools we have, but how wisely we choose to use them, that shapes meaningful outcomes.

IBM launched Watson for Oncology with the ambitious goal of revolutionizing cancer treatment through AI. Developed in partnership with Memorial Sloan Kettering Cancer Center, the system was designed to assist doctors by recommending evidence-based cancer treatments. However, the project fell short of expectations and was eventually scaled back.

One of the key reasons for the struggle, was leadership overreach and cultural misalignment. Senior executives at IBM aggressively marketed Watson as a transformative AI solution before it was fully validated in clinical settings. This created unrealistic expectations among hospital partners and patients. Meanwhile, frontline clinicians, those expected to use the system, often distrusted its recommendations, finding them either irrelevant or inconsistent with local practice. Many doctors felt the system lacked the nuanced judgment required in complex medical decisions.

The leadership focused on rapid commercialization but failed to foster a culture of co-creation, trust, and iterative learning between AI developers and healthcare professionals. Additionally, insufficient attention was paid to change management; clinicians were neither fully trained nor included in shaping how AI fit into their daily workflows.

Ultimately, Watson for Oncology serves as a cautionary tale: *Even the most advanced AI cannot succeed without grounded leadership, cultural alignment, and a deep understanding of the human context.*

10.1 Transformation Required

AI has moved from experimentation to enterprise-wide deployment. Success is no longer defined solely by data quality or algorithmic performance. What increasingly determines the value, and risk, of AI is the strength of an organization's governance structures, the culture in which AI operates, and the leadership guiding its adoption.

Business leaders need to understand that AI is not just a technical tool; it is a strategic capability that reshapes how decisions are made, how work is structured, and how value is created. This

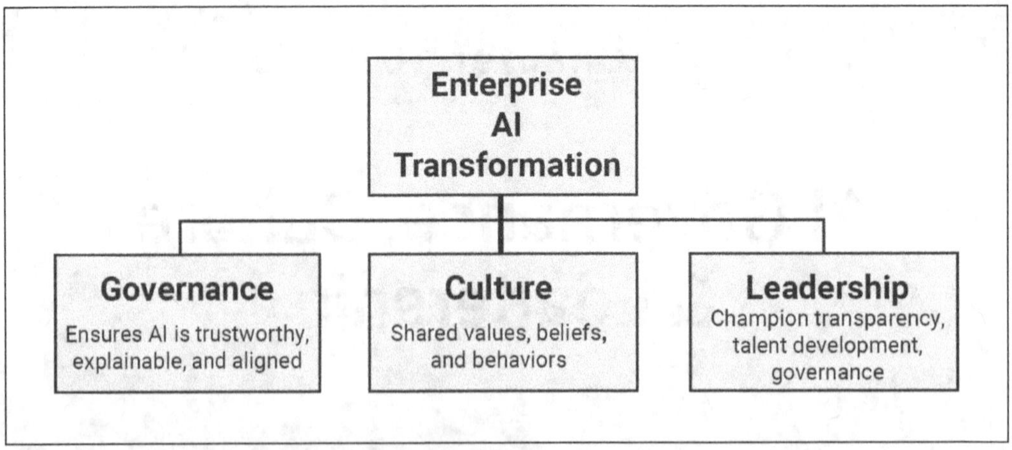

Figure 10.1 Foundations for AI Transformation: Governance, Culture, and Leadership

transformation requires clear frameworks for accountability, transparency, and ethical decision making. This is broadly called **AI governance.** Governance ensures that AI systems are trustworthy, explainable, and aligned with business goals and societal norms. Without it, organizations risk model failures, bias, regulatory penalties, and reputational damage. Simply put, without governance, AI initiatives will fall apart.

But governance alone is not enough. Organizations must also foster a culture of learning, adaptability, and digital confidence. AI often introduces uncertainty and challenges established norms. If the culture is risk-averse, siloed, or resistant to change, AI initiatives will stall, even with the best tools and plans. A culture that promotes experimentation, cross-functional collaboration, and responsible innovation becomes the foundation for scaling AI responsibly and sustainably.

At the center of all this is leadership. The role of AI leadership is not only to approve technology investments but also to set a vision, create alignment, and navigate the ethical and organizational implications of intelligent systems. Effective leaders champion transparency, prioritize talent development, and embed governance into the decision-making fabric of the enterprise. They recognize that AI success is less about machines and more about managing people, change, and trust.

In short, governance, culture, and leadership are not peripheral to AI—they are foundational. Together, they enable organizations to move from isolated pilot projects to responsible, enterprise-wide transformation that delivers long-term value.

Chapter 8, "The AI Roadmap," presented the DEFINE framework that outlines an organizational path of implementing AI. However, the implementation of the roadmap cannot happen without instituting organizational change. AI, although exciting and full of possibilities, has created a rapidly changing environment that requires adaptation and continuous improvement. Implementing the roadmap requires organizational change.

The following sections show you how to lead such a transformation. Let's first look at an example of why such a transformation is needed.

COMPANY EXAMPLE: Schneider Electric

Schneider Electric is a global leader in energy management and industrial automation, operating in more than 100 countries. By the early 2010s, the company faced mounting complexity from its vast geographic footprint, multiple manufacturing sites, and an increasingly digital customer base. In response, Schneider initiated a comprehensive transformation strategy: to digitize its supply chain and embed artificial intelligence across its operations.

1. STRATEGIC VISION AND ROADMAP

The transformation began with a clear strategic vision: to create a Tailored, Sustainable, Connected, Supply Chain 4.0. This initiative was part of a broader digital transformation roadmap aligned with the company's sustainability and customer-centric goals.

Key technologies implemented included the following:

- AI-powered control towers for predictive decision making
- IoT-enabled sensors for real-time data capture on factory floors
- Digital twins to simulate scenarios across manufacturing and logistics
- Machine-learning algorithms to optimize demand forecasting, inventory levels, and energy efficiency

The implementation was rolled out in phases across global regions, ensuring that local teams were involved, and systems were gradually integrated.

2. CULTURE CHANGE AND EMPLOYEE ENGAGEMENT

Cultural transformation was central to success. Schneider Electric recognized that digital tools would fail without workforce buy-in and a shift in mind-set. Management implemented a number of key actions to upskill workforce and get buy-in. It included the following:

- Training and reskilling programs were introduced across departments to build AI literacy, targeting not just IT staff but also operations, supply chain, and leadership.
- Digital ambassadors were appointed within business units to evangelize AI benefits and support change from within.
- A culture of data democratization was promoted, making data accessible and understandable to non-technical employees through dashboards and self-service analytics tools.
- A shift from traditional siloed operations to a collaborative, cross-functional mind-set was reinforced through integrated project teams.

This cultural investment helped reduce fear around automation, fostered experimentation, and cultivated a mind-set of continuous improvement.

3. LEADERSHIP COMMITMENT

Leadership played a pivotal role in the success of the transformation. CEO Jean-Pascal Tricoire publicly championed digital innovation as a core growth driver. One of the key leadership practices was modeling digital behaviors at the top. Executives engaged directly with data platforms and showcased use cases during town halls.

The company further worked to align KPIs and incentive structures with digital success metrics.

They established a Digital Transformation Office reporting directly to the executive committee to coordinate initiatives, ensure alignment, and track ROI. Partnerships with external tech firms (e.g., Microsoft) helped accelerate innovation and provided access to world-class AI infrastructure.

The resulting operational benefits were significant. The company saw a 25 percent reduction in forecasting error due to AI-enhanced demand sensing. They increased supply-chain resilience during the COVID-19 crisis, enabling real-time scenario modeling and faster supplier pivots. AI enabled smart factories that resulted in a 30 percent increase in productivity and reduced downtime through predictive maintenance. Last, they saw energy efficiency gains of 10 to 15 percent across automated facilities, supporting the company's sustainability agenda.

Schneider also experienced a significant cultural shift. Employee sentiment surveys indicated a growing sense of empowerment around data usage. AI was no longer viewed as a threat but rather as an enabler of better decision making. Cross-functional collaboration improved, as digital tools broke down departmental silos.

Other successes followed as well. Schneider was named a Global Lighthouse by the World Economic Forum for its smart factory in Batam, Indonesia. It was also recognized in the Gartner Supply Chain Top 25 for multiple years in a row due to digital excellence.

The lesson learned in this example is that transformation is far more than tech. AI systems alone cannot succeed without a cultural shift that embraces data and experimentation. Leadership must lead from the front. Executives must actively sponsor, use, and communicate the value of AI initiatives. Employee inclusion is critical in developing the right culture. Training, transparency, and cross-functional engagement are essential to reduce resistance and build momentum.

Remember that digital maturity is an ongoing journey. Schneider continues to evolve, using AI not just for efficiency, but also for innovation and sustainability.

QUESTIONS

1. How did Schneider Electric align its AI transformation with broader strategic goals such as sustainability and customer responsiveness, and why is this alignment critical for success?
2. What specific cultural barriers might companies face when introducing AI, and how did Schneider Electric address these through its digital ambassadors and training programs?
3. In what ways did leadership behavior influence employee buy-in and trust during Schneider Electric's AI transformation journey? Provide examples.
4. How did Schneider Electric's phased, cross-functional approach to AI implementation contribute to risk mitigation and organizational learning?
5. What lessons from Schneider Electric's AI journey can be applied to smaller firms with fewer resources, especially in terms of culture, leadership, and incremental transformation?

10.2 AI Governance

AI governance is a specialized subset of corporate governance that focuses on the responsible development, deployment, and oversight of artificial intelligence systems. While corporate governance broadly addresses the structures and processes for overall organizational accountability and ethical management, AI governance zeros in on issues like algorithmic transparency, bias mitigation, model validation, data privacy, and regulatory compliance in AI applications. It ensures that AI systems align with corporate values and societal norms, while corporate governance provides the broader framework within which AI governance operates.

More specifically, governance in the context of AI implementation refers to the framework of policies, procedures, roles, responsibilities, and oversight mechanisms that guide the ethical, effective, and accountable use of AI across an organization. It ensures that AI systems are aligned with organizational goals, comply with regulations, manage risks, and promote transparency and fairness throughout their life cycle, from design and development to deployment and monitoring.

10.2.1 Why Governance is Important

Governance ensures that AI initiatives are selected and designed to support the organization's strategic goals. As we discussed earlier, there must be **alignment** between AI initiatives and organizational strategy. Without alignment, AI projects risk becoming isolated tech experiments with limited business impact. Consider a retail company that may prioritize AI investments to engage in customer personalization. It is then up to the governance structures to validate alignment of the investment and ROI with the company's growth plans and profitability goals. A mere desire to enhance customer personalization is not enough.

Another important issue governance oversees is **risk management**. AI introduces new risks, including model bias, data privacy breaches, explainability gaps, and system failures. A governance framework identifies, assesses, and mitigates these risks. For example, financial institutions implement model risk governance to regularly audit AI credit scoring tools for fairness and regulatory compliance. Without governance, risks could easily lead to significant failure and losses, as there is no oversight.

Governance enables **ethical oversight**. Governance provides a mechanism to uphold ethical standards—such as fairness, accountability, and non-discrimination—in AI system design and deployment. Consider a health-tech company that must ensure that its diagnostic algorithms do not disproportionately misdiagnose based on race or gender. An ethics committee within the governance framework reviews models before deployment. Governance also provides oversight of regulatory compliance. Many jurisdictions now impose legal obligations on AI systems, including explainability, auditability, and user consent. Governance ensures adherence to these evolving laws (e.g., EU AI Act, General Data Protection Regulation [GDPR]). For example, an AI-powered HR screening tool must comply with employment law and privacy regulations, which is enforced through governance processes.

Governance also ensures transparency and, therefore, accountability. Without accountability initiatives easily get lost. Governance ensures there is clarity around who is responsible for AI outcomes, which fosters trust among internal stakeholders and external users. One example is establishing clear documentation protocols for AI model decisions. These protocols can help explain outcomes to customers and regulators if challenged. They also document a standardized and fair process.

Last, governance ensures ongoing continuous improvement and life cycle management. Remember that AI systems are not static. They require continuous monitoring, updating, and auditing. Today's technologies are tomorrow's legacy systems. Governance provides the structure for model maintenance, retraining, and retirement and replacement. Consider a supply-chain forecasting model that is reviewed quarterly under governance protocols to incorporate new data and changing demand patterns, as well as evaluation of new systems and technological capabilities.

AI governance is not just a compliance function; it is a strategic enabler that maximizes the value of AI while minimizing risks. It instills discipline, trust, and accountability, which are essential for scaling AI responsibly across an organization. Without governance, AI implementation can become fragmented, ethically questionable, or even legally problematic. As AI becomes embedded in enterprise operations and global supply chains, strong governance is essential to ensure that AI is implemented responsibly, strategically, and at scale. Governance is the system of rules, roles, and processes that guide how AI is selected, built, deployed, and monitored. Without a structured governance approach, organizations risk deploying AI in fragmented or unsafe ways, leading to poor outcomes, regulatory exposure, and loss of trust among partners and customers.

Let's look at an example of AI governance at Microsoft, showcasing how they implemented AI governance and the specific mechanisms they used.

COMPANY EXAMPLE: Microsoft (Part I)

As AI becomes increasingly embedded in enterprise functions, organizations face the growing challenge of ensuring that their AI systems are ethical, transparent, fair, and accountable. Microsoft stands out as a leading example of how to successfully implement AI governance in a way that is both structured and adaptable across business functions.

Microsoft's approach to AI governance has been both proactive and strategic. Recognizing early the potential risks associated with advanced AI systems—such as bias, lack of explainability, and unintended societal consequences—Microsoft formalized its AI governance around a core set of responsible AI principles: fairness, reliability and safety, privacy and security, inclusiveness, transparency, and accountability.

To operationalize these values, Microsoft created a multitiered governance structure:

1. Office of Responsible AI (ORA): This centralized function sets company-wide rules for responsible AI and ensures adherence across product teams.
2. The Aether Committee: A cross-disciplinary advisory body that evaluates sensitive AI use cases and provides guidance on ethical challenges.
3. Responsible AI Champions: Designated individuals embedded within product teams to ensure AI governance is translated into everyday decision making and product development.

One of the most notable successes of Microsoft's AI governance model was seen in its development of the AI behind Azure Cognitive Services and GitHub Copilot. Each product was subjected to a rigorous review process involving both technical assessments (e.g., model validation, robustness testing, and bias audits) and ethical considerations (e.g., impact assessments, stakeholder consultation, and scenario planning).

In the case of GitHub Copilot, a generative AI tool that assists programmers, the company paused its broader release until governance concerns were addressed. This included mitigating risks of code plagiarism, evaluating model training data sources, and ensuring output did not replicate biased or harmful patterns.

Several key factors contributed to Microsoft's successful AI governance implementation:

1. **Early commitment.** Microsoft began developing its responsible AI framework in 2016, giving it a significant head start in building institutional knowledge and trust.
2. **Cross-functional integration.** Governance was not siloed. By integrating legal, engineering, design, ethics, and business functions, Microsoft ensured that governance was practical, not just aspirational.
3. **Transparency and learning.** Microsoft regularly publishes updates, principles, and case studies about its responsible AI practices, promoting transparency and public accountability. This has helped improve its reputation and trust with enterprise clients and regulators alike.
4. **Scalable oversight.** Rather than relying solely on top-down enforcement, Microsoft empowered local product teams through Responsible AI Champions, ensuring governance was embedded at the team level.

Microsoft's experience demonstrates that AI governance can be both rigorous and flexible when it is tied to a company's strategic goals and ethical commitments. Their model has become a benchmark in the industry and has influenced the development of external AI governance guidelines, including contributions to the Organization for Economic Cooperation and Development and National Institute of Standards and Technology AI frameworks.

Microsoft's AI governance journey highlights the importance of clear principles, organizational structure, and continuous learning. By balancing innovation with responsibility, Microsoft not only mitigated risk but also enhanced product quality and trust, proving that good governance is not a constraint but a competitive advantage in the AI era.

QUESTIONS

1. What role did cross-functional collaboration play in the success of Microsoft's AI governance strategy, and how can other companies replicate this model?
2. How does Microsoft's approach to embedding Responsible AI Champions within product teams help bridge the gap between policy and practice?
3. Why is early commitment to AI governance, before products are fully deployed, critical for long-term success and trust?
4. How might Microsoft's governance structure (e.g., ORA and Aether Committee) need to evolve as AI technologies become more autonomous and generative?
5. To what extent can Microsoft's governance framework be applied to smaller organizations with limited resources? What adaptations would be necessary?

10.2.2 Governance Across Supply Chains

Unlike siloed IT initiatives, supply chains are inherently cross-functional, spanning procurement, manufacturing, logistics, planning, customer service, and finance. These functions rely on shared data and interdependent systems, making the coordination and oversight of AI applications far more complex. Governance ensures that AI use cases (e.g., demand forecasting, inventory optimization, supplier risk assessment) are aligned with enterprise objectives, built on consistent data, and subject to clear accountability mechanisms. It also provides a framework to manage risks related to bias, model drift, data privacy, and unintended consequences.

Further, supply chains are increasingly digitized and automated, requiring continuous oversight of autonomous decision-making tools. Without governance, companies may face operational inconsistencies, noncompliance with regulatory standards (e.g., GDPR, ESG reporting), or failures due to unmonitored algorithms.

There are steps in establishing AI governance. Although they can vary, here are some basics:

1. Establish a Cross-Functional AI Governance Board.

The first step is to form an AI governance board or steering committee composed of representatives from the following departments:

- IT and Data Science
- Supply Chain Operations
- Legal and Compliance
- Finance and Risk
- Business Units and Strategy
- Ethics and Sustainability Officers

This board should have the authority to approve AI projects, review model performance, assess ethical implications, and resolve conflicts across departments. Regular meetings and escalation procedures should be built into its charter. IBM exemplifies a significant approach to AI governance within its own operations, as well as supply chain and in client engagements. As part of its Watson Supply Chain platform, IBM established a governance model that includes a cross-functional AI board responsible for reviewing use cases, data quality, and compliance risks. Each AI deployment, such as predictive demand sensing or supplier risk analytics, undergoes internal validation and ethical review before full rollout. Data stewards monitor data health across global locations, and model performance is tracked against standardized metrics. Governance is baked into the platform itself, allowing IBM clients to adopt similar oversight practices. By investing in governance early, IBM accelerated adoption, reduced operational risk, and built stakeholder confidence, proving that governance is not a barrier to innovation but an enabler of responsible scale.

Another example is Unilever that established an AI Governance Council to oversee its use of AI in supply-chain forecasting and supplier management. The council included leaders from data science, supply chain, and compliance. It enforced model validation, ethical use standards, and audit trails to ensure fairness and transparency. Regular reviews aligned AI outputs with business goals. This structure enabled Unilever to scale AI responsibly while maintaining control and accountability.

2. Define Policies, Principles, and Standards.

The governance board should create a formal AI policy framework that includes Guidelines for AI use cases (e.g., acceptable versus prohibited applications):

- Data standards and interoperability protocols
- Model development, validation, and documentation requirements
- Bias detection and fairness assessments
- Privacy and cybersecurity protocols
- Guidelines for human oversight ("human-in-the-loop")

These policies should be embedded into enterprise AI workflows to ensure every deployment adheres to common standards, regardless of department or geography. For example, as part of its AI rollout in supply-chain operations, Schneider Electric created a set of clear AI policies and principles. These included a focus on **fairness**, where AI tools used for supplier scoring must be explainable and avoid bias. They also considered **transparency**, where all forecasting models had to provide interpretable outputs for planners. They also included **accountability**, where each AI system had an assigned owner responsible for monitoring performance and ethical compliance.

These standards were documented and embedded into procurement and logistics processes, ensuring consistent and responsible AI use across global operations.

3. Assign Roles and Responsibilities.

Clear roles must be defined for all parties involved in AI implementation:

- **AI Product Owner:** Oversees development and business alignment
- **Model Validator:** Reviews algorithms for accuracy, explainability, and bias
- **Data Steward:** Ensures data quality, lineage, and compliance
- **Process Owner:** Ensures that AI outcomes integrate into business operations
- **Ethics Adviser or Ombudsperson:** Flags ethical concerns and unintended impacts

These roles help distribute accountability while ensuring checks and balances. For example, when implementing AI for route optimization, the company Maersk assigned clear roles: data scientists built the models, supply-chain managers validated outputs, and a newly appointed AI operations lead ensured alignment with business goals and compliance standards. This role clarity accelerated adoption and minimized confusion during deployment.

4. Implement Monitoring and Life Cycle Management.

As soon as AI models are deployed, governance structures must ensure ongoing performance monitoring. This involves checking KPIs and monitoring for model drift, conducting regular audits and recalibrations, and creating incident response procedures for AI failures. This includes sunset or retirement plans for outdated models. Supply chain disruptions, changes in customer behavior, or external shocks (e.g., COVID-19) can make models obsolete quickly. Governance ensures these changes are addressed in a controlled and transparent way.

Consider the example of Dell. Dell implemented AI to optimize inventory across its global supply chain. After deployment, the company established a monitoring protocol to track model performance weekly. If accuracy dropped below a set threshold, a cross-functional team was alerted to retrain or adjust the model. Dell also set a review cycle every six months to reassess business alignment, data quality, and compliance. This life cycle management approach ensured the AI remained effective, ethical, and up to date.

5. Build Transparency and Communication Channels.

Transparency is key to trust, both internally and with external stakeholders. This means maintaining explainability of AI decisions, especially in procurement or logistics decisions that affect partners and creating reporting tools and dashboards to show AI's impact on supply-chain KPIs. Another important aspect is communicating ethical guardrails and success stories to build enterprise-wide confidence in AI.

A good example here is when Walmart rolled out its AI-driven demand-forecasting platform. The company prioritized transparency by creating dashboards that showed how forecasts were generated and updated. Planners across regions could see model logic, data inputs, and confidence levels. Regular cross-team meetings and a dedicated AI communication portal allowed users to ask questions, flag concerns, and suggest improvements. This openness builds trust and improved adoption across the supply chain.

COMPANY EXAMPLE: Amazon

Amazon's push into predictive logistics through its anticipatory shipping model offers a powerful example of both the potential and complexity of AI-led innovation. The initiative, patented in 2014, proposed a radical idea: ship products toward customers before they actually place an order. The concept relied on using customer data (search history, purchase patterns, wish lists, and even cursor activity) to forecast demand and proactively move inventory to regional fulfillment centers.

On paper, the benefits were compelling: faster delivery, higher customer satisfaction, and a competitive edge over brick-and-mortar retailers. However, Amazon's journey reveals that technical capability alone is not enough. The real challenge lies in governance—how data is used, how decisions are made, and how organizational systems align around AI.

Data Governance and Ethical Boundaries. At the heart of anticipatory shipping is a sophisticated data ecosystem. But turning behavioral signals into actionable forecasts raised profound governance questions:

- What level of surveillance is acceptable?
- How transparent must Amazon be about the use of customer data?
- How should consent and privacy be managed, especially across different jurisdictions?

Although Amazon has invested heavily in data security and compliance, anticipatory shipping tests the limits of what customers consider ethical. Without clear frameworks for data usage and algorithmic accountability, the risk of public backlash or regulatory intervention looms large.

Algorithmic Accountability and Explainability. Predictive models that decide where to ship products need to be explainable, auditable, and aligned with business objectives. Amazon had to ensure that

- models were free of systemic bias (e.g., not disproportionately favoring wealthy ZIP codes),
- models can be explained to internal stakeholders (why this product to this location?), and
- there are protocols in place for when predictions go wrong.

Governance structures around model validation, oversight, and feedback loops became essential, not just to reduce error but also to sustain trust across the organization and with customers.

Cross-Functional Coordination and Ownership. Operationalizing anticipatory shipping required breaking down silos among data science, supply chain, legal, compliance, and customer service teams. Governance failures, such as misalignment over who owns decision rights or failure to coordinate responses to prediction errors, could lead to costly misfires. Amazon's experimentation underscored the need for a dedicated AI oversight board or governance committee, Clear protocols for exception handling and continuous learning, and a unified KPI frameworks to track both performance and unintended consequences.

Balancing Innovation and Risk. Anticipatory shipping challenged traditional notions of inventory management by placing Amazon in a position of assuming demand risk without a confirmed order. Governance mechanisms were needed to assess and balance

- forecast confidence thresholds (When is it "safe enough" to ship?),
- financial exposure from pre-shipments that fail to convert, and
- legal risk from unsolicited deliveries.

In response, Amazon adopted a phased deployment strategy, applying anticipatory shipping first to low-cost, high-frequency products and fine-tuning models using controlled feedback from test markets.

Lessons in AI Governance: Amazon's anticipatory shipping model illustrates that implementing predictive AI at scale is as much about governance maturity as it is about data science. The case reveals five governance imperatives:

- Data ethics and transparency must be embedded into design and deployment.
- Cross-functional governance boards should oversee AI initiatives to manage risk and alignment.
- Model explainability and auditability are essential to prevent bias and ensure trust.
- Dynamic policy frameworks are needed to adjust models and strategies as the environment evolves.
- Innovation governance must weigh opportunity against operational and reputational risk.

While anticipatory shipping has not yet been fully commercialized in its most radical form, it remains a testament to Amazon's forward-thinking culture and a reminder that successful AI implementation hinges not just on technological sophistication but also on strong governance, ethical clarity, and organizational readiness.

QUESTIONS

1. Amazon's anticipatory shipping model relies on detailed behavioral data (e.g., cursor movement, browsing patterns). What ethical considerations arise from using this type of data to predict consumer behavior? How should companies balance personalization with privacy?
2. Predictive AI systems can influence logistics, inventory, and customer experience. What kind of governance structures (e.g., AI oversight board, ethics committee) should be in place to oversee such systems? Who should be included?
3. AI models might favor certain regions or demographics based on data availability or purchase patterns. How can companies ensure fairness in how anticipatory logistics systems serve customers? What measures can prevent regional or socioeconomic bias?
4. Predictive models need to be understood and trusted internally. How important is explainability in AI-driven logistics systems? What steps should Amazon take to ensure that its internal teams understand and can audit these models?
5. Shipping products before an order is placed carries financial and reputational risk. How should companies govern the trade-offs between innovation speed and operational risk in AI implementation?
6. Anticipatory shipping requires coordination among data science, operations, compliance, and customer service. What challenges might arise in aligning these functions? How can governance frameworks support effective cross-functional collaboration?
7. Privacy laws differ significantly across jurisdictions (e.g., GDPR in Europe versus more lenient US policies). How should a global company like Amazon adapt its AI governance practices across different regulatory environments?
8. Even if anticipatory shipping is efficient, customers might see it as intrusive. How should companies govern the public rollout of controversial AI capabilities? Should customers be informed or able to opt out?

AI governance is the backbone of successful, scalable AI implementation across enterprises and supply chains. It brings structure to complexity, accountability to innovation, and clarity to uncertainty. From forming a governing board to instituting policies, roles, and monitoring protocols, governance ensures that AI serves strategic objectives while minimizing risk. As supply chains become more data-driven and autonomous, companies that embed governance into their AI roadmaps will be best positioned to unlock long-term value with integrity and resilience.

10.2.3 Governing the AI Roadmap

As organizations increasingly integrate AI into their operations and decision-making processes, the importance of AI governance cannot be overstated. **AI governance** are the structures, policies, processes, and controls that guide the ethical and responsible development, deployment, and use of AI systems. Embedding governance into the AI roadmap is not a discrete step but rather a continuous thread that must be woven into every phase, from strategic alignment to system implementation and performance evaluation. Four foundational pillars of AI governance are particularly critical—model validation, fairness, explainability, and auditability:

1. **Model Validation.** Model validation ensures that AI systems perform as intended, both technically and functionally. It involves testing for accuracy, robustness, reliability, and alignment with business goals. Validation should be conducted before deployment and periodically thereafter, using realistic data sets and stress tests to uncover unintended behaviors under varying conditions. One good governance practice is to establish a formal model validation protocol as part of the AI roadmap. This includes independent review processes, cross-functional validation teams, and scenario-based testing. Also important is to create checkpoints in the roadmap. For example, during the implementation and measure phases, validation criteria should be revisited continuously as models evolve or retrain on new data.

2. **Fairness.** Fairness addresses the risk of bias in AI outputs, especially those affecting hiring, credit scoring, pricing, or resource allocation. Bias can emerge from historical data, algorithmic design, or operational context. AI governance must prioritize fairness to uphold legal compliance and preserve stakeholder trust. One governance practice here is to define fairness metrics that are aligned with the organization's values and regulatory requirements. Equally important is to implement bias detection tools and mitigation strategies early in model development. Good checkpoints in the roadmap are "fairness audits." These should be integrated into the roadmap particularly during data collection and AI development.
3. **Explainability.** Explainability refers to the degree to which AI decision-making processes can be understood and interpreted by humans. Without explainability, stakeholders—including regulators, customers, and employees—may struggle to trust, or even challenge, AI-driven outcomes. There are many models or post hoc explanation techniques (e.g., LIME [Local Interpretable Model-Agnostic Explanations] and SHAP [SHapley Additive exPlanations]) that provide clear insight into how decisions are made. They are beyond the scope of this discussion but there are many that can be used by the AI team. It is important to ensure that explanations are accessible to both technical and non-technical stakeholders. Also, a good checkpoint is to require that all AI systems have documented explanation mechanisms tailored to end users' roles and needs.
4. **Audit Trails.** Auditability ensures traceability of AI decisions and system changes. Maintaining audit trails supports transparency, regulatory compliance, and error diagnosis. One good governance practice is to create immutable logs of data sources, model parameters, versioning, and decision outcomes. Then, set up internal audit schedules and escalation procedures. A good checkpoint for audit readiness should be included in KPI reviews, with detailed records maintained throughout the life cycle of the AI application.

Let's look at a business example of these steps in governing the AI roadmap.

COMPANY EXAMPLE: JPMorgan Chase

JPMorgan Chase provides an example of embedding AI governance into the core of its enterprise roadmap. As one of the world's largest financial institutions, Chase relies heavily on AI to enhance fraud detection, automate customer interactions, improve operational efficiency, and optimize credit scoring. However, the bank's leadership recognized early on that these high-stakes applications carried significant regulatory, ethical, and reputational risks. As a result, JPMorgan developed a comprehensive Model Risk Management framework to govern the full AI life cycle, from model design to deployment and monitoring. Let's look at these steps.

1. Model Validation. JPMorgan mandates that all AI models undergo independent validation by teams that are structurally separate from those who develop the models. This ensures objectivity and reduces the risk of unchecked bias. One of the major hurdles here is the shortage of qualified AI validators, individuals with both domain expertise and deep understanding of machine learning. Training, recruiting, and retaining such talent is a continuous organizational priority for JP Morgan.

2. Fairness. Particularly in credit risk models, JPMorgan integrates fairness metrics aligned with the US Equal Credit Opportunity Act (ECOA) requirements, aiming to prevent algorithmic discrimination. A key issue, however, has been bias embedded in historical data, which can unintentionally penalize minority groups. Correcting for these requires delicate recalibration of models and continuous regulatory consultation.

3. Explainability. The bank uses explainable AI techniques, such as SHAP, to clarify how inputs influence decisions like loan approvals. SHAP is a method for explaining the output of ML models. It assigns each feature (input variable) a value that represents its contribution to a specific prediction, based on principles from cooperative game theory. SHAP helps make black box AI models more interpretable by showing how each input drives the final output. JPMorgan uses SHAP to increase model transparency and regulatory compliance, especially in high-stakes applications like credit scoring, fraud detection, and loan approvals.

How does this help?

As an example, when a customer is denied a loan, SHAP can explain which factors (e.g., income level, credit utilization, or payment history) contributed most to the decision. This allows JPMorgan to justify decisions to clients and regulators clearly. It can detect and mitigate bias by reviewing SHAP value distributions across demographic groups. It also builds trust by ensuring that AI-driven outcomes are explainable, auditable, and fair.

4. Audit Trails. JPMorgan maintains a centralized model and data inventory, including version control, model lineage, and automated logging. The challenge, however, has been ensuring consistent documentation and traceability across thousands of models in a federated enterprise system remains a challenge. Standardization and automation have been key strategies.

Below is a timeline of AI governance implementation at JPMorgan. Notice the gradual steps involved and that full implementation took close to ten years. This timeline is important to illustrate that there is a process that is followed and takes time.

Year	Milestone
2016	Identified need for governance standards.
2017	Formed the Model Risk Office to oversee AI governance.
2018	Piloted explainable AI tools for credit and compliance models.
2019	Rolled out centralized Model and Data Inventory platform to track metadata, usage, and ensure version control.
2020	Aligned AI fairness protocols with ECOA and began routine fairness assessments in lending and underwriting.
2021	Integrated real-time audit logging and automated documentation workflows to support internal and external audits.
2022	Created a cross-functional AI Governance Council to align technical, legal, and ethical perspectives across business units.
2023–Present	Focus on scaling governance frameworks globally, expanding tools, enhancing employee training on AI.

JPMorgan's experience underscores that AI governance is not a bolt-on control mechanism but rather a strategic pillar that must evolve with the AI systems it supports. In industries like finance, where trust, compliance, and risk are foundationa, governance ensures that innovation does not come at the cost of fairness or transparency.

Key challenges such as talent shortages, explainability trade-offs, and biased training data require not only technical solutions but organizational change and cross-disciplinary collaboration. By embedding governance into its AI roadmap early and proactively, JPMorgan has created a replicable model for others to follow.

QUESTIONS

1. Why is independent model validation critical in financial institutions like JPMorgan, and what challenges might arise in separating validation from development teams?
2. How does JPMorgan's use of SHAP contribute to both regulatory compliance and customer trust? Can this level of explainability be expected in all industries using AI?
3. What role does centralized data and model inventory play in ensuring effective AI governance? How might this system break down in a decentralized organization?
4. JPMorgan aligned its fairness metrics with US regulatory standards like ECOA. What are the limitations of using legal compliance alone as a measure of fairness in AI?
5. Given the timeline of JPMorgan's AI governance implementation, what lessons can other companies learn about phasing in governance practices as AI maturity evolves?

10.3 Building an AI Culture

As AI reshapes industries, organizations are recognizing that successful AI adoption is not solely a technological challenge; it is a cultural one. The development of a strong AI culture is essential for ensuring that AI tools are not just implemented but also embraced, trusted, and embedded into the fabric of how work is done. Without the right cultural foundation, even the most advanced AI systems risk rejection, underuse, or failure to deliver value.

10.3.1 Why AI Culture Matters

AI culture refers to the shared values, beliefs, and behaviors within an organization that support the ethical, responsible, and effective use of AI. A healthy AI culture encourages curiosity, cross-functional collaboration, responsible experimentation, data-drive decision making, and a balanced relationship between human judgment and machine intelligence. Importantly, it also includes awareness of bias, fairness, and accountability, particularly when AI is used to influence decisions about people or customers.

Organizations lacking an AI culture often encounter resistance. Employees may mistrust algorithms, fear job displacement, or fail to see the relevance of AI in their roles. Leaders may struggle to align AI initiatives with strategic goals. In contrast, companies that foster a positive AI culture are more likely to accelerate learning curves, increase employee engagement, and achieve meaningful business outcomes.

Building an AI culture requires intentional effort across multiple levels of the organization:

Leadership Commitment. Senior leaders must champion AI not just as a tool but also as a catalyst for transformation. This involves setting a clear vision, allocating resources, and modeling a learning mind-set. Leaders should communicate not only the benefits of AI but also the responsibilities that come with it.

Democratizing AI Literacy. AI culture thrives when people across the organization understand what AI is, and what it is not. Providing accessible training, case studies, and forums for experimentation helps employees feel empowered rather than threatened. Initiatives such as AI bootcamps, lunch-and-learns, and citizen data science programs can spark engagement.

1. **Cross-Functional Collaboration.** AI should not live in a silo. Collaboration among technical experts, business leaders, domain experts, and frontline employees ensures that AI solutions are relevant, human-centered, and context-aware. Culture is reinforced when diverse voices contribute to model design, deployment, and feedback.
2. **Responsible AI Practices.** Embedding ethical principles, such as transparency, fairness, and accountability, into AI development reinforces cultural trust. Establishing governance structures and audit mechanisms helps maintain confidence and prevent reputational or regulatory risks.
3. **Recognizing and Rewarding AI Champions.** Recognizing early adopters, problem solvers, and innovators fosters a sense of ownership and pride. Creating communities of practice around AI encourages knowledge sharing and continuous improvement. Fostering AI culture inevitably involves change, and change can be difficult. Change management provides the structured approach needed to guide people through the transition. Key components include the following:
    - **Assess Readiness:** Before launching AI initiatives, organizations should evaluate their cultural, technological, and leadership readiness.
    - **Stakeholder Engagement:** Actively involving employees in the design and implementation process increases buy-in and surfaces concerns early.
    - **Communication Strategy:** Transparent, consistent messaging about the "why," "what," and "how" of AI helps counter misinformation and anxiety.

- **Support Mechanisms:** Providing coaching, peer support, and resources ensures that employees are not left behind.
- **Measure Progress:** Tracking adoption, usage patterns, sentiment, and outcomes allows leaders to course-correct and celebrate milestones.

AI will not succeed in a vacuum. Its transformative potential depends not only on algorithms and data but also on the culture that surrounds it. Organizations that invest in fostering a strong AI culture, supported by clear leadership, inclusive practices, responsible governance, and effective change management, position themselves to unlock long-term value and build a workforce that sees AI as a partner in progress, not a threat to be feared.

We already discussed Microsoft's AI roadmap. Let's again look at Microsoft.

COMPANY EXAMPLE:: Microsoft (Part II)

Over the past decade, Microsoft has undergone a significant cultural transformation under the leadership of CEO Satya Nadella, particularly in how it approaches AI. Rather than treating AI as a purely technical endeavor, Microsoft embedded it into its company-wide culture and business strategy, focusing on democratization, ethics, and continuous learning.

When Nadella took over in 2014, he emphasized a growth mind-set across the organization. This philosophy aligned perfectly with AI innovation, encouraging teams to experiment, fail fast, and continuously learn. Microsoft trained tens of thousands of employees, not just engineers, on AI concepts through internal programs such as AI Business School and AI 101. These initiatives aimed to foster a shared understanding of how AI could improve customer solutions, streamline operations, and open new markets.

AI was not confined to R&D. Teams across marketing, HR, finance, and operations were invited to propose AI-driven projects. For instance, Microsoft HR used AI to improve employee engagement and predict turnover risks, while finance used machine learning models to forecast revenue more accurately. By involving non-technical teams, Microsoft promoted a culture of inclusion and broke down the myth that AI belonged only to data scientists.

Microsoft also took a leading role in AI ethics by establishing the Aether Committee. This group includes experts from law, engineering, social science, and product teams, and it reviews high-risk projects to ensure they align with Microsoft's principles around fairness, transparency, privacy, and accountability.

Throughout this cultural transformation, Microsoft applied change management techniques to ensure successful AI integration. Leaders communicated the purpose of AI initiatives clearly and frequently. They invested in upskilling programs and encouraged innovation hubs internally. Resistance was addressed through inclusion and education, and early adopters were celebrated. Progress was tracked using metrics such as AI project success rates, employee engagement scores, and business impact.

Microsoft's cultural and organizational approach to AI allowed it to shift from a tools provider to a leader in enterprise AI solutions. The cultural foundation helped accelerate innovations like Azure AI services, GitHub Copilot, and integration of generative AI tools across Office products. Crucially, the company did not just implement AI. It made AI part of how things are done across all business units.

10.4 Organizational Change

10.4.1 Transformational versus Incremental Change

Organizational change is any substantive modification to an aspect of an organization. It may consist of a change in the workforce, technologies, structure, work processes, culture, values, or strategic vision. Yet change is difficult. Half to two-thirds of these changes fail to achieve their expected results. In a rapidly changing environment, the knowledge that is most useful to organizations is that which helps them change and adapt. Leaders must understand when change is needed and how to guide their organizations through it. This implementation requires a major organizational change: to become an AI-driven organization. This section explains how to guide this change process.

The way that change is implemented throughout the organization depends on the type of change involved. There are two types of organizational change. Change can be either transformational or incremental. **Transformational change** occurs when an organization changes its strategic direction or reengineers its culture or operations in response to dramatic change, such as a technological breakthrough or a merger. Transformational changes are typically a reaction to a major problem or challenge in the competitive environment. These changes require leaders and all members of the organization to embrace learning and innovation in order to succeed. They represent a major shift for the organization.

Incremental change occurs when an organization takes relatively smaller steps toward its goal. This may involve restructuring to be more efficient or expanding its line of products or services to promote more growth. These changes occur incrementally or gradually. The organization exists in a state of equilibrium. The focus with incremental change is on fine-tuning existing practices.

A great example of transformational change in action is what happened at Rakuten Inc., Japan's online shopping equivalent of Amazon. In 2012, the founder and CEO of Rakuten, Hiroshi Mikitani, made a transformational change to his large company when he instituted an English only policy for all company communications. This change required that all meetings, training sessions, and even emails would be in English. His goal was to globalize the firm, broaden the mind-set of current employees, and move away from the Japan-centric thinking of his employees. Mikitani's actions were dramatic but necessary if the company was to become a global competitor. His actions were intended to lead his organization through a transformational change.

As happens with most changes of this magnitude, his requests were met with a bit of resistance. Some employees resigned in protest, while others struggled to meet the new expectations. Originally, Mikitani expected employees to adapt to the change without the organization's help. However, after feedback came in regarding employees experiencing apprehension and anxiety, the company began offering free English classes and time on the job to study. Mikitani has made it clear that speaking English is a requirement for all jobs. The organization is leading transformational change, and it is working. Currently, more than half of the company's Japanese employees use English for all internal communications.

The type of transformational change that has taken place at Rakuten is the kind of change required to implement big data analytics and create an AI-driven, decision-making culture. Learning to make decisions based on data and information, rather than guess work, is akin to forcing an organization to communicate in an entirely different language. For this to be successful, there are certain steps that need to be followed in the change process.

Another notable example of transformational change is Microsoft, which we already discussed, under the leadership of Satya Nadella. When Nadella assumed the role of CEO in 2014, the company faced significant challenges in adapting to the evolving technology landscape, particularly in mobile and cloud computing. Nadella spearheaded a cultural transformation, shifting the organization's focus from a Windows-centric approach to one that embraced innovation, agility, and customer-centricity. He encouraged employees to take risks and fostered a more open, collaborative environment. This approach underscores the profound impact of transformational change in repositioning an organization to thrive in a competitive landscape. His leadership demonstrates the necessity of embracing new realities, fostering innovation, and aligning an organization's culture with strategic goals, an approach critical for organizations aiming to implement big data analytics and build a data-driven culture.

10.4.2 Stages in Transformational Change

Transformational change is change of a large magnitude. It requires a major shift in the organization. This process of organizational change has been outlined in the seminal work of Kurt Lewin (1951). Lewin outlined four steps that organizations go through in the change process,

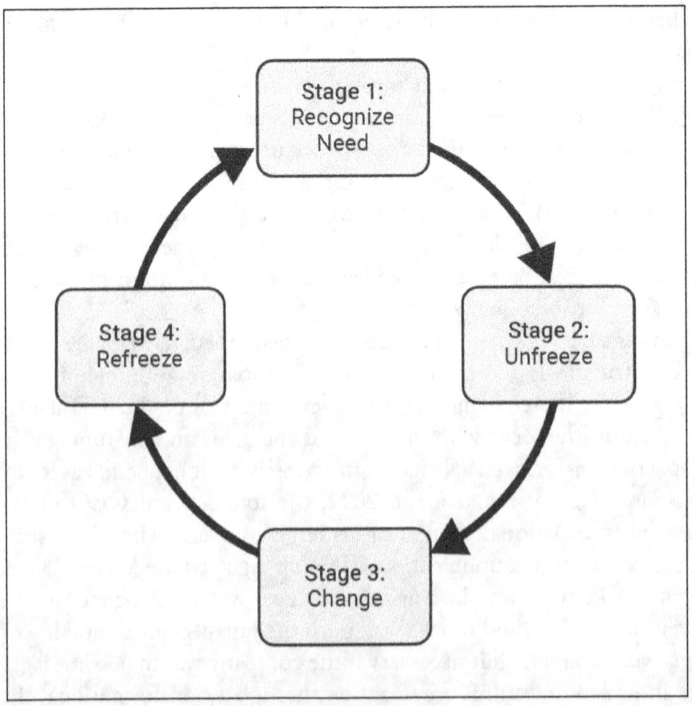

Figure 10.2 Steps in the Change Process

if transformational change is to occur and last. These are steps every organization goes through. Although approaches differ in the descriptions of how to perform each step, there is consensus on the need for all four steps.

The steps represent a continuum, shown in Figure 10.2. They involve recognizing a need, unfreezing, changing, and refreezing. The unfreezing and refreezing steps are especially important. For meaningful change to occur it cannot just be forced on the organization. The organization must be prepared through the unfreezing stage, where old habits thaw. After change is implemented, it must be solidified, in the refreezing stage. Otherwise, the organization will fall back into its old habits.

A notable example of transformational change guided by Lewin's model is IBM's turnaround in the early 1990s under CEO Lou Gerstner. Facing declining market share, financial losses, and inefficiencies, Gerstner initiated the unfreezing stage by challenging IBM's identity as solely a hardware company and shifting its focus to value added services.

During the changing stage, Gerstner implemented strategic initiatives to streamline operations, divest non-core businesses, and invest in emerging technologies like the internet and software services. These efforts refocused the company on its core competencies and operational efficiency. In the refreezing stage, Gerstner reinforced these changes by embedding them into IBM's culture. He championed a customer-centric approach, empowered decision making, and fostered innovation and collaboration.

This transformation helped IBM reclaim its leadership position in the technology industry, achieving improved profitability and market competitiveness. The success of IBM's turnaround demonstrates how Lewin's model provides a structured approach to implementing and sustaining transformational change.

10.4.3 Top-Down or Bottom-Up

The traditional view of organizational change and learning has been largely top-down: senior leaders set the agenda, craft a vision, and rely on managers and consultants to drive implementation. In this model, expertise and decision-making authority are concentrated at the top. Leaders analyze the external environment, identify what must change, and then work to persuade the rest of the organization to adopt new practices, structures, or cultures. They manage change through oversight, communication, and enforcement of alignment.

However, this approach increasingly falls short in fast-moving, digital environments where agility, innovation, and employee engagement are crucial. A more sustainable and adaptive approach to change is bottom-up, where employees at all levels participate in shaping and implementing change initiatives. This model is core to the Japanese management style seen in companies like Toyota and Sony, and it aligns with modern servant leadership philosophies that emphasize empowerment, collaboration, and respect for individual contributions.

Today, companies like Haier (China) and Spotify (Sweden) have pushed this concept even further. Haier operates with a microenterprise model, where frontline teams function as semi-autonomous units, making decisions based on customer needs. Similarly, Spotify's squad model empowers cross-functional teams to drive product innovation independently, while leaders provide overall direction and values rather than micromanaging execution.

Importantly, a bottom-up approach does not mean leaderless change. Senior leaders still define the vision and strategic intent, but they avoid prescribing tactical details. Employees closest to the work are trusted to interpret the vision, devise local plans, and adapt processes in their own contexts. These plans are then shared upward for alignment and support. This decentralized decision making builds ownership, accountability, and innovation at every level.

A notable historical example is Nobuyuki Idei's leadership at Sony. He used the motto Digital Dream Kids not as a specific directive but rather as a loose vision to inspire interpretation across the organization. Although the phrase puzzled many Western analysts, it empowered teams to reflect on what the digital future meant in their context, and act accordingly. This participatory mind-set helped Sony navigate a period of digital transition.

Research continues to show that participatory change processes are more effective than top-down mandates. They foster dialogue, increase buy-in, and build a shared commitment to long-term goals. Studies also confirm that organizations that adopt this approach are more likely to cultivate sustainable innovation and resilience.

Modern technology strengthens this participatory approach. Enterprise platforms, AI tools, and data visualization systems (e.g., SAP S/4HANA, Microsoft Power BI, or Snowflake) give frontline employees access to insights previously reserved for executives. This democratization of data enables better local decision making and empowers staff to act in alignment with organizational strategy.

In today's digital age, successful transformation requires more than tools. It requires culture. Becoming a data-driven organization is not just about adopting AI or analytics; it is about engaging people at all levels to question assumptions, adapt quickly, and contribute meaningfully. As companies grow more analytical, they must also grow more participatory.

With that in mind, let's now explore how the process of change unfolds in this modern context.

10.5 The Four-Step Change Process

10.5.1 Step 1: Recognize Need

The first step in any successful change process is recognizing the need for change, an insight that most often originates from leadership. Change opportunities emerge from a variety of internal and external sources, including market shifts, regulatory changes, technological advancements, customer behavior, or workforce dynamics. These signals may be formal, such as data trends or shareholder demands, or informal, such as emerging competitor activity or grassroots employee input. Effective leaders are attuned to these indicators and respond with a strategic vision for adaptation.

For instance, as the energy sector pivots away from fossil fuels, many traditional power companies are recognizing the unsustainability of depending on coal or oil. This shift has spurred global players like Ørsted, formerly Danish Oil and Natural Gas, to completely transform its core business from fossil fuels to renewable energy. By 2023, more than 90 percent of its energy generation came from wind and solar, a radical shift enabled by recognizing the environmental, regulatory, and consumer-driven imperatives for clean energy.

In parallel, internal capabilities can also trigger the need for change. A growing talent shortage in data science and AI capabilities has pushed companies to reexamine their workforce development strategies. For example, Walmart launched its "Live Better U" education program to upskill employees in data analytics, cloud computing, and machine learning, recognizing that future competitiveness depends on having a data-literate workforce.

External changes in consumer behavior also serve as powerful catalysts. Social media analytics, for example, have evolved into critical tools for understanding customer sentiment in real time. T-Mobile leverages AI to analyze customer sentiment across social channels, identifying churn risk patterns and proactively offering retention incentives, an evolution rooted in its recognition of the strategic value of unstructured social data.

A hallmark case of recognizing transformational change is Netflix. Originally a DVD-by-mail service, the company pivoted sharply in the early 2010s as digital streaming emerged. Under the leadership of Reed Hastings, Netflix foresaw the decline of physical media and invested heavily in its streaming infrastructure and original content. Today, Netflix is not just a content distributor but also a global content creator, operating in more than 190 countries with more than 260 million subscribers by 2025. The bold pivot was triggered by recognizing both technological change and shifting consumer expectations.

Technological innovation also pushes companies to rethink their business models. Rolls-Royce, the jet engine manufacturer, exemplifies this through its servitization model. Instead of just selling engines, it now provides real-time monitoring services using AI and IoT technologies, allowing predictive maintenance and usage-based billing. The company monitors more than 10,000 engines in operation and earns more than 50 percent of its civil aviation revenue from services—a complete reinvention of its business model driven by data.

Another standout example is Domino's Pizza, which reframed itself as a tech-enabled delivery company. After years of struggling with customer satisfaction and declining market share, the company recognized the need to overhaul its operations and digital strategy. It invested in data analytics, GPS tracking, AI-powered order prediction, and mobile ordering. The result? Domino's now processes more than 75 percent of its orders through digital channels and has consistently outperformed many fast-food chains in growth and customer retention.

Similarly, CEMEX, a global cement manufacturer, redefined its value proposition by focusing not on cement but instead on reliable and timely delivery. By embedding GPS and predictive analytics into its logistics systems, it began offering customers guaranteed delivery windows, a major differentiator in the construction industry. As a result, CEMEX moved from selling a low-margin commodity to offering a high-value logistics service, commanding a premium price.

In each of these cases, recognizing the need to become a data-driven organization was the inflection point. The value of big data goes beyond improving operational efficiency. It enables entirely new products, services, and business models. Recognizing that potential is a leadership imperative. Companies that fail to see these shifts, or hesitate to act, risk being left behind in a rapidly evolving digital landscape.

10.5.2. Step 2: Unfreeze

Unfreezing is the critical first stage of any successful organizational change effort. It involves preparing the organization to embrace change, particularly transformational shifts such as becoming a data-driven enterprise. At this stage, leaders must reduce resistance, increase awareness of why the change is necessary, and cultivate a collective readiness to act. Without this psychological and cultural preparation, even well-designed change initiatives can falter due to passive resistance or even active sabotage.

Research consistently shows that employees are more likely to support change when they perceive it as an opportunity rather than a threat. Conversely, when change is viewed as a risk to status, security, or identity, a phenomenon known as the threat-rigidity response often emerges. In such cases, employees cling to familiar routines and behaviors, effectively freezing the organization in place. Leaders must proactively "unfreeze" these behaviors by creating awareness, trust, and momentum toward the desired future.

A modern and compelling example of unfreezing can be seen, yet again, at Microsoft under Satya Nadella. When Nadella became CEO in 2014, Microsoft was stagnating under a rigid, siloed culture that resisted collaboration and failed to capitalize on emerging cloud technologies. Nadella immediately focused on unfreezing this culture. He replaced the long-standing internal rivalry with a growth mind-set, emphasizing learning, humility, and cross-functional collaboration. He famously asked employees to shift from a culture of know-it-alls to learn-it-alls. This cultural unfreezing set the stage for Microsoft's massive pivot to Azure cloud services and AI, contributing to its dramatic resurgence and trillion-dollar valuation.

Unfreezing Through Opportunity versus Threat. Unfreezing can be approached in two key ways by highlighting a threat that demands change or by painting a picture of the opportunity that awaits. Both are intended to break the psychological inertia of the status quo, but research suggests that opportunity-driven approaches often lead to more lasting and enthusiastic adoption.

1. **The Threat-Based Approach.** A threat-based approach draws attention to potential risks or negative consequences of inaction. For example: "If we do not modernize our operations with AI, we will lose market share, fall behind competitors, and face significant layoffs."

   This method can generate urgency, but it often only triggers the minimal compliance necessary to mitigate the threat, not full commitment. Worse, if employees perceive the change itself as a threat, such as risking job security or eroding work relationships, they may disengage or leave altogether.

   An example of this approach can be seen in traditional retailers like Sears, which acknowledged the rise of e-commerce as a threat but responded too slowly and without engaging employees in a positive vision for transformation. The result was a halfhearted effort that failed to mobilize internal support and ultimately led to bankruptcy.
2. **The Opportunity-Based Approach.** An opportunity-based strategy highlights the benefits of change, both organizational and personal. For example: "Implementing AI and analytics will help us improve our decision-making, reduce routine tasks, increase efficiency, grow our stock value, and give everyone the skills they need for tomorrow's workforce."

   This optimistic, future-oriented framing is more likely to inspire commitment and creativity. A prime example is Unilever's digital transformation initiative, which framed its AI strategy around sustainability, innovation, and employee empowerment. By positioning AI as a tool for achieving its purpose-driven mission, Unilever engaged employees at all levels, building a strong internal coalition around change.

A classic example of opportunity-based unfreezing is Apple under Steve Jobs, but it is worth contextualizing for today's digital environment. When Jobs returned in 1997, Apple was on the brink of collapse. Rather than resorting to fear-based messaging, Jobs offered a compelling vision of innovation and design excellence. He ignited internal excitement with products like the iMac, iPod, and iPhone, emphasizing not just technical features but also how they would change the world. Jobs' approach to unfreezing was built on inspiration, clarity of purpose, and an uncompromising focus on excellence, which catalyzed Apple's transformation into a global icon.

Participatory Unfreezing: Building Readiness from the Ground Up. While top-down leadership can jumpstart unfreezing, the most sustainable and effective change happens when organizations also use bottom-up participation. By involving employees in the process of diagnosing the current state and imagining the future, leaders build ownership and engagement.

A modern example is ING Bank's agile transformation, which involved flattening hierarchies, creating self-organizing squads, and including employees in redesigning work processes. Instead of imposing new structures from above, ING engaged frontline workers in shaping the transformation. This participatory approach not only reduced resistance but also unlocked innovation and accelerated adoption.

10.5.3 Step 3: Change

Once an organization has successfully unfreezed, it is ready to enter the second phase of the change process: implementation. This stage involves transforming vision into reality through targeted interventions across **technology, processes, and people**. These three pillars are not just barriers to overcome. They represent levers for lasting, systemic transformation.

Remember that change is not a one-time event—it is a progression toward capability. Organizations that integrate technology, redesign processes, and invest in people are best positioned to become truly data-driven. These efforts must be coordinated through a unified vision and supported by leadership, clear governance, and ongoing feedback loops.

Successful change requires not just the right tools or talent, but also the right mind-set. As companies face unprecedented complexity and competition, aligning the three pillars of transformation, discussed in Chapter 9, is the foundation. This is shown in Figure 10.3

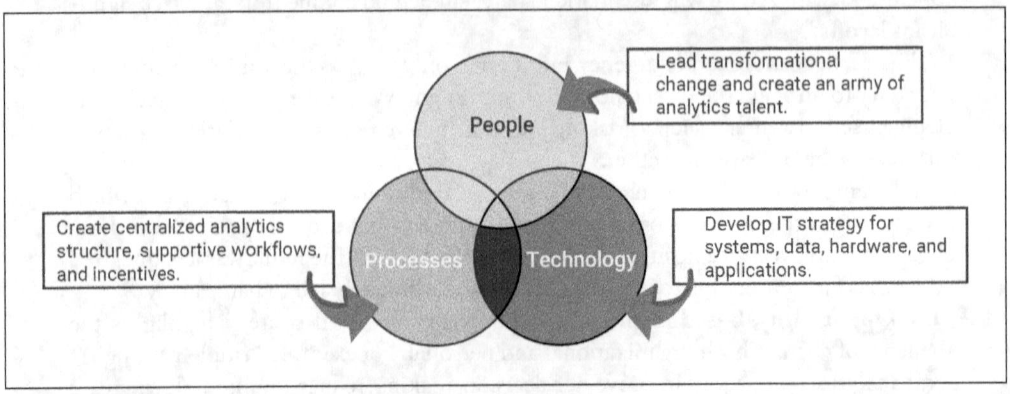

Figure 10.3 Points for Intervention: People, Processes, and Technology

1. PEOPLE: Developing Data-Driven Talent.

People are the ultimate drivers of change. Even the most sophisticated systems are ineffective without employees who understand how to use them. A data-driven organization must invest in talent development, cultural transformation, and leadership alignment.

Walmart has implemented comprehensive training programs to equip employees at all levels with digital and analytical skills. Through its Walmart Academy, the company offers modules on machine learning, data visualization, and real-time analytics. This approach helps shift the mind-set from gut-based decisions to data-augmented decision making across functions, from merchandising to logistics.

There are a few additional considerations in changing people:

- **Behavioral Change Tools:** This involves implementing tools to enhance and gamify the change process. One innovative example is Schneider National's gamified simulation used in dispatch planning. Employees engage in a game-based platform that simulates load planning and routing using analytics dashboards, helping them understand the benefits of data in real-time decisions. These tools foster learning, acceptance, and curiosity.

- **Decision Support Systems:** These are tools that enhance human decision making, not replace it. Introducing these to people helps their acceptance and engagement of the change process. One example is a tool like IBM's Watson Health, which illustrates how analytics can augment rather than replace human judgment. Watson provides physicians with diagnostic probabilities and treatment recommendations, empowering them to make more informed decisions. This model exemplifies the synergy between machine intelligence and human expertise.

2. PROCESSES: Redesigning Workflows for Agility.

Process change interventions target the structure and routines that shape how work is done. This may include decision rights, approval chains, compensation systems, or even physical workspaces.

An excellent example here is Schneider Electric's global supply-chain transformation. It showcases how digital technologies and agile processes can be synchronized. By integrating IoT, AI, and predictive analytics into its Supply Chain Control Towers, Schneider gained real-time visibility across operations, from demand forecasting to logistics. This required rethinking how data was shared across departments and partners, as well as shifting to a cross-functional decision-making model.

A few additional considerations are important in changing processes:

- **Physical Environment:** The role of physical space in collaboration and innovation is also critical. Companies like Salesforce have adopted flex teams and reimagined workspaces with open design, shared hubs, and digital collaboration tools. This approach fosters communication, creativity, and speed—key drivers in a data-enabled environment.
- **Organizational Structure Shift:** Many companies are also moving from hierarchical structures to networked, agile teams. For example, ING Bank transitioned to a squad-based model inspired by tech companies, empowering cross-functional teams to operate autonomously with their own data tools and goals.

3. TECHNOLOGY: Building the Analytical Backbone.

Technological interventions are the backbone of becoming a data-driven enterprise. These interventions enhance the way organizations gather, process, and act on data. They influence workflows, decision-making systems, and the infrastructure required to support data at scale.

A modern example is Amazon that continues to set the standard for technological transformation. Through AI-driven demand forecasting, real-time inventory tracking, and robotics-enhanced warehouses, Amazon has redefined customer expectations. Its logistics engine, powered by technologies like Kiva robots and predictive analytics, ensures faster, cheaper, and more reliable deliveries. In 2024, Amazon introduced generative AI tools to optimize customer service queries and product descriptions, dramatically increasing efficiency and personalization.

To implement such systems, companies must address legacy infrastructure, siloed data systems, and gaps in interoperability. A strategic IT roadmap is essential (deciding whether to overhaul, integrate, or migrate systems) and determining whether to build or buy solutions. They must also secure access to data, whether from internal sources, external partnerships, or data marketplaces.

A key consideration here is that technology should not be implemented for its own sake. Organizations must ensure that the analytics systems they adopt are aligned with business goals, easy to integrate into workflows, and scalable over time.

Most companies make the mistake of addressing technology without the other two pillars. Lasting change occurs when interventions across people, processes, and technology are aligned. Misalignment would be, for example, implementing an AI tool without training staff or integrating it into workflows, which leads to underperformance or outright failure.

UPS's ORION system is a standout success in integrated change. Powered by advanced analytics, ORION calculates the most efficient delivery routes, saving the company millions annually. Yet its true success lies in simplicity. Drivers do not need to understand algorithms; they simply trust a user-friendly device that improves their day-to-day work.

To encourage adoption, organizations are increasingly turning to analytical apps: lightweight, intuitive tools that serve specific business decisions like inventory management, supplier performance, or transportation analytics. These tools lower the barrier to entry, reduce resistance, and expand the reach of analytics beyond specialists.

10.5.4 Step 4: Refreezing (or Re-Slushing)

Refreezing (or Re-Slushing) is reinforcing and sustaining organizational change. Once the implementation phase of change is complete, the final challenge is ensuring that the new behaviors, systems, and mind-sets become embedded in the organization. This stage is traditionally referred to as refreezing, the process of stabilizing the organization after change by institutionalizing new ways of working and thinking. Without this reinforcement, organizations risk sliding back into familiar habits, eroding the progress they have made.

Refreezing involves more than just maintaining momentum; it requires reshaping organizational structures, routines, incentives, and culture to support the new reality. It means codifying the change into the DNA of the organization through policies, metrics, training, and norms.

Under CEO Satya Nadella, Microsoft executed one of the most successful cultural transformations of the past decade. Nadella's emphasis on a growth mind-set, a culture of curiosity, collaboration, and resilience—was not just aspirational. To refreeze these values, Microsoft redesigned systems to reinforce them.

The company launched the One Microsoft strategy to break down silos, restructured teams around shared goals, and revamped its performance review system to reward collaboration over individual competition. Employee development programs were expanded to support ongoing learning in cloud computing and AI, helping employees adapt to new technologies and work practices. Most critically, leadership modeled the behaviors they sought to instill (empathy, humility, and continuous learning) making cultural change tangible, not just rhetorical.

Embedding Change Through Design. In the refreezing stage, leaders must make conscious decisions about people, processes, and technology to sustain change:

People. Reinforce desired behaviors through aligned incentive systems, revised job descriptions, and ongoing training. Promote employees who embody the new values and create mentorship opportunities that transfer these behaviors to others.

Processes. Modify workflows, reporting structures, and standard operating procedures to reflect the new way of working. Integrate analytics into decision-making pipelines, redefine how cross-functional teams collaborate, and formalize review and feedback loops.

Technology. Ensure that new systems are fully integrated into daily operations and provide ongoing support for users. Data dashboards, collaboration platforms, and automation tools should be intuitive and widely accessible.

For example, Adobe, after transitioning to a cloud-based subscription model, embedded change by developing new success metrics (such as monthly recurring revenue and customer retention) and reengineering sales and support processes to align with long-term customer engagement rather than one-time transactions. The company also revamped its internal tools and compensation structures to reinforce these shifts, helping the change take root.

From Refreezing to Re-Slushing: Embracing Continuous Adaptation. While the idea of refreezing emphasizes stability, many organizations today are adopting a more fluid mind-set, known as "re-slushing." This term acknowledges that in a fast-paced, technology-driven environment, change is not a one-time event; it is continuous.

Re-slushing means designing systems and cultures that are inherently adaptive. It places emphasis on agility, experimentation, and learning. Rather than hardening around one set of practices, organizations build flexible infrastructures that evolve with customer needs, market trends, and technological advances.

Spotify exemplifies this approach. Its agile team structure, comprised of squads, tribes, and chapters, enables rapid experimentation and iterative improvements without needing to refreeze the entire organization. Teams are empowered to test, learn, and adapt autonomously within a loose but aligned framework. This structural flexibility has allowed Spotify to stay responsive in the competitive and fast-evolving music streaming market.

Practices for Sustained Change. Whether organizations lean toward refreezing or re-slushing, several principles remain essential for sustaining transformation:

- Align incentives with the new behaviors: Recognition and reward systems should reinforce what the organization wants to see repeated.
- Communicate ongoing success: Share wins and learnings to maintain momentum and highlight the value of change.
- Model the way: Leadership behavior must reflect the new norms, consistently and visibly.
- Invest in continuous learning: Build learning platforms and pathways that support upskilling and curiosity.
- Institutionalize flexibility: Create systems that can flex, scale, and evolve as circumstances change.

Beyond Stabilization to Sustainable Growth. The final phase of the change process, whether called refreezing or re-slushing, is about turning change into capability. It is about making sure that what was once new becomes normal, and what is normal continues to evolve. This is especially crucial in data-driven organizations, where agility, learning, and technology must coexist in a dynamic ecosystem.

Change that sticks is not just embedded; it is enabled, reinforced, and designed to grow.

10.6 AI — Leadership

Only senior leadership can spearhead the transformational change required to build and apply analytical and AI capabilities across their organizations. This transformation cannot be outsourced or delegated to middle managers, professional analysts, or external consultants alone. It requires top executives to take ownership—not only in vision but also in daily execution, prioritization, and culture shaping.

Organizations that have successfully put AI and analytics at the center of their strategy do so because their leaders have made a visible, sustained commitment. As Sundar Pichai, CEO of Alphabet, recently emphasized at Google, "AI is the most profound technology we are working on today. Leaders must make sure that innovation does not happen in isolation but is embedded across every layer of the organization." This message is echoed across industries; transformation is not a technical initiative; it is a leadership imperative.

For example, Lenovo's Group President, Luca Rossi, has stressed that scenario modeling and predictive analytics are how the company stays resilient in uncertain times. At Amazon, the culture of data obsession has continued under CEO Andy Jassy, who has stressed that invention alone is not enough and that what is needed is a culture that measures, tests, and then scales relentlessly. This follows the tradition established by Jeff Bezos, who famously removed teams or products that failed to align with Amazon's metrics-driven ethos.

Leadership commitment must go beyond slogans. It must shape decisions, budgets, priorities, and team structures. This is especially true during periods of change, when the role of leadership expands from setting direction to becoming change agents, individuals who actively drive and manage the transformation process. These leaders do not merely support change, they catalyze it. They create the vision, allocate resources, redesign roles, and establish systems that enable shifts in technology, process, and culture. They also ensure that their workforce is prepared. Satya Nadella, Microsoft CEO, has emphasized that AI will change every job and every industry and that it is up to leaders to ensure that people are ready. That also means upskilling, transparent communication, and building trust.

This recalls the example of Hiroshi Mikitani, CEO of Rakuten, who recognized that radical transformation (in his case, mandating English fluency across the company) requires education, empathy, and sustained leadership. Equally important are the idea champions, individuals within the organization who actively and enthusiastically support innovation. When leaders empower these internal champions, they accelerate adoption, overcome resistance, and build a coalition for change. Change agents and champions together create an environment where innovation thrives, not just in pilots but also in full-scale enterprise impact.

For transformational change to take root, there must be organization-wide commitment. Such commitment is nurtured through clear communication, aligned incentives, shared purpose, and the personal engagement of top leadership. Only when senior executives consistently embody these values can a company fully harness the power of analytics and AI to reimagine its future.

10.6.1 What Should Leaders Do?

Organizations that seek a competitive advantage from AI must take an enterprise-wide perspective in their applications. Leaders need to ensure that analytics takes a cross-functional, cross-product, cross-customer, cross-supply-chain approach. This is the only way analytics will make a difference in the organization. Many analytics applications are currently embedded in a set of organizational silos. Leaders need to ensure that silos are broken down and that analytics applications connect detailed information on product availability with the models that predict demand, so the interaction between demand and supply can be better understood. However, this is not enough for a competitive advantage. Business leaders need to know how to take advantage of the big data capability they are creating. Leaders must have some understanding and appreciation of analytical capability and techniques. Big data analytics has demonstrated its potential. It is now up to leaders to lead their organizations through the transformation.

As supply chains grow more complex, volatile, and interdependent, leadership in this domain is undergoing a fundamental shift. Traditional supply-chain leadership focused on efficiency, cost minimization, and lean operations. But the infusion of AI is redefining what it means to lead. AI leadership in supply-chain management now demands a vision that integrates digital capabilities, fosters cross-functional collaboration, and cultivates a culture of innovation and adaptability.

- **Set a Strategic AI Vision:** Leaders must articulate how AI will support business objectives, be it demand forecasting, inventory optimization, transportation efficiency, or risk mitigation. The roadmap should align AI capabilities with measurable outcomes such as reduced lead times, improved service levels, and increased resilience.
- **Champion Cross-Functional Collaboration:** AI initiatives in the supply chain span procurement, logistics, finance, and IT. Leaders need to break down silos and promote collaboration across departments to ensure clean data, shared goals, and coordinated execution.
- **Develop and Attract AI Talent**: AI success requires talent with expertise in data science, analytics, and supply-chain operations. Leaders must invest in upskilling existing staff, attract external talent, and create hybrid roles that bridge analytics and operations.

- **Foster a Culture of Experimentation and Trust:** AI systems often challenge human intuition. Leaders must build trust in AI-driven insights, encourage data-driven decision making, and create safe spaces for testing and learning without fear of failure.
- **Ensure Governance and Ethics:** As AI tools influence critical supply decisions, such as sourcing, allocation, and pricing, leaders must ensure ethical use, fairness, and transparency. This involves establishing clear governance policies for data usage, model validation, and accountability.
- **Measure What Matters:** Effective leadership uses the right metrics to track performance, such as forecast accuracy, order fulfillment rates, and AI model ROI. Leaders must tie these KPIs to strategic business goals and continuously refine them as the organization evolves.

COMPANY EXAMPLE: Stitch Fix

Stitch Fix, founded in 2011 by Katrina Lake, is a small-to-mid-size fashion retail company that has become a standout in the application of AI for personalization, inventory management, and styling recommendations. With under 10,000 employees and a focused business model, Stitch Fix offers a compelling case of how a small company can lead with AI not just by investing in technology, but by fostering a culture of experimentation, cross-functional collaboration, and customer-centric innovation.

From the beginning, Katrina Lake had a vision: to use data science to reinvent personal shopping. Unlike many startups, Stitch Fix did not bolt on AI as an afterthought; it was foundational to the company's value proposition. Lake's leadership centered on making data-driven decision making core to every function, from merchandising to marketing to logistics. She also took a bold early step: hiring Eric Colson, former Netflix VP of Data Science, as Chief Algorithms Officer. This move signaled that AI leadership at Stitch Fix would be at the executive table and not buried in a back-office analytics team.

Stitch Fix pioneered what it calls a human-in-the-loop model, where AI algorithms suggest clothing items based on customer data, but stylists make the final curation. This collaboration between machine and human is a hallmark of thoughtful AI leadership, balancing efficiency with the empathy and creativity only humans provide.

Leaders at Stitch Fix trained stylists on how to interpret and enhance AI outputs, fostering mutual respect between technical teams and customer-facing employees. Rather than fearing AI, employees saw it as an augmentation tool, helping them to better understand customer preferences and deliver personalized service.

One of Stitch Fix's smartest leadership decisions was to embed data scientists within business teams. Instead of creating a centralized data science function, they built small, agile pods that paired data scientists with inventory planners, buyers, and stylists. This made it easier to identify real-world problems and rapidly co-develop solutions using machine learning.

As a small company, Stitch Fix could not compete with Google or Amazon on salary. But it attracted top AI talent by offering meaningful, high-impact problems, a collaborative culture, and leadership that truly understood and valued data science. Stitch Fix developed proprietary algorithms to predict inventory needs based on customer profiles, regional preferences, and return behavior. These AI models helped reduce excess inventory and improve fulfillment accuracy, which is critical for a small company with tight margins. For example, their Style Shuffle app, which gamifies preference collection, became a rich data source. Leaders ensured that insights from this tool fed directly into AI models for trend forecasting, sizing optimization, and supplier negotiations.

AI can feel invasive in the fashion world, but Stitch Fix leadership communicated clearly how data was used, offering transparency and control to customers. Clients could give feedback on every shipment, which not only improved recommendations but also reinforced a sense of agency and trust. Lake and her team often emphasized ethical data use and customer value in investor calls and public messaging, reinforcing that AI was a tool for empowerment, not manipulation.

Stitch Fix achieved profitability early—rare in the e-commerce space—due in large part to its efficient, AI-enhanced operations. More than 80 percent of items shipped are selected using

AI-driven models, leading to higher retention and satisfaction rates. The blend of AI and human stylists led to a high Net Promoter Score and deep customer loyalty, despite being a niche player.

Why does this example matter?

Stitch Fix shows that AI leadership is not about size. It is about mind-set. By placing AI at the center of strategy, embedding it in culture, and balancing automation with human creativity, the company built a sustainable competitive edge. Leadership's clarity of vision, ethical stance, and investment in people over raw tech are lessons any small company can learn from.

QUESTIONS

1. How did Stitch Fix's early leadership decisions, such as hiring a chief algorithms officer, set the foundation for long-term AI integration? What message did this send internally and externally about the role of data science?
2. Stitch Fix's human-in-the-loop approach blends AI with human stylists. What are the advantages and potential trade-offs of this model, especially in maintaining personalization and customer trust?
3. In what ways did Stitch Fix's organizational structure (embedding data scientists in cross-functional teams) support rapid innovation and AI adoption? How might this differ from more traditional centralized analytics models?
4. Given that Stitch Fix is a relatively small company, how did it compete for top AI talent without the resources of tech giants like Google or Amazon? What can other small or mid-size businesses learn from this talent strategy?
5. Stitch Fix emphasized transparency and ethical data use in a sensitive industry like fashion. Why is this particularly important in AI applications that involve personal identity and taste, and how did the company build customer trust?
6. How did tools like the Style Shuffle app both enhance customer experience and fuel AI model development? What does this tell us about designing data collection with dual-purpose value (for both user and system)?
7. What broader leadership lessons can be drawn from Stitch Fix's success with AI, especially for companies outside the tech or fashion sectors? Is AI leadership more about technical capability or cultural mind-set?

10.6.2 Lesson in Leadership – IBM

Research has shown that employees of an organization will be more committed to change if they have confidence in the competencies of their leaders. There are three ways this confidence can develop. *First*, leaders need to lead by example and exhibit appropriate skills. A leader who wants to lead an AI-driven organization needs to exhibit at least a rudimentary level of analytical literacy. Leaders certainly do not have to be experts but just demonstrate that they value fact-based decisions. *Second*, leaders need to make sure that expectations are clear and that everyone has the right training. *Third*, leaders need to ensure there are both extrinsic and intrinsic rewards. Last, the organization needs to maintain continuity.

A powerful example of leadership-driven change through analytics is the transformation of IBM, first under Ginni Rometty and continuing under her successor, Arvind Krishna. Together, their leadership illustrates how sustained executive commitment can embed analytics into the core fabric of an enterprise and drive strategic renewal. Let's look at this.

Rometty's Era: Laying the Foundation for a Data-Driven Enterprise

During her tenure as CEO from 2012 to 2020, Ginni Rometty led IBM through a bold shift from legacy IT infrastructure toward a future rooted in cloud, AI, and data analytics. Recognizing that analytics would be central to value creation, she repositioned IBM around cognitive computing, particularly through the development and promotion of IBM Watson.

Rometty's leadership was deeply hands-on. She articulated a clear vision of a data-driven IBM, reinforced it through internal communications, and aligned it with company strategy. Under her direction

- IBM acquired companies such as the Weather Company to enhance Watson's predictive capabilities;
- employees were reskilled through large-scale learning platforms like Your Learning, fostering a company-wide culture of data literacy;
- performance metrics were tied to the successful adoption of analytics tools; and
- innovation in analytics was recognized and rewarded, creating strong intrinsic and extrinsic motivation.

These moves embedded analytics into IBM's structure, building both capability and cultural alignment around data-driven decision-making.

Krishna's Era: Scaling and Operationalizing AI

When Arvind Krishna assumed the CEO role in 2020, he doubled down on Rometty's foundation but shifted the emphasis toward operationalizing AI at scale, especially in hybrid cloud environments. Having led IBM's Cloud and Cognitive Software division, Krishna brought a more technical focus and accelerated IBM's push into enterprise AI solutions. Under his leadership

- IBM spun off its legacy managed infrastructure business Kyndryl, allowing the company to focus more sharply on AI, cloud, and quantum computing;
- the acquisition of Red Hat was fully leveraged to integrate AI into open hybrid cloud systems; and
- IBM enhanced its Watsonx platform, launched in 2023, to provide enterprise-ready foundation models and tools for governance, transparency, and performance management—key concerns for businesses scaling AI.

Krishna also prioritized responsible AI governance, emphasizing tools for explainability, fairness, and traceability. His leadership reflects a more mature stage of AI integration in which analytics are not just strategic differentiators but also operational necessities across the enterprise.

A Continuum of Leadership

The evolution from Rometty to Krishna illustrates a continuum of AI leadership: from vision-setting and cultural transformation to technical execution and scale. Both leaders understood that analytics would redefine IBM's value proposition, and they each took complementary approaches: Rometty by championing the cultural shift and foundational capabilities, and Krishna by industrializing and embedding AI into business operations. This transition showcases the critical role of sustained, adaptive leadership in achieving true AI transformation.

Questions

1. How did Ginni Rometty's leadership style contribute to fostering a data-driven culture at IBM, and what specific actions helped ensure company-wide adoption of analytics?
2. In what ways did Arvind Krishna's approach to AI and analytics differ from Rometty's, and how did his technical background influence IBM's strategic direction post-2020?
3. What are the challenges of transitioning from a vision-oriented phase of AI adoption (like under Rometty) to an operational and scalable phase (like under Krishna)? How can leaders manage that shift effectively?
4. How can organizations ensure continuity in AI strategy and culture during leadership transitions? What lessons from IBM's leadership evolution can be applied to other companies undergoing digital transformation?

Chapter Review

Key Points

1. Companies are rapidly becoming data-intensive, taking advantage of the large volumes of data to better operate and manage their businesses.
2. Every organization goes through four stages during the process of organizational change: recognizing a need, unfreezing, changing, and refreezing.
3. Only senior leadership can spearhead the transformational change required to build and apply analytical capabilities across their organizations.
4. Leaders need to ensure that analytics takes a cross-functional, cross-product, cross-customer, cross-supply-chain approach.

Case Study

Unilever's AI Leadership and Change Management Strategy

Unilever, a global consumer goods company with more than 400 brands sold in 190 countries, is a standout example of how a multinational organization can embrace AI not as a short-term efficiency play but as a long-term enabler of resilience, sustainability, and innovation. Faced with growing operational complexity, global supply-chain disruptions, and fast-changing consumer behavior, Unilever used AI to lead change across people, processes, and technology, anchored by clear leadership and ethical stewardship.

Strategic Vision from the Top

Under former CEO Alan Jope (2019–2023) and continuing with Hein Schumacher, Unilever embedded AI within its broader Connected 4 Growth digital transformation initiative. This vision emphasized simplification, agility, and the intelligent use of data to empower decisions across the organization. AI was not confined to IT or analytics teams; it became a strategic capability woven into the fabric of the business.

Key areas of transformation included

- predictive demand forecasting,
- marketing personalization,
- sustainable procurement, and
- intelligent hiring and workforce planning.

Leadership communicated AI not just as a tool but also as a way to live Unilever's purpose: "to make sustainable living commonplace." This alignment between technology and values helped build internal trust and stakeholder confidence.

Ethical AI Governance and Transparency. To manage AI responsibly, Unilever created a Data and AI Ethics Committee composed of cross-functional leaders from data science, legal, HR, and compliance. This group reviews AI use cases, especially those affecting consumers, employees, or suppliers, to ensure adherence to transparency, fairness, and privacy principles.

Key ethical safeguards included

- adoption of the OECD AI Principles for internal governance;
- mandatory bias and fairness audits for AI used in hiring, pricing, and sourcing; and
- collaboration with vendors to ensure explainability in systems like automated video interviews.

Unilever publicly shares how it uses AI in its operations, including disclosures on its website and investor communications, positioning transparency as a competitive differentiator in a trust-driven world.

Cultural Transformation and Talent Enablement

Rather than outsourcing AI, Unilever focused on building internal capability and digital fluency. It launched a Data and Analytics Academy, which provided tailored training across departments, from marketing to supply chain to HR.

Key initiatives included

- hands-on workshops in AI literacy, data visualization, and ethics;
- role-specific upskilling programs (e.g., for procurement or logistics planners); and
- executive education in digital strategy and AI implications for decision makers.

This approach democratized AI, breaking it out of technical silos and empowering employees to understand, challenge, and collaborate with AI tools. Leaders were expected to actively participate in AI strategy, not passively defer to data teams.

AI in Action: The People Data Center

One of the most transformative AI platforms Unilever deployed is the People Data Center (PDC), a multilingual, real-time analytics hub that captures consumer sentiment from social media, e-commerce reviews, forums, and retail data across more than 30 languages.

PDC allowed Unilever to move from lagging insights to predictive marketing intelligence. They were able to launch hyper-localized campaigns tailored to evolving consumer behavior. They were also able to accelerate product innovation cycles by months. This AI system empowered marketers to identify emerging trends and sentiment shifts, enabling agile response to changing market conditions while maintaining alignment with brand values and local sensitivities.

AI-Driven Supply-Chain Leadership

Unilever's supply-chain transformation exemplifies how AI can create agility, resilience, and sustainability at scale. Under

Chief Supply Chain Officer Reginaldo Ecclissato, the company embraced AI and automation to reimagine its global operations.

Key capabilities included

- AI-powered forecasting using real-time weather, demand signals, and social trends;
- digital twin technology to simulate production and logistics scenarios;
- AI dashboards in control towers to monitor supplier risk, bottlenecks, an emissions; and
- cross-functional collaboration with Microsoft, SAP, and Google Cloud to deploy scalable tools.

During the COVID-19 pandemic, Unilever's AI recommended reallocating hygiene product production. Local managers, using contextual judgment, adapted these recommendations in underserved markets with limited infrastructure, an example of human–AI collaboration in action.

Results and Impact

Unilever's AI initiatives produced measurable, enterprise-wide benefits. The company saw a 15 percent reduction in inventory in key markets. There was double-digit improvement in campaign engagement through personalization. Further, forecast accuracy was improved by more than 10 percent, enhancing service levels and reducing waste. They also saw increased internal adoption and trust, thanks to training, ethics, and transparency.

By anchoring AI in leadership, culture, and purpose, Unilever moved beyond the hype to create real value, demonstrating that AI leadership is not about machines; it is about guiding people through change.

Lessons from Unilever

- Leadership sets the tone. Unilever's success began with C-suite commitment to AI as a strategic enabler—not a technical initiative.
- AI must reflect values. Ethics, fairness, and transparency were embedded into governance structures from the outset.
- Culture is far more important than technology. Broad training, executive education, and employee empowerment made AI part of everyone's role.
- Human–AI collaboration is key. Decisions are informed by algorithms but grounded in human judgment and context.
- AI is a journey, not a destination. Unilever embraced AI as a tool to deliver on its long-term purpose, not just near-term KPIs.

Questions

1. How did Unilever's leadership frame AI adoption as more than just a technology initiative? What role did purpose, culture, and ethical governance play in driving employee engagement and stakeholder trust?
2. What are the advantages and potential challenges of building internal AI capabilities (e.g., Data and Analytics Academy) instead of outsourcing to vendors? How might this approach affect scalability, innovation, and organizational learning?
3. Unilever created an AI Ethics Committee to evaluate the effect of AI across business functions. In your view, what are the essential elements of effective AI governance, and how should companies balance innovation with regulation?
4. What does the People Data Center (PDC) illustrate about the shift from reactive to predictive decision making? How can organizations ensure that such real-time insights are used responsibly and effectively?
5. In what ways does Unilever's supply-chain AI implementation demonstrate the principle of human–AI collaboration? Why is it important that final decisions remained with local planners, even when algorithms generated recommendations?
6. Unilever's executive training in digital fluency empowered leaders to engage critically with AI outputs. How important is executive-level digital literacy in shaping effective AI transformation strategies?
7. Unilever's approach reflects the concept of "refreezing" or even "re-slushing" to sustain change. What mechanisms did the company put in place to make AI adoption stick, and how could these be applied in other industries?

Index

www.ingramcontent.com/pod-product-compliance
Lightning Source LLC
Chambersburg PA
CBHW082132100726
47617CB00008B/4